MARKET OPERATIONS IN ELECTRIC POWER SYSTEMS

MARKET OPERATIONS IN ELECTRIC POWER SYSTEMS
Forecasting, Scheduling, and Risk Management

Mohammad Shahidehpour, Ph.D.
Electrical and Computer Engineering Department
Illinois Institute of Technology
Chicago, Illinois

Hatim Yamin, Ph.D.
Energy Information System Department
ABB Information System
Raleigh, North Carolina

Zuyi Li, Ph.D.
Research and Development Department
Global Energy Markets Solutions (GEMS)
Minneapolis, Minnesota

The Institute of Electrical and Electronics Engineers, Inc., New York

A JOHN WILEY & SONS, INC., PUBLICATION

To:

Jamie, Maha, and Xuping

Copyright © 2002 by John Wiley & Sons, Inc., New York. All rights reserved.

Published simultaneously in Canada.

No part of this publication may be reproduced, stored in a retrieval system or transmitted in any form or by any means, electronic, mechanical, photocopying, recording, scanning or otherwise, except as permitted under Sections 107 or 108 of the 1976 United States Copyright Act, without either the prior written permission of the Publisher, or authorization through payment of the appropriate per-copy fee to the Copyright Clearance Center, 222 Rosewood Drive, Danvers, MA 01923, (978) 750-8400, fax (978) 750-4470. Requests to the Publisher for permission should be addressed to the Permissions Department, John Wiley & Sons, Inc., 111 River Street, Hoboken, NJ 07030, (201) 748-6011, fax (201) 748-6008.

For ordering and customer service, call 1-800-CALL-WILEY.

Library of Congress Cataloging-in-Publication Data:

Library of Congress Cataloging-in-Publication Data is available.
ISBN 978-0-471-44337-7

Contents

Preface XIII

CHAPTER

1 Market Overview in Electric Power Systems 1
    1.1 Introduction 1
    1.2 Market Structure and Operation 2
        1.2.1 Objective of Market Operation 2
        1.2.2 Electricity Market Models 4
        1.2.3 Market Structure 5
        1.2.4 Power Market Types 9
        1.2.5 Market Power 13
        1.2.6 Key Components in Market Operation 14
    1.3 Overview of the Book 15
        1.3.1 Information Forecasting 15
        1.3.2 Unit Commitment in Restructured Markets 17
        1.3.3 Arbitrage in Electricity Markets 18
        1.3.4 Market Power and Gaming 19
        1.3.5 Asset Valuation and Risk Management 19
        1.3.6 Ancillary Services Auction 19
        1.3.7 Transmission Congestion Management and Pricing . . . 19

2 Short-Term Load Forecasting 21
    2.1 Introduction 21

	2.1.1 Applications of Load Forecasting	21
	2.1.2 Factors Affecting Load Patterns	22
	2.1.3 Load Forecasting Categories	23
2.2	Short-Term Load Forecasting with ANN	25
	2.2.1 Introduction to ANN	25
	2.2.2 Application of ANN to STLF	29
	2.2.3 STLF using MATLAB'S ANN Toolbox	31
2.3	ANN Architecture for STLF	33
	2.3.1 Proposed ANN Architecture	33
	2.3.2 Seasonal ANN	34
	2.3.3 Adaptive Weight	36
	2.3.4 Multiple-Day Forecast	37
2.4	Numerical Results	38
	2.4.1 Training and Test Data	38
	2.4.2 Stopping Criteria for Training Process	42
	2.4.3 ANN Models for Comparison	43
	2.4.4 Performance of One-Day Forecast	45
	2.4.5 Performance of Multiple-Day Forecast	51
2.5	Sensitivity Analysis	53
	2.4.1 Possible Models	53
	2.4.2 Sensitivity to Input Factors	54
	2.4.3 Inclusion of Temperature Implicitly	55

3 Electricity Price Forecasting — 57

3.1	Introduction	57
3.2	Issues of Electricity Pricing and Forecasting	60
	3.2.1 Electricity Price Basics	60
	3.2.2 Electricity Price Volatility	61
	3.2.3 Categorization of Price Forecasting	63
	3.2.4 Factors Considered in Price Forecasting	64
3.3	Electricity Price Simulation Module	65
	3.3.1 A Sample of Simulation Strategies	66
	3.3.2 Simulation Example	67
3.4	Price Forecasting Module based on ANN	69
	3.4.1 ANN Factors in Price Forecasting	70
	3.4.2 118-Bus System Price Forecasting with ANN	72
3.5	Performance Evaluation of Price Forecasting	77

		3.5.1 Alternative Methods	77
		3.5.2 Alternative MAPE Definition	78
	3.6	Practical Case Studies	81
		3.6.1 Impact of Data Pre-Processing	82
		3.6.2 Impact of Quantity of Training Vectors	84
		3.6.3 Impact of Quantity of Input Factors	86
		3.6.4 Impact of Adaptive Forecasting	89
		3.6.5 Comparison of ANN Method with Alternative Methods	90
	3.7	Price Volatility Analysis Module	91
		3.7.1 Price Spikes Analysis	91
		3.7.2 Probability Distribution of Electricity Price	105
	3.8	Applications of Price Forecasting	111
		3.8.1 Application of Point Price Forecast to Making Generation Schedule	111
		3.8.2 Application of Probability Distribution of Price to Asset Valuation and Risk Analysis	112
		3.8.3 Application of Probability Distribution of Price to Options Valuation	112
		3.8.4 Application of Conditional Probability Distribution of Price on Load to Forward Price Forecasting	112
4	Price-Based Unit Commitment		115
	4.1	Introduction	115
	4.2	PBUC Formulation	117
		4.2.1 System Constraints	118
		4.2.2 Unit Constraints	118
	4.3	PBUC Solution	119
		4.3.1 Solution without Emission or Fuel Constraints	120
		4.3.2 Solution with Emission and Fuel Constraints	129
	4.4	Discussion on Solution Methodology	134
		4.4.1 Energy Purchase	134
		4.4.2 Derivation of Steps for Updating Multipliers	134
		4.4.3 Optimality Condition	137
	4.5	Additional Features of PBUC	139
		4.5.1 Different Prices among Buses	139
		4.5.2 Variable Fuel Price as a Function of Fuel Consumption	140
		4.5.3 Application of Lagrangian Augmentation	141
		4.5.4 Bidding Strategy based on PBUC	145

	4.6	Case Studies 150
		4.5.1 Case Study of 5-Unit System 150
		4.5.2 Case Study of 36-Unit System 154
	4.7	Conclusions 160

5 Arbitrage in Electricity Markets 161

5.1 Introduction 161
5.2 Concept of Arbitrage 161
    5.2.1 What is Arbitrage 161
    5.2.2 Usefulness of Arbitrage 162
5.3 Arbitrage in a Power Market 163
    5.3.1 Same-Commodity Arbitrage 163
    5.3.2 Cross-Commodity Arbitrage 164
    5.3.3 Spark Spread and Arbitrage 164
    5.3.4 Applications of Arbitrage Based on PBUC 165
5.4 Arbitrage Examples in Power Market 166
    5.4.1 Arbitrage between Energy and Ancillary Service .. 166
    5.4.2 Arbitrage of Bilateral Contract 171
    5.4.3 Arbitrage between Gas and Power 174
    5.4.4 Arbitrage of Emission Allowance 182
    5.4.5 Arbitrage between Steam and Power 186
5.5 Conclusions 188

6 Market Power Analysis Based on Game Theory 191

6.1 Introduction 191
6.2 Game Theory 192
    6.2.1 An Instructive Example 192
    6.2.2 Game Methods in Power Systems 195
6.3 Power Transactions Game 195
    6.3.1 Coalitions among Participants 197
    6.3.2 Generation Cost for Participants 198
    6.3.3 Participant's Objective 201
6.4 Nash Bargaining Problem 202
    6.4.1 Nash Bargaining Model for Transaction Analysis .. 203
    6.4.2 Two-Participant Problem Analysis 204
    6.4.3 Discussion on Optimal Transaction and Its Price . 206

CONTENTS

        6.4.4 Test Results 207

  6.5 Market Competition with Incomplete Information 215
        6.5.1 Participants and Bidding Information 215
        6.5.2 Basic Probability Distribution of the Game 216
        6.5.3 Conditional Probabilities and Expected Payoff 217
        6.5.4 Gaming Methodology 218

  6.6 Market Competition for Multiple Electricity Products . . 222
        6.6.1 Solution Methodology 222
        6.6.2 Study System 223
        6.6.3 Gaming Methodology 225

  6.7 Conclusions 230

7 Generation Asset Valuation and Risk Analysis 233

  7.1 Introduction 233
        7.1.1 Asset Valuation 233
        7.1.2 Value at Risk (VaR) 234
        7.1.3 Application of VaR to Asset Valuation in Power Markets 235

  7.2 VaR for Generation Asset Valuation 236
        7.2.1 Framework of the VaR Calculation 236
        7.2.2 Spot Market Price Simulation 238
        7.2.3 A Numerical Example 240
        7.2.4 A Practical Example 246
        7.2.5 Sensitivity Analysis 258

  7.3 Generation Capacity Valuation 267
        7.3.1 Framework of VaR Calculation 268
        7.3.2 An Example 268
        7.3.3 Sensitivity Analysis 270

  7.4 Conclusions 273

8 Security-Constrained Unit Commitment 275

  8.1 Introduction 275

  8.2 SCUC Problem Formulation 276
        8.2.1 Discussion on Ramping Constraints 280

  8.3 Benders Decomposition Solution of SCUC 285
        8.3.1 Benders Decomposition 286
        8.3.2 Application of Benders Decomposition to SCUC 287

		8.3.3 Master Problem Formulation 287
	8.4	SCUC to Minimize Network Violation 290
		8.4.1 Linearization of Network Constraints 290
		8.4.2 Subproblem Formulation 293
		8.4.3 Benders Cuts Formulation 296
		8.4.4 Case Study 296
	8.5	SCUC Application to Minimize EUE - Impact of Reliability 303
		8.5.1 Subproblem Formulation and Solution 303
		8.5.2 Case Study 306
	8.6	Conclusions 310
9	Ancillary Services Auction Market Design 311	
	9.1	Introduction 311
	9.2	Ancillary Services for Restructuring 313
	9.3	Forward Ancillary Services Auction – Sequential Approach 315
		9.3.1 Two Alternatives in Sequential Ancillary Services Auction 317
		9.3.2 Ancillary Services Scheduling 318
		9.3.3 Design of the Ancillary Services Auction Market 320
		9.3.4 Case Study 322
		9.3.5 Discussions 334
	9.4	Forward Ancillary Services Auction – Simultaneous Approach 334
		9.4.1 Design Options for Simultaneous Auction of Ancillary Services 336
		9.4.2 Rational Buyer Auction 338
		9.4.3 Marginal Pricing Auction 347
		9.4.4 Discussions 354
	9.5	Automatic Generation Control (AGC) 354
		9.5.1 AGC Functions 354
		9.5.2 AGC Response 356
		9.5.3 AGC Units Revenue Adequacy 357
		9.5.4 AGC Pricing 358
		9.5.5 Discussions 366
	9.6	Conclusions 367

CONTENTS

10 Transmission Congestion Management and Pricing 369
    10.1 Introduction 369
    10.2 Transmission Cost Allocation Methods 372
        10.2.1 Postage-Stamp Rate Method 372
        10.2.2 Contract Path Method 373
        10.2.3 MW-Mile Method 373
        10.2.4 Unused Transmission Capacity Method 374
        10.2.5 MVA-Mile Method 376
        10.2.6 Counter-Flow Method 376
        10.2.7 Distribution Factors Method 376
        10.2.8 AC Power Flow Method 379
        10.2.9 Tracing Methods 379
        10.2.10 Comparison of Cost Allocation Methods ... 386
    10.3 Examples for Transmission Cost Allocation Methods ... 387
        10.3.1 Cost Allocation Using Distribution Factors Method ... 388
        10.3.2 Cost Allocation Using Bialek's Tracing Method 389
        10.3.3 Cost Allocation Using Kirschen's Tracing Method ... 391
        10.3.4 Comparing the Three Cost Allocation Methods 392
    10.4 LMP, FTR, and Congestion Management 393
        10.4.1 Locational Marginal Price (LMP) 393
        10.4.2 LMP Application in Determining Zonal Boundaries .. 405
        10.4.3 Firm Transmission Right (FTR) 408
        10.4.4 FTR Auction 412
        10.4.5 Zonal Congestion Management 421
    10.5 A Comprehensive Transmission Pricing Scheme ... 431
        10.5.1 Outline of the Proposed Transmission Pricing Scheme 432
        10.5.2 Prioritization of Transmission Dispatch 434
        10.5.3 Calculation of Transmission Usage and Congestion Charges and FTR Credits 439
        10.5.4 Numerical Example 443
    10.6 Conclusions 453

APPENDIX

A List of Symbols 455

B	Mathematical Derivation		461
	B.1	Derivation of Probability Distribution	461
	B.2	Lagrangian Augmentation with Inequality Constraints	462
C	RTS Load Data		467
D	Example Systems Data		469
	D.1	5-Unit System	469
	D.2	36-Unit System	472
	D.3	6-Unit System	476
	D.4	Modified IEEE 30-Bus System	477
	D.5	118-Bus System	479
E	Game Theory Concepts		483
	E.1	Equilibrium in Non-Cooperative Games	483
	E.2	Characteristics Function	484
	E.3	N-Players Cooperative Games	485
	E.4	Games with Incomplete Information	486
F	Congestion Charges Calculation		489
	F.1	Calculations of Congestion Charges using Contributions of Generators	489
	F.2	Calculations of Congestion Charges using Contributions of Loads	493

References 495

Index 509

Preface

During the last five years, Illinois Institute of Technology in Chicago has been offering a master's degree program in electricity markets which is a joint venture between the College of Engineering and the School of Business. The subject of this book is currently offered as a required course for students majoring in the master of electricity markets.

We believe that the subject of this book will be of interest to power engineering faculty and students, consultants, vendors, manufacturers, researchers, designers, and electricity marketer, who will find a detailed discussion of electricity market tools throughout the book with numerous examples. We assume that the readers have a fundamental knowledge of power system operation and control.

Much of the topics in this book are based on the presumption that there are two major objectives in establishing an electricity market: ensuring a secure operation and facilitating an economical operation. Security is the most important aspect of the power system operation be it a regulated operation or a restructured power market. In a restructured power system, security could be ensured by utilizing the diverse services available to the market. The economical operation facilitated by the electricity market is believed to help reduce the cost of electricity utilization, which is a primary motive for restructuring and a way to enhance the security of a power system through its economics. To accomplish these objectives, proper market tools must be devised and efficient market strategies must be employed by participants based on power system requirements.

The topics covered by this book discuss certain tools and procedures that are utilized by the ISO as well as GENCOs and TRANSCOs. These topics include electricity load and price forecasting, security-constrained unit commitment and price-based unit commitment, market power and monitoring, arbitrage in electricity markets, generation asset valuation and risk analysis, auction market design for energy and ancillary services, as well as transmission congestion management and pricing. For instance, chapters that discuss price forecasting, price-based unit commitment, market power, arbitrage, and asset valuation and risk analysis, present market tools that can be utilized by GENCOs for analyzing electricity market risks, valuation of GENCO's assets and formulation of their strategies for maximizing profits. The chapters that discuss load forecasting, gaming methods, security-constrained unit commitment, ancillary services auction, and transmission congestion management and pricing present market tools that can be utilized by certain market coordinators (such as the ISOs). In addition, the chapter that discusses transmission congestion management and pricing present the role of TRANSCOs in restructured electric power systems.

We have intended to preserve the generality in discussing the structure and the operation of electricity markets so that the proposed tools can be applied to various alternatives in analyzing the electricity markets.

We take this opportunity to acknowledge the important contributions of Professor Muwaffaq Alomoush of the Yarmouk University to our book. He provided much of the presentation in Chapter 10 on transmission congestion management and pricing. We thank Dr. Ebrahim Vaahedi (Perot Systems) and Professor Noel Schulz (Mississippi State University) who reviewed an earlier version of this book and provided several constructive comments. This book could not have been completed without the unconditional support of our respective families. We thank them for their sacrifice and understanding.

Mohammad Shahidehpour
Hatim Yamin
Zuyi Li

Chapter 1

Market Overview in Electric Power Systems

1.1 INTRODUCTION

This book discusses the hierarchy, structure, and operation of electricity markets in a general sense. The generality will allow readers to apply the presented tools to various alternatives in analyzing electricity markets. These tools will help electricity market participants apply the market rules efficiently, and maximize their individual revenues by enhancing their position in competitive electricity markets.

The electricity industry throughout the world, which has long been dominated by vertically integrated utilities, is undergoing enormous changes. The electricity industry is evolving into a distributed and competitive industry in which market forces drive the price of electricity and reduce the net cost through increased competition.

Restructuring has necessitated the decomposition of the three components of electric power industry: generation, transmission, and distribution. Indeed, the separation of transmission ownership from transmission control is the best application of *pro forma* tariff. An independent operational control of transmission grid in a restructured industry would facilitate a competitive market for power generation and direct retail access. However, the independent operation of the grid cannot be guaranteed without an independent entity such as the independent system operator (ISO).

The ISO is required to be independent of individual market participants, such as transmission owners, generators, distribution companies, and end-users. In order to operate the competitive market efficiently while ensuring the reliability of a power system, the ISO, as the market operator, must establish sound rules on energy and ancillary services markets, manage the transmission system in a fair and non-discriminatory fashion, facilitate hedging tools against market risks, and monitor the market to ensure that it is free from market power. The ISO must be equipped with powerful computational tools, involving market monitoring, ancillary services auctions, and congestion management, for example, in order to fulfill its responsibility.

The Federal Energy Regulatory Commission (FERC) Order No. 888 mandated the establishment of unbundled electricity markets in the newly restructured electricity industry. Energy and ancillary services were offered as unbundled services, and generating companies (GENCOs) could compete for selling energy to customers by submitting competitive bids to the electricity market. They could maximize their profits regardless of the systemwide profit. In this market, GENCOs would no longer be controlled by entities that control the transmission system and could choose to acquire computational tools, such as price forecasting, unit commitment, arbitrage and risk management to make sound decisions in this volatile market. Figure 1.1 depicts such a possible alternative electricity market. However, the design is general and could encompass other alternatives. The market components presented in this design are discussed throughout this book.

1.2 MARKET STRUCTURE AND OPERATION

1.2.1 Objectives of Market Operation

There are two objectives for establishing an electricity market: ensuring a secure operation and facilitating an economical operation.

Security is the most important aspect of the power system operation be it a regulated operation or a restructured power market. In a restructured environment, security could be facilitated by utilizing the diverse services available to the market. The economical operation of the electricity market would reduce the cost of electricity utilization. This is a primary motive for restructuring, and a way to enhance the security of a power system through its economics. To do this, proper strategies must be designed in the markets based on power system requirements. For example, financial instruments such as contracts for differences (CFDs), transmission congestion contracts (TCCs) and firm transmission rights (FTRs) could be considered in hedging volatility risks. Besides, monitoring tools are being devised in several markets to avoid a possible market power.

MARKET OVERVIEW IN ELECTRIC POWER SYSTEMS 3

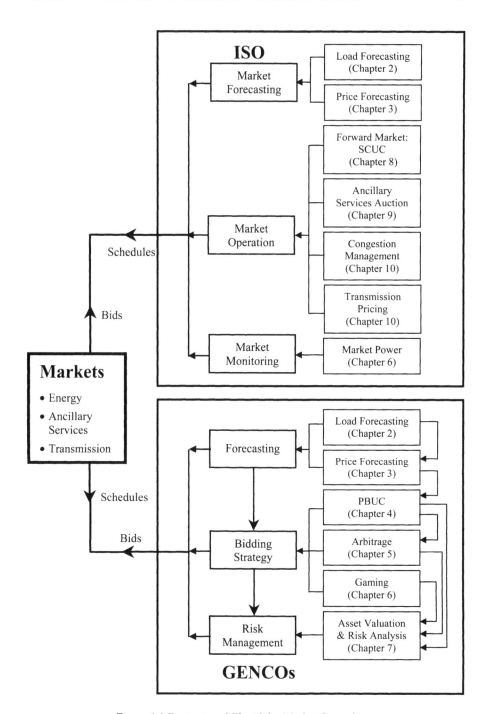

Figure 1.1 Restructured Electricity Market Operation

1.2.2 Electricity Market Models

In order to achieve electricity market goals, several models for the market structure have been considered. Three basic models are outlined as follows.

PoolCo Model. A PoolCo is defined as a centralized marketplace that clears the market for buyers and sellers. Electric power sellers/buyers submit bids to the pool for the amounts of power that they are willing to trade in the market. Sellers in a power market would compete for the right to supply energy to the grid, and not for specific customers. If a market participant bids too high, it may not be able to sell. On the other hand, buyers compete for buying power, and if their bids are too low, they may not be able to purchase. In this market, low cost generators would essentially be rewarded. An ISO within a PoolCo would implement the economic dispatch and produce a single (spot) price for electricity, giving participants a clear signal for consumption and investment decisions. The market dynamics in the electricity market would drive the spot price to a competitive level that is equal to the marginal cost of most efficient bidders. In this market, winning bidders are paid the spot price that is equal to the highest bid of the winners.

Bilateral Contracts Model. Bilateral contracts are negotiable agreements on delivery and receipt of power between two traders. These contracts set the terms and conditions of agreements independent of the ISO. However, in this model the ISO would verify that a sufficient transmission capacity exists to complete the transactions and maintain the transmission security. The bilateral contract model is very flexible as trading parties specify their desired contract terms. However, its disadvantages stem from the high cost of negotiating and writing contracts, and the risk of the creditworthiness of counterparties.

Hybrid Model. The hybrid model combines various features of the previous two models. In the hybrid model, the utilization of a PoolCo is not obligatory, and any customer would be allowed to negotiate a power supply agreement directly with suppliers or choose to accept power at the spot market price. In this model, PoolCo would serve all participants (buyers and sellers) who choose not to sign bilateral contracts. However, allowing customers to negotiate power purchase arrangements with suppliers would offer a true customer choice and an impetus for the creation of a wide variety of services and pricing options to best meet individual customer needs. In our discussion of market structure, we assume the use of a hybrid model.

1.2.3 Market Structure

In this section, we initiate a discussion on a possible market structure[1] encompassing market entities (i.e., entities that take part in a market) and market types (e.g., energy and ancillary services). In addition, we discuss issues related to market power.

1.2.3.1 Key Market Entities

The restructuring of electricity has changed the role of traditional entities in a vertically integrated utility and created new entities that can function independently. Here, we categorize market entities into market operator (ISO) and market participants. The ISO is the leading entity in a power market and its functions determine market rules. The key market entities discussed here include GENCOs and TRANSCOs. Other market entities include DISCOs, RETAILCOs, aggregators, brokers, marketers, and customers.

ISO. A competitive electricity market would necessitate an independent operational control of the grid. The control of the grid cannot be guaranteed without establishing the ISO. The ISO administers transmission tariffs, maintains the system security, coordinates maintenance scheduling, and has a role in coordinating long-term planning. The ISO should function independent of any market participants, such as transmission owners, generators, distribution companies, and end-users, and should provide non-discriminatory open access to all transmission system users.

The ISO has the authority to commit and dispatch some or all system resources and to curtail loads for maintaining the system security (i.e., remove transmission violations, balance supply and demand, and maintain the acceptable system frequency). Also, the ISO ensures that proper economic signals are sent to all market participants, which in turn, should encourage efficient use and motivate investment in resources capable of alleviating constraints.

In general, there are two possible structures for an ISO, and the choice of structure depends on the ISO's objectives and authority. The first structure (MinISO) is mainly concerned with maintaining transmission security in the operations of the power market to the extent that the ISO is

---

[1] This is also referred to as "market architecture".

able to schedule transfers in a constrained transmission system. This structure of the ISO is based on the coordinated multilateral trade model [Var97], and the ISO has no market role. Its objective is restricted to security, and its authority is modest. The California ISO is an example of this structure in which the ISO has no jurisdiction over forward energy markets and very limited control over actual generating unit schedules.

The second structure for an ISO (MaxISO) includes a power exchange (PX) that is integral to the ISO's operation. The PX is an independent, non-government and non-profit entity that ensures a competitive marketplace by running an auction for electricity trades. The PX calculates the market-clearing price (MCP) based on the highest price bid in the market. In some market structures, the ISO and the PX are separate entities, although the PX functions within the same organization as the ISO. This second structure for an ISO is based on an optimal power flow dispatch model. Market participants must provide extensive data, such as cost data for every generator, and daily demand for every consumer or load. With these extensive data, the ISO obtains the unit commitment and dispatch that maximizes social welfare, and sets transmission congestion prices (as the Lagrange or dual variables corresponding to the transmission capacity constraint in the optimal power flow program). The PJM ISO and the National Grid Company (NGC) in the United Kingdom are examples of this structure having a wide-ranging of authority and control.

In this book, we consider both structures. We assume that the ISO has the authority to operate an ancillary services market and manage a transmission network. We also discuss the tools needed for an ISO to operate a constrained electricity market.

GENCOs. A GENCO operates and maintains existing generating plants. GENCOs are formed once the generation of electric power is segregated from the existing utilities. A GENCO may own generating plants or interact on behalf of plant owners with the short-term market (power exchange, power pool, or spot market). GENCOs have the opportunity to sell electricity to entities with whom they have negotiated sales contracts. GENCOs may also opt to sell electricity to the PX from which large customers such as DISCOs and aggregators may purchase electricity to meet their needs. In addition to real power, GENCOs may trade reactive power and operating reserves. GENCOs are not affiliated with the ISO or TRANSCOs. A GENCO may offer electric power at several locations that will ultimately be delivered through TRANSCOs and DISCOs to customers.

MARKET OVERVIEW IN ELECTRIC POWER SYSTEMS

GENCOs include IPPs, QFs, exempt wholesale generators (EWGs) created under EPAct, foreign utilities, and others. Its generating assets include power-producing facilities and power purchase contracts. Since GENCOs are not in a vertically integrated structure, their prices are not regulated. In addition, GENCOs cannot discriminate against other market participants (e.g., DISCOs and RETAILCOs), fix prices, or use bilateral contracts to exercise market power. GENCOs may be entitled to funds collected for the stranded power costs recovery. GENCOs will communicate generating unit outages for maintenance to the ISO within a certain time (usually declared by the ISO) prior to the start of the outage. The ISO then informs the GENCOs of all approved outages.

In the restructured power market, the objective of GENCOs is to maximize profits. To do so, GENCOs may choose to take part in whatever markets (energy and ancillary services markets) and take whatever actions (arbitraging and gaming). It is a GENCO's own responsibility to consider possible risks.

TRANSCOs. The transmission system is the most crucial element in electricity markets. The secure and efficient operation of the transmission system is the key to the efficiency in these markets.

A TRANSCO transmits electricity using a high-voltage, bulk transport system from GENCOs to DISCOs for delivery to customers. It is composed of an integrated network that is shared by all participants and radial connections that join generating units and large customers to the network. The use of TRANSCO assets will be under the control of the regional ISO, although the ownership continues to be held by original owners in the vertically integrated structure. TRANSCOs are regulated to provide non-discriminatory connections and comparable service for cost recovery.

A TRANSCO has the role of building, owning, maintaining, and operating the transmission system in a certain geographical region to provide services for maintaining the overall reliability of the electrical system. TRANSCOs provide the wholesale transmission of electricity, offer open access, and have no common ownership or affiliation with other market participants (e.g., GENCOs and RETAILCOs). Authorities at the state and federal levels regulate TRANSCOs, and they recover their investment and operating costs of transmission facilities using access charges (which are usually paid by every user within the area/region), transmission usage charges (based on line flows contributed by each user), and congestion revenues collected by the ISO.

1.2.3.2 Other Market Entities

DISCOs. A DISCO distributes the electricity, through its facilities, to customers in a certain geographical region. A DISCO is a regulated (by state regulatory agencies) electric utility that constructs and maintains distribution wires connecting the transmission grid to end-use customers. A DISCO is responsible for building and operating its electric system to maintain a certain degree of reliability and availability. DISCOs have the responsibility of responding to distribution network outages and power quality concerns. DISCOs are also responsible for maintenance and voltage support as well as ancillary services.

RETAILCOs. A RETAILCO is a newly created entity in this competitive industry. It obtains legal approval to sell retail electricity. A RETAILCO takes title to the available electric power and re-sells it in the retail customer market. A retailer buys electric power and other services necessary to provide electricity to its customers and may combine electricity products and services in various packages for sale. A retailer may deal indirectly with end-use customers through aggregators.

Aggregators. An aggregator is an entity or a firm that combines customers into a buying group. The group buys large blocks of electric power and other services at cheaper prices. The aggregator may act as an agent (broker) between customers and retailers. When an aggregator purchases power and re-sells it to customers, it acts as a retailer and should initially qualify as a retailer.

Brokers. A broker of electric energy services is an entity or firm that acts as a middleman in a marketplace in which those services are priced, purchased, and traded. A broker does not take title on available transactions, and does not generate, purchase, or sell electric energy but facilitates transactions between buyers and sellers. If a broker is interested in acquiring a title on electric energy transactions, then it is classified as a generator or a marketer. A broker may act as an agent between a GENCO, or an aggregation of generating companies, and marketers.

Marketers. A marketer is an entity or a firm that buys and re-sells electric power but does not own generating facilities. A marketer takes title, and is approved by FERC, to market electric energy services. A marketer performs as a wholesaler and acquires transmission services. A marketer may handle both marketing and retailing functions.

MARKET OVERVIEW IN ELECTRIC POWER SYSTEMS

Customers. A customer is the end-user of electricity with certain facilities connected to the distribution system, in the case of small customers, and connected to transmission system, in the case of bulk customers. In a vertically integrated structure, a user obtains electric energy services from a utility that has legal rights to provide those services in the service territory where the customer is located. In a restructured system, customers are no longer obligated to purchase any services from their local utility company. Customers would have direct access to generators or contracts with other providers of power, and choose packages of services (e.g., the level of reliability) with the best overall value that meets customers' needs. For instance, customers may choose providers that would render the option of shifting customer loads to off-peak hours with lower rates.

1.2.4 Power Market Types

This book will cover the operation of a market from the ISO's perspective, and discuss the algorithms for maximizing a GENCO's profit. Based on trading, the market types include the energy market, ancillary services market, and transmission market. Furthermore, markets are classified as forward market (day-ahead or hour-ahead) and real-time market. It is important to note that markets are not independent but interrelated. In the following, we will learn how these market types are organized.

1.2.4.1 Energy, Ancillary Services, and Transmission Markets

Energy Market. The energy market is where the competitive trading of electricity occurs. The energy market is a centralized mechanism that facilitates energy trading between buyers and sellers. The energy market's prices are reliable prices indicators, not only for market participants but for other financial markets and consumers of electricity as well. The energy market has a neutral and independent clearing and settlement function. In general, the ISO or the PX operates the energy market.

In the MinISO model, the ISO (or PX) accepts demand and generation bids (a price and quantity pair) from the market participants, and determines the market-clearing price (MCP) at which energy is bought and sold. In general, the way to determine the MCP is as follows: Aggregate the supply bids into a supply curve and aggregate the demand bids into a demand curve. The intersection point of the supply curve and demand curve is the MCP. In time periods of congestion, a corresponding adjustment would be made. In California, the adjustment is implemented in the form of congestion charge (or usage charge) for each congested transmission path. In the electricity markets of England and Wales, the MCP is adjusted in the form of a capacity charge, which includes the loss of load probability (LOLP) and the value of lost load (VOLL). In the MinISO model, it is not the ISO (or PX) but the GENCOs who are responsible for unit commitment.

In the MaxISO model, the market participants must submit extensive information similar to that required by a regulated industry, such as energy offer, start-up cost, no-load cost, ramp rates, and minimum ON/OFF time. From these data, the ISO implements security-constrained unit commitments that maximize social welfare. The ISO will either set transmission congestion prices as dual variables corresponding to the transmission capacity constraints or obtain locational marginal prices (LMPs) as the dual variables corresponding to the load balance constraints as in the PJM market. In this book, we will fully discuss the unit commitment problem in the MaxISO model.

Ancillary Services Market. Ancillary services are needed for the power system to operate reliably. In the regulated industry, ancillary services are bundled with energy. In the restructured industry, ancillary services are mandated to be unbundled from energy. Ancillary services are procured through the market competitively. In the United States, competitive ancillary services markets are operated in California, New York, and New England.

In general, ancillary services bids submitted by market participants consist of two parts: a capacity bid and an energy bid. Usually, ancillary services bids are cleared in terms of capacity bids. The energy bid represents the participants' willingness to be paid if the energy is actually delivered.

Different ancillary services in the market could be cleared sequentially or simultaneously. In the sequential approach, a market is cleared for the highest quality service first, then the next highest, and so on. For example, suppose that four types of ancillary services are traded, including regulation, spinning reserve, non-spinning reserve, and replacement reserve, which are from the highest quality to the lowest quality. The market would be cleared first for regulation, then spinning, non-spinning, and replacement reserves. In each round, market participants would be allowed to rebid their unfulfilled resources in the previous rounds. For example, if a participant's regulation bid is not accepted in the regulation clearing round, the participant could bid it again as spinning reserve. The participant could modify the bid in a new round before resubmitting it.

In the simultaneous approach, market participants would submit all bids for ancillary services at once, and the ISO (or PX) would clear the ancillary services market simultaneously by solving an optimization problem. The objective of the optimization problem would depend on the market and could include the minimization of social cost, the minimization of procurement cost, and so on. In the optimizing process, the ISO (or PX) could also consider the substitutability of ancillary services, which refers to substituting a higher quality reserve for a lower quality one. In this book, we will discuss an efficient ancillary services market operation that includes the substitutability of reserves.

Transmission Market. In a restructured power system, the transmission network is where competition occurs among suppliers in meeting the demands of large users and distribution companies. The commodity traded in the transmission market is a transmission right. This may be the right to transfer power, the right to inject power into the network, or the right to extract power from the network. The holder of a transmission right can either physically exercise the right by transferring power or be compensated financially for transferring the right for using the transmission network to others. The importance of the transmission right is mostly observed when congestion occurs in the transmission market. In holding certain transmission rights, participants can hedge congestion charges through congestion credits.

The transmission right auction would represent a centralized auction in which market participants submit their bids for purchase and sale of transmission right. The auction is conducted by the ISO or an auctioneer appointed by the ISO, and its objective is to determine bids that would be feasible in terms of transmission constraints and that would maximize revenues for the transmission network use. A buyer of a transmission right is required to provide the maximum amount of transmission right that the buyer is willing to trade, in addition to buying price and points of injection and extraction. A seller of a transmission right is required to provide the maximum amount of transmission right that the seller is willing to trade, selling price and points of injection and extraction. Transmission rights could be obtained initially from an annual primary auction as in the CAISO case, through the purchase of network transmission services based on their anticipated peak loads for load serving entities (LSEs), or through the purchase of firm point-to-point transmission services as in the PJM case.

More significant is the secondary auction market for transmission rights, since it would facilitate a more robust and liquid market for transmission rights and facilitate energy trading markets. The secondary auction could be monthly, weekly or daily.

We will discuss the issue of transmission pricing in this book, and include issues like how to manage the transmission market efficiently.

1.2.4.2 Forward and Real-time Markets

Forward Market. In most electricity markets, a day-ahead forward market is for scheduling resources at each hour of the following day. An hour-ahead forward market is a market for deviations from the day-ahead schedule. Both energy and ancillary services can be traded in forward markets.

In general, the forward energy market is cleared first. Then, bids for ancillary services are submitted, which could be cleared sequentially or simultaneously as discussed before. Whenever energy schedules in a forward market can be accommodated without congestion management, the ISO would procure ancillary services through a systemwide auction. However, if a congestion exists somewhere in the system, the auction for ancillary services would be implemented on a zonal basis.

Real-Time Market. To ensure the reliability of power systems, the production and consumption of electric power must be balanced in real-time. However, real-time values of load, generation, and transmission system can differ from forward market schedules. Therefore, the real-time market is established to meet the balancing requirement[2].

The real-time market is usually operated by the ISO. Available resources for accommodating real-time energy imbalances can be classified according to their response time, including that of automatic generation control (AGC) which could respond within few seconds, and spinning, non-spinning, and supplemental reserves which could be available within minutes of the ISO's dispatch instruction based on ramping considerations.

The ISO aggregates energy bids into a systemwide bid curve for incremental energy. However, if there are congestions in the real-time market, then prices are set on a zonal basis. The ISO would dispatch units

---

[2] It is also called a balancing market.

in real-time starting from the unit with the lowest energy bid, subject to its prevailing constraints. If supply exceeds demand in real-time, decremental adjustment bids for generation can be utilized. When supply exceeds demand, the ISO would call on the highest priced decremental bid to restore the balanced price as the price of the last unit that was called upon to adjust its schedule. In the case of an undersupply, this would be the highest incremental bid taken.

The balancing energy price is usually calculated at 10-minute or 5-minute intervals. Suppliers who have committed a capacity for supplying energy, except regulation, to one of the ancillary services markets receive payments for energy supply in addition to capacity payment.

In this book, we will see that AGC is essential to the operation of a real-time market.

1.2.5 Market Power

Non-competitive practices in the electric power industry, especially in the generation sector, mainly concerns market power. When an owner of a generation facility is able to exert a significant influence (monopoly) on pricing or on the availability of electricity, a market power is manifested. Market power could prevent the competition and the customer choice in a restructured power system.

Market power may be defined as owning the ability by a seller, or a group of sellers, to drive the spot price over a competitive level, control the total output, or exclude competitors from a relevant market for a significant period of time. A market power could hamper the competition in power production, service quality, and technological innovation. The net result of the existence of market power is a transfer of wealth from buyers to sellers through a misallocation of resources.

Market power may be exercised intentionally or accidentally. For example, in the generation sector, market power could arise from offering an excessive amount of generation to a market (intentional), by committing costly generating units for maintaining reliability while other units could have been less expensive (intentional), or by transmission constraints that could limit the transfer capability in a certain area (accidental). Transmission constraints could prohibit certain generating units from supplying power and persuade dominant providers to drive market prices up by offering more costly units to the market. Another example is when hourly metering is unavailable in customer sites. Hourly price information could encourage customers to manage their loads (elastic loads) as prices go up in peak periods. The lack of hourly information could persuade generating companies to drive market prices up, when resources are scarce, to their own benefit. In the earlier restructuring era, transmission owners could exercise market power by offering pertinent transmission information to their affiliated generating companies and withholding it from other competitors.

Authorities in the electricity industry must identify and correct situations in which some companies possess market power. Some of the tools for identifying market power will be discussed later in this book.

1.2.6 Key Components in Market Operation

In this section, we identify the key components in market operation, as are discussed in this book.

The responsibilities of the ISO are to operate the market securely and efficiently, and to monitor the market free from market power. Thus, first, the ISO needs to forecast the system load accurately to guarantee that there is enough energy to satisfy the load and enough ancillary services to ensure the reliability of the physical power system. Second, the operational responsibilities of the ISO include the energy market, the ancillary services market, and the transmission market. The ISO must be equipped with powerful tools to fulfill those responsibilities, such as through security-constrained unit commitment (SCUC), the ancillary services auction, and transmission pricing. Third, the ISO must be equipped to monitor the market to suppress the market power and protect the market participants.

GENCOs are key players in the power market. The sole objective of a GENCO is to maximize its profit. In order to do so, first, the GENCO must make an accurate forecast about the system, including its load and its price. In most situations, load forecasting is the basis for price forecasting since the load is the most important price driver. Price forecasting is most important for the GENCO in the restructured power industry, since the price reflects the market situation. Price is a signal that should lead every action the GENCO may take. Second, to achieve the maximum profit, the GENCO should have a good bidding strategy based on the forecasted system information. In the restructured power market, the price-based unit commitment (PBUC), replacing the traditional unit commitment, would be the basis for a good bidding strategy. In addition, identifying arbitrage opportunities in the market and exploiting those opportunities to achieve maximum profit should be one of the capabilities of the GENCO. In most

MARKET OVERVIEW IN ELECTRIC POWER SYSTEMS 15

cases, the identification of arbitrage opportunities depends on PBUC. Because of the uncertainty and the competitiveness of the market, a game strategy would be an indispensable tool for the GENCO. Third, enough attention must be paid to risk management, and the various risk factors. Asset valuation is an important issue in risk management, and this would utilize PBUC, arbitrage, and gaming. The reader is referred to Figure 1.1, on Page 3, which shows the details of a market design discussed in this book.

1.3 OVERVIEW OF THE BOOK

1.3.1 Information Forecasting

Two important sets of information that are forecasted in the restructured power market are the load and the electricity price.

1.3.1.1 Load Forecasting (Chapter 2)

In a restructured power system, a GENCO would have to forecast the system demand and the corresponding price in order to make an appropriate market decision. For the ISO, load forecasting has several applications, including generation scheduling, prediction of power system security, generation reserve of the system, providing information to the dispatcher, and market operation. In this chapter, we mainly discuss short-term load forecasting (STLF).

A proper non-linear mathematical model should take into account load and other data such as temperature, wind, and humidity. The lack of a mathematical model for the inclusion of all the prevailing factors was the main problem with the previous work on this subject. Another method to represent non-linear functions can be obtained in using artificial neural networks (ANNs). ANNs are capable of sufficiently representing any nonlinear functions. In this chapter, we use ANN architecture to design the STLF, which features a seasonal network, an adaptive weight update, and a multiple-day forecast. The considered scheme is compared with two alternative models and it shows better performance.

One of the keys to a good architecture in ANN is choosing appropriate input variables. We apply a sensitivity analysis in our study of this issue. Possible input variables include historical load and weather information. The sensitivity analysis in this chapter shows that the previous day's load has the largest impact, and the consideration of weather factors can improve the performance of load forecasting. If humidity and wind weather information are included among the direct inputs to ANN, the performance is better than that with temperature alone, although it might take longer to train the ANN. To speed up the ANN training, we could add humidity and wind weather information indirectly by the introduction of an effective temperature. By doing so, we find that speed is significantly increased, but the forecasting performance is a bit compromised.

1.3.1.2 Price Forecasting (Chapter 3)

In the restructuring of the power industry, the price of electricity has been the motive behind all activities. In this chapter, we mainly describe short-term price forecasting (STPF) of electricity in restructured power markets. We consider a comprehensive framework for price forecasting, denoted as *ForePrice*, which has four functional modules: price simulation, price forecasting by ANN, performance analysis and volatility analysis.

In the price simulation module where the actual system dispatch includes the system's operating requirements and constraints, *ForePrice* can provide insights on the price curve. Potential price drivers such as line limits, line outages, generator outages, load patterns, and bidding patterns can be identified in *ForePrice* through a sensitivity analysis.

In the price forecasting module, which is based on price simulation, *ForePrice* can select the significant price drivers and establish the relationship between these price drivers and electricity price using ANN. *ForePrice* uses an adaptive scheme to adjust the parameters of ANN with the latest available data. *ForePrice* employs different data pre-processing techniques to improve the quality of the available data for ANN. *ForePrice* automatically decides how much historical information is necessary to achieve the best forecasting accuracy.

In the performance analysis module, the results of the price forecasting module are compared based on ANN and alternative techniques. The alternative techniques involve linear interpolation and similar brute force methods for forecasting the electricity price. *ForePrice* allows for a more reasonable error analysis index to be used in evaluating the forecasting performance.

The volatility analysis module is the most distinctive feature of *ForePrice*. The probability of price spikes is analyzed based on different

MARKET OVERVIEW IN ELECTRIC POWER SYSTEMS 17

load levels and different price forecast levels. In addition, the volatility analysis module analyzes the probability distribution of electricity price using both statistical and ANN methods.

From the price forecasted by *ForePrice*, a GENCO can obtain its price-based unit commitment and optimize its generation resources for achieving the maximum profit. In utilizing the probability distribution of price and its spikes, engineers and marketers can perform generation asset valuation, risk management, and option valuation.

1.3.2 Unit Commitment in Restructured Markets

In the restructured power markets, different entities may be responsible for executing unit commitment (UC). In California and New England, GENCOs will run the unit commitment, which is called price-based unit commitment. In PJM and New York, the ISO runs the transmission security and the voltage-constrained unit commitment, which is called a security-constrained unit commitment.

1.3.2.1 Price-Based Unit Commitment (Chapter 4)

In the restructured power markets, unit commitment is used by individual GENCOs for maximizing its profit in scheduling generation resources. This is referred to as price-based unit commitment to emphasize the importance of the price signal. The most distinctive feature of price-based unit commitment is that all market information is reflected in the market price.

In this chapter, we consider the formulation and the solution methodology for the PBUC problem in a restructured market structure. Distinct features of our scheme include handling different prices among buses, variable fuel prices as a function of fuel consumption, and bidding strategies based on PBUC.

1.3.2.2 Security-Constrained Unit Commitment (Chapter 8)

In some restructured markets, including the PJM interconnection, the New York market, and the U.K. Power Pool, the ISO plans the day-ahead schedule using security-constrained unit commitment. The ISO collects detailed information on each generating unit including characteristics such as start-up and no-load costs, minimum start-up and shut-down times, minimum and maximum unit outputs, and bids representing incremental heat rate. The ISO also obtains information from TRANSCOs via the OASIS on transmission line capability and availability. Then, the ISO uses the SCUC model to determine the optimal allocation of generation resources.

An efficient algorithm is considered in this chapter for a network-constrained unit commitment that takes into account unit generation, phase shifter and tap transformer controls. The methodology is based on Benders decomposition by which the network-constrained unit commitment is divided into a master problem and a subproblem. The master problem is formulated to solve unit commitment with all its prevailing constraints–except transmission security, voltage and reliability constraints–by an augmented Lagrangian relaxation method. Given the unit commitment schedule, the subproblem minimizes the network (i.e., transmission and voltage) violations or the expected unserved energy (EUE). A Benders cut is generated if any violation is detected after subproblems are solved. With Benders cuts, the unit commitment is solved iteratively to provide a minimum cost generation schedule while satisfying all constraints. Since the decomposed problem is easier to solve and requires less complicated and smaller computing capabilities, the generating scheduling is more accurate and faster.

1.3.3 Arbitrage in Electricity Markets (Chapter 5)

Arbitrage refers to making profit by a simultaneous purchase and sale of the same or equivalent commodity with net-zero-investment and without any risk. The usage of arbitrage also includes any activity that attempts to buy a relatively under-priced commodity and to sell a similar and relatively over-priced commodity for profit. In the restructured power industry, there exist many inconsistencies in electricity pricing, which could provide opportunities for arbitrage.

In this chapter, we consider applying PBUC to identify arbitrage opportunities in power markets, including arbitrage between energy and ancillary services, arbitrage of bilateral contracts, arbitrage between gas and energy, arbitrage of emission allowances, arbitrage between steam and energy, and arbitrage between leasing an existing plant and building a new plant.

MARKET OVERVIEW IN ELECTRIC POWER SYSTEMS

1.3.4 Market Power and Gaming (Chapter 6)

Power market authorities must identify and correct situations in which some companies possess market power. In this chapter, we consider methodologies based on game theory, which can be used to identify non-competitive situations in the restructured energy marketplaces (transaction analysis from the market coordinator's point of view) and to provide support for minimizing risks involved in price decisions (transaction analysis from a participant's point of view).

1.3.5 Asset Valuation and Risk Management (Chapter 7)

Asset valuation for generating units is an important issue in the restructured power market. In this chapter, we consider two types of valuation for generating units. One is the valuation based on the daily scheduled generation, and the other is the valuation based on the available capacity of generating units. Since the value of generating units depends on market prices which could be very uncertain in restructured power market, in this chapter, we consider applying the concept of value at risk (VaR) to value generation assets and assess the risk of generation capacity profitability. Frameworks for application of VaR to both types of asset valuation are considered.

1.3.6 Ancillary Services Auction (Chapter 9)

Ancillary services are necessary to support the transmission of power from sellers to buyers given the obligation of control areas and transmission utilities to maintain a reliable operation of the interconnected transmission system. In the restructured power market, ancillary services should be procured competitively through market auctions. In this chapter, we discuss two different approaches that can be used to implement the ancillary services auction: sequential and simultaneous. We also discuss an AGC operation and its pricing as a major component of the ancillary services auction.

1.3.7 Transmission Congestion Management and Pricing (Chapter 10)

The transmission network plays a vital role in competitive electricity markets. In a restructured power system, the transmission network is the key mechanism for generators to compete in supplying large users and distribution companies. A proper transmission pricing scheme that considers transmission congestion could motivate investors to build new transmission and/or generating capacity for improving the efficiency. In a competitive environment, proper transmission pricing could meet revenue expectations, promote an efficient operation of electricity markets, encourage investment in optimal locations of generation and transmission lines, and adequately reimburse owners of transmission assets.

In this chapter, we consider a comprehensive transmission pricing scheme. This scheme can be used by the ISO to modify preferred schedules, trace participants' contributions and allocate transmission usage and congestion charges. By this scheme, the ISO would adjust preferred schedules on a non-discriminatory basis to keep the system within its limits and apply curtailment priority according to the participants' willingness to avoid curtailing transactions. In this scheme, transmission congestion and losses are calculated based on LMPs. A flow-based tracing method is utilized to allocate transmission charges. FTR holders' credits are calculated based on line flow calculations and LMPs.

Chapter 2

Short-Term Load Forecasting

2.1 INTRODUCTION

The principal objective of short-term load forecasting (STLF) is to provide the load prediction for basic generation scheduling functions, for assessing the security of system operation, and for timely dispatcher information [Gro87]. It is well known that STLF plays an important role in the traditional monopolistic power systems. In a restructured power system, a GENCO would have to forecast the system demand and its corresponding price in order to make an appropriate market decision.

Different forecasting models have been employed in power systems for achieving forecasting accuracy. Among the models are regression, statistical and state-space methods. In addition, artificial intelligence-based algorithms have been introduced based on expert system, evolutionary programming, fuzzy system, artificial neural network (ANN), and a combination of these algorithms. Among these algorithms, ANN has received more attention because of its clear model, easy implementation, and good performance.

2.1.1 Applications of Load Forecasting

Short-term load forecasting in power systems operation has several functions, namely:

Generation Scheduling. Scheduling is the main purpose of short-term load forecasting. For hydro generating systems, forecasting would determine the flow of water from reservoirs. For thermal systems, forecasting is used for unit commitment, in calculating start-up and shutdown of each plant. For hydrothermal systems, forecasting is required for balancing the amount of power produced by hydro and thermal plants to achieve an economical operation.

Power System Security. STLF could lead to more secure operation of the system. Using load forecasting, the effect of scheduled operation on power system security can be predicted, and preventive and corrective actions can be prescribed, before the occurrence of contingencies.

Generation Reserve of the System. Power generation reserves could overcome shortcomings caused by sudden load increases and plant failures. The appropriate amount of reserve can be determined based on load forecasting.

Providing Information to Dispatchers. The dispatcher would need the real time information regarding very short-term loads in order to operate the system most economically.

Market Operation. With deregulation of the power industry, load forecasting is becoming even more important, not only for system operators but also for market operators, transmission owners, and other market participants, so that adequate energy transactions can be scheduled, and appropriate operational plans and bidding strategies can be established [Che01].

2.1.2 Factors Affecting Load Patterns

The first step to make a proper load forecasting is to identify factors that would affect load patterns. Some of these factors are:

Economical Factors. An economical condition in one area could affect the load shape. This condition could cover issues like the type of customers, demographic conditions, industrial activities, and population. These conditions would mainly affect the long-term load forecasting.

Time Factors. Time factors include seasonal, weekly, and holiday effects. Examples for the seasonal effect include the number of daylight hours in one season, which affects the load pattern. Industrial load on weekdays will be higher than that of weekends. Holidays will have much effect on the load pattern as loads decrease below normal.

Weather Factors. Temperature is the most influential weather factor in load forecasting. The temperature change could impact the amount of power needed for heating in the winter and air conditioning in the summer.

SHORT-TERM LOAD FORECASTING

Other weather factors that affect load forecasting include humidity especially in hot and humid areas, precipitation, thunderstorms, and the wind and light intensity of the day.

Random Disturbances. Large industrial consumers, like steel mills, may cause sudden load changes. In addition, certain events and conditions can cause sudden load changes such as popular TV shows or the shutdown of an industrial operation.

Price Factors. In electricity markets, electricity price, which is volatile and could present a complicated relationship with system load, is becoming an important factor in load forecasting.

Other Factors. A load shape may be different due to geographical conditions. For example, the load shape for rural areas is different from that of urban areas. The load shape may also depend on the type of consumer. For instance, the residential load shape could be different from that of commercial and industrial consumers.

2.1.3 Load Forecasting Categories

There are two distinct categories of load forecasting for power systems planning and operation. The distinction is based on the forecasting duration.

- In power systems planning, load forecasting is for a span of several months to one year. This type of forecasting is mainly for fuel scheduling. Longer terms in load forecasting could be from one to ten years, which is used for determining the economical location, type, and size of future power plants.

- In power systems operation, load forecasting is mostly for a span of few minutes to 168 hours. There are two main categories of load forecasting in power systems operation: i.e., very short- and short-term load forecasting. Very short-term load forecasting is for minutes ahead and is used for automatic generation control (AGC). Short-term load forecasting is for one to 168 hours ahead. Short-term load forecasting results are mainly used for generation scheduling purposes.

In general, short-term load forecast ought to be available every morning before 7 o'clock for the next 40 hours. Friday's forecast includes weekend and next Monday's forecasts. But when the Monday forecast becomes crucial, the forecast could be made on Sunday.

Load forecasting could also be categorized into peak load and load shape forecasting [Gro87]. This is based on the way loads are modeled. Different models of load would suggest different load forecasting techniques. In the peak load model, load is a function of weather and independent of time. A typical form of the peak load model is peak load = base load + weather-dependent component. The advantage of the peak load model lies in its simplicity, which would require a small amount of data. The disadvantage of this model stems from its time independence.

In the load shape model, load is a function of time. Load forecasting here is for 30 minutes to one hour while the measured quantity is the consumed energy in that period. There are two types of load shape model, time of day model and dynamic model. In the time of day model, the load is defined by a time series at each discrete sampling time of the forecasting period. The advantage of this model is its simple structure which can be updated rapidly. The disadvantage of this model is that it misrepresents the relation between load and weather. The dynamic model recognizes the fact that the load is not only a function of the time of the day but also of its most recent behavior as well as the weather as a random input.

There are two basic types of dynamic models, autoregressive moving average (ARMA) model and state-space model. The ARMA model's common form is

$$z(t) = y_p(t) + y(t) \tag{2.1}$$

where $y_p(t)$ is a component dependent on the time and weather for a particular day, and $y(t)$ is an additive load residual term describing influences due to weather pattern deviations from normal and random correlation. The ARMA model would include weather as an explicit input. However, the model would have to adjust several parameters in areas with a lot of climatic changes. The most important weather input is based on the temperature deviation. This input is described as a non-linear function of the difference between normal and actual temperatures.

The state space model has a general form of

$$z(t) = c^T x(t) \tag{2.2}$$

where

$$x(t+1) = A\, x(t) + B\, u(t) + w(t) \tag{2.3}$$

$x(t)$ is state vector at time t, $u(t)$ is a vector of base weather input, and $w(t)$ is the vector of white noise input. Matrices A and B are considered constant. The main difference between ARMA and state space models is

SHORT-TERM LOAD FORECASTING

that the state space model assumes that the parameters that define the periodic component of the load are random processes.

There are no certain advantages of the state-space model over the ARMA model. However, one possible area where the state-space model may be more advantageous is in the development of bus load forecasting where bus loads exhibit a high degree of correlation.

2.2 SHORT-TERM LOAD FORECASTING WITH ANN

The main problem with the previous works was to make proper non-linear mathematical models for load and other data such as temperature and humidity. Another method to represent non-linear functions is by using ANN. ANN is capable of sufficiently representing any non-linear functions.

2.2.1 Introduction to ANN

ANN is a computer information processing system that simulates the function of human brain. Human brain consists of billions of interconnected cells called neurons. Neurons have four different parts: soma or cell body, dendrite, axon and synapses, as shown in Figure 2.1.

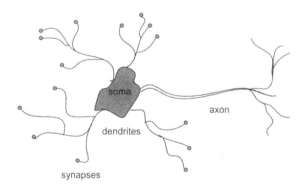

Figure 2.1 Illustration of Biological Neuron

Dendrites receive electric potential from other neurons. These potentials were given weight by synapses. The weight given by synapses can be either excitatory or inhibitory. The excitatory weight increases the voltage of potential, and the inhibitory weight decreases the voltage of potential. The soma sums all potentials given by the dendrites. If the sum of all potentials exceeds a certain value (threshold), the soma will fire an action potential through an axon. The axon will deliver this action potential to another neuron. After firing the action potential, the soma will reset the voltage to the resting potential, and it has to wait some time until it can fire another potential (refractory period).

Biological form of a neuron can be modeled as shown in Figure 2.2. Dendrites are modeled as an input *vector* that gathers the information from an outside neuron. The weight vector represents synapses that put weight in the information. The adder is a representation of the soma that sums all of the incoming information. The transfer function represents a certain value that controls neuron fire, and finally the axon can be represented as an output vector.

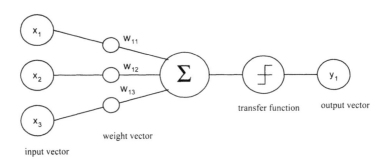

Figure 2.2 Mathematical Model for Neuron

Every ANN model can be classified by its architecture, processing, and training. The architecture describes the neural connections. Processing describes how networks produce output for every input and weight. The training algorithm describes how ANN adapts its weight for every training vector.

In general, the ANN architecture consists of three parts: input layer, hidden layer, and output layer, as shown in Figure 2.3. The input layer is a layer with connection to the outside world. The input layer will receive information from the outside world. The hidden layer does not have connection to the outside world; it only connects to the input layer and the output layer. The output layer will give the ANN output to the outside world after the incoming information is processed by the network.

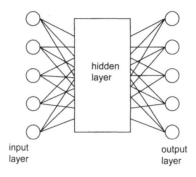

Figure 2.3 General ANN Model

There are four kinds of ANN architecture that are commonly used, namely, a single layer network, a multi layer perceptron, a Hopfield network, and a Kohonen network. The single-layer network is an ANN architecture that does not have hidden layer, as shown in Figure 2.4. Since the input layer sometimes is not counted as a layer, this architecture is called a single-layer network. This network can be categorized as a feedforward ANN, because information flows in one direction, that is, to the output layer.

The multi-layer perceptron is the most common architecture. Unlike the single-layer perceptron, the multi-layer perceptron always has a hidden layer, as shown in Figure 2.4. The simplest form of the multi-layer perceptron will have three layers, an input layer, a hidden layer and an output layer. There is still no best method in determining the number of hidden layer and neurons for each hidden layer. This number is usually found by using heuristics.

The Hopfield network can be categorized as a feedback ANN, as shown in Figure 2.5, since in this network a layer not only receives information from previous layer but also from previous output and bias. Every neuron connected with each other will have two output values, -1

(OFF) or +1 (ON). Output for every neuron is dependent on the previous activation.

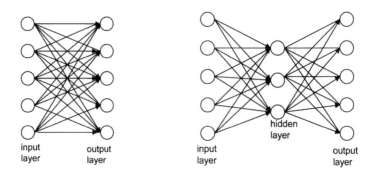

Figure 2.4 Single- and Multi-Layer Network

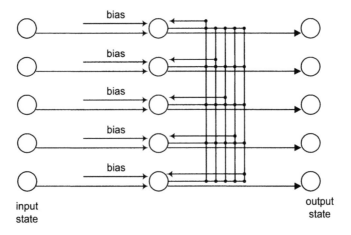

Figure 2.5 Hopfield Network

The Kohonen network consists of feedforward input units and a lateral layer, as shown Figure 2.6. The lateral layer has several neurons, which are laterally connected to their neighbors. The Kohonen network can organize itself and can cause the neighboring unit to react the same way.

SHORT-TERM LOAD FORECASTING

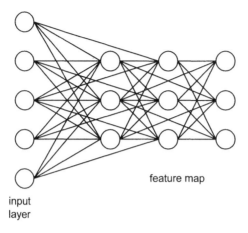

Figure 2.6 Kohonen Network

2.2.2 Application of ANN to STLF

The first ANN architecture for short-term load forecasting (STLF) was a simple feedforward multi-layered perceptron. The output produced by this ANN was an integrated load value, including peak, valley, and mean daily load values. The ANN input was the prospective hour and day.

The use of ANN on STLF can be broken down into two groups based on learning strategies: supervised and unsupervised learning. In supervised learning, the network produces output based on the given input and compares this output with pre-tested outputs. If the output produced by the network is not correct, the network will adjust its weights. In unsupervised learning, only the input has to be given to the network without a feedback from the corrected output. The network will adjust itself to satisfy the inputs. Most applications use a supervised learning ANN. In a supervised leaning ANN, a feedforward multi-layered perceptron ANN is widely used, and many enhancements have been explored.

The partitioning method is one of the enhancements. It was developed because of differences in the load shape for every season and every day. The partitioning method divides the network into several sub networks. For example, in [Els93] the network was divided into the following groups: Monday, Tuesday, Wednesday through Friday, Saturday, and Sunday. In [Cha93] and [Kha93] one week was divided into three groups: Saturday, Sundays, and weekdays. In [Ker93] the network was divided into seven ANN, each with 24 outputs. However, [Bak95] pointed out that one ANN for every day of each week was better than the partitioning method.

Another way to implement partitioning is to divide the input based on hours or seasons. In [Bac92], 24 networks were used to represent every hour in a day. In [Kha92], 2 models was used, one for summer and the other for winter. In [Kho95], a short-term load forecasting was divided into three modules to represent weekly, daily, and hourly modules. The modular outputs were combined using an adaptive linear element. Another method for partitioning [Moh95] divides a year into four seasons (summer, winter, and transitions), and every season divided into three different kinds of day (Monday, weekday and weekend), and each day divided into five periods (1am, 6 am, 11 am, 1 pm, 9pm).

Another supervised learning method that is used for STLF is the recurrent ANN [Dju93]. The difference between a recurrent ANN and a multi-layered perceptron is in the use of context layer. The function of the context layer is to preserve historical data. The output of hidden layers will propagate not only to the output layer but also to the context layer.

[Bau93] is an example of unsupervised learning in STLF. In this study, seven Kohonen maps are used to process data without the weather effect. Each map represents a day in the week. During the training process, every neuron will have a load profile. For one-year data, the neurons store the load profile in the form of 6×6 Kohonen map. For considering the weather effect, two layers of ANN are used. The basic idea is to relate errors produced by the Kohonen network to weather data. Other techniques are combined with ANN, such as fuzzy logic, machine learning, and a genetic algorithm, to improve the forecasting performance.

Many papers report the use of fuzzy logic with ANN on STLF. The first use of fuzzy logic in STLF was in a fuzzy-expert system algorithm subset as a pre-processing layer. The basic idea for the use of fuzzy logic is to divide the variables into subsets, defined as low, medium, and high. The membership degree for each subset is processed by ANN. In [Tor91] and [Kim95] fuzzy-expert system is used to improve the forecast done by ANN.

Another study uses fuzzy as a post-processor [Kim93]. In that study, multi-layered feedforward ANN is used to make a provisional forecast. The input data include load data from an hour before, two hours before and some load data from a week and two weeks before. By observing these

SHORT-TERM LOAD FORECASTING

input data, the ANN would forecast the load without any weather data. The fuzzy-expert system is then used to capture the effect of weather. The final forecast is obtained by summing the ANN forecast and the effect of temperature changes introduced by the fuzzy system.

In [Sri95] a fuzzy system was used to overcome the ANN's inability to forecast on weekends, because of the lack of historical data. Another interesting subject in that paper is a study made in Singapore, which has a tropical climate. So the effect of temperature was not as large as that in a four-season country. Despite this, the study uses maximum, minimum, and daily average temperatures. Rain forecasts were also used for this study since rain has a bigger effect than temperature. The input data further included special events to cover holiday patterns. A fuzzy system in the study was used as a front-end processor. Input data were fuzzified and supplied as input to the multi-layer perceptron ANN. The ANN output from this ANN was then defuzzified to produce the 24 hour load forecast.

Fuzzy was not the only system that was combined with ANN to make load forecasts. Another hybrid system is based on using a machine learning approach with ANN [Rah93]. The machine learning approach is developed to complement ANN for enhancing the reliability of forecast and improving the overall accuracy. In the case of maximum load forecast, the proposed method displays better accuracy and a smaller variance than ANN.

Genetic algorithm is also combined with ANN for load forecasting [Erk97]. In this method, STLF consists of three modules. The first module uses a modified Kohonen network to cluster the daily load curve. A genetic algorithm is used in the second module to determine the most appropriate supervised ANN topology and initial weights for each cluster. In the third module, a three-layer back propagation ANN is used for the hourly load forecast.

2.2.3 STLF Using MATLAB'S ANN Toolbox

MATLAB is a high-performance language for technical computing. It integrates computation, visualization, and programming in an easy-to-use environment [Mat99a]. MATLAB toolboxes are collections of m-files that extend MATLAB's capabilities to a number of technical fields such as control system, signal processing, optimization, and ANN. In the ANN Toolbox version 3.0 [Mat99b], MATLAB provides 12 training functions, many of which are highly efficient. Also provided in the toolbox is the modular network representation, which allows a great deal of flexibility for the design of one's own custom network.

Since MATLAB has provided a variety of tools, it is natural to apply MATLAB's ANN Toolbox to load forecasting. A commonly used 3-layer back propagation ANN is employed. Some discussion of input layer, hidden layer, and output layer follows.

Input Layer. It is important to note that the input layer can accommodate any factors affecting the load pattern directly or implicitly, including types of day, load, temperature, humidity, and wind.

Hidden Layer. The number of neurons in the hidden layer has to be determined by heuristics, since there is no general method available to determine the exact number of neurons in the hidden layer. If the number is too small, the network cannot find the complex relationship between input and output and may have difficulty in convergence during training. If the number is too large, the training process would take longer and could harm the capability of ANN.

The number of neurons in the hidden layer would vary for different applications and could usually depend on the size of the training set and the number of input variables. A few heuristic rules are given as follows:

- The number of hidden layer neurons is equal to two times the number of input layer neurons plus one, or
- The number of hidden layer neurons is equal to the number of input layer neurons plus number of output layer neurons, or
- The number of hidden layer neurons is equal to sum of the number of input layer neurons and the number of output layer neurons divided by two.

In practice, it would be better to make several tries before determining an appropriate number. Fortunately, using MATLAB, it would be a very easy and quick task to compare the impact of the number of hidden layer neurons on the performance of ANN.

Output Layer. As for the output layer, it is relatively easy to set up as compared to the input and hidden layers. We would need to provide target loads for training at the training stage and produce forecasting results at the forecasting stage.

# SHORT-TERM LOAD FORECASTING

2.3 ANN ARCHITECTURE FOR STLF

2.3.1 Proposed ANN Architecture

One of the keys for designing a good architecture in ANN is choosing appropriate input variables. In the case of short-term load forecasting, these inputs can be divided into time, electrical load and weather information. The time information may include the type of season, days of a week, and hours of a day. The load information may include previous loads. The weather information may include previous and future temperatures, cloud cover, thunderstorm, humidity, and rain.

Until now, there have been no general regulations on input types in designing the ANN for STLF. However, as a matter of principle, historical load and temperature represent the most important inputs. For a normal climate area, these two inputs and other related inputs (e.g., time) would be enough to make a good short-time load forecasting. However, for extreme weather conditions in humid areas or in areas with many thunderstorms, additional weather factors should be included for forecasting.

In the proposed architecture, ANN is designed based on previous loads, type of season, type of day, hours of a day, previous day's temperature and temperature forecast. Only two weather factors are used in this architecture, since the forecasted load is assumed to be a normal climate area.

A block diagram for the proposed ANN architecture is shown in Figure 2.7. There are a total of 73 neurons in the input layer. The first neuron is used to define the day of the forecast. A day of the week would be assigned to a number ranging from 1 to 7, Monday as 1, Tuesday as 2, Wednesday as 3, and so on. for this neuron. We could represent the type of day in a week by using seven neurons; however, that would mean adding more neurons and slow down the training process. Another possibility is to try a separate network for every day of the week, which would end up with too many networks to train.

The next 24 input neurons represent the previous day's hourly load. We could have included the previous week's load data corresponding to the same day. However, that would increase the length of the input data with only a minor improvement in performance.

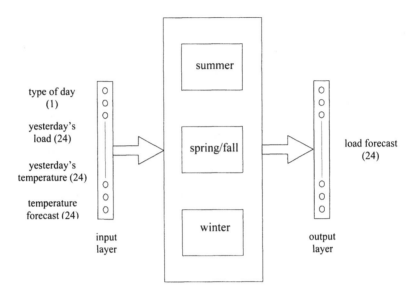

Figure 2.7 Block Diagram

The next 48 neurons are used to capture the effect of temperature. The first 24 neurons are used for previous day's temperature and another 24 neurons for the next day's temperature forecast. In the hidden layer, three seasonal networks are used, which will be discussed in the next section. The output layer of all seasonal networks comprises 24 neurons. These neurons represent 24 hours in a day of forecast.

2.3.2 Seasonal ANN

Figure 2.8 depicts the weekly loads for each season. The highest load would occur in winter, and the second highest in summer. The loads in spring and fall have slight differences. The temperature would also differ in every season, that is, winter would have the lowest temperature and summer the highest temperature.

With this in mind, it would be better to differentiate between the seasons by using different ANN modules. Accordingly, the training would be easier and there is a chance to have better results. We consider three ANN modules for summer, winter, and spring/fall. Winter and summer seasons are represented by different networks. Spring and fall seasons are represented by one network, because of the similarity in their loads and temperatures. In the training process, every network is only supplied by data on that particular season.

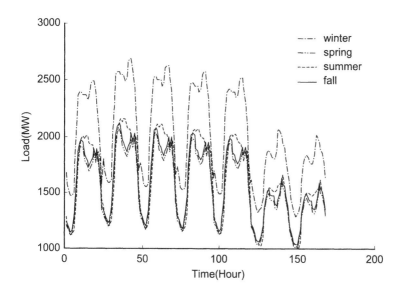

Figure 2.8 Weekly Loads for Each Season

Three seasonal networks have the same architecture, which is a three-layer feedforward ANN. It has 73 neurons in the input layer, 90 neurons in the hidden layer, and 24 neurons in the output layer. The seasonal network model is shown in Figure 2.9.

The number of neurons in the input or the output layer is already fixed, based on the input and output data chosen. The number of neurons in the hidden layer is determined as follows. We begin with a small number and then increase the number gradually until the training process is matured. Accordingly, an appropriate number for neurons in the hidden layer is chosen as 90, based on training time and training results.

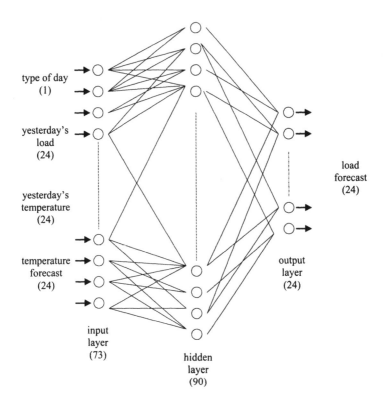

Figure 2.9 Seasonal Network Architecture

2.3.3 Adaptive Weight

In engineering applications of ANN, the training process could be done once, and the weights could be treated as fixed values. However, in order to correlate load changes with time and weather, ANN weights for load forecasting are updated regularly, based on the latest information.

Suppose that the weight adaptation is done daily based on back-propagation. Everyday, ANN forecasts the load for the following day and stores this information. Once the actual load is made available, the difference between actual and forecasted loads would be calculated and propagated back to the weights, using the same method as in the training process. Figure 2.10 illustrates the weight adaptation procedure. This procedure would implicate the latest information for training.

SHORT-TERM LOAD FORECASTING

2.3.4 Multiple-Day Forecast

The proposed ANN could be expanded to forecast loads for several days ahead with the following strategy, which is illustrated in Figure 2.11:

- Load forecast for the d day acts as the previous day load for the $d + 1$ day forecast, and

- Temperature forecast for the d day acts as the previous day temperature for the $d + 1$ day forecast.

As long as the temperature forecast for the $d + k$ day is available, this procedure can be repeated $d + k$ times.

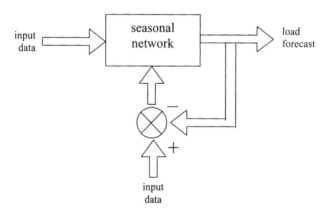

Figure 2.10 Weight Adaptation Procedure

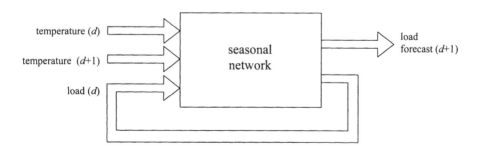

Figure 2.11 Multiple Days Forecast

Although a forecast for more than seven days can be done, it is generally not recommended because of cumulative errors. Since each forecast will have an error with regard to the actual load, a load forecast for a wider time span will have cumulative errors, which make the forecast no longer precise. Another reason for not doing a forecast beyond seven days is the availability of a temperature forecast. Hourly temperature forecasts for more than seven days are rarely made by weather stations. Even if the data were available, the forecast would not reliable.

2.4 NUMERICAL RESULTS

This section will first describe training and test data used to test the proposed ANN model for short-term load forecasting. Both load and temperature data will be presented. Stopping criteria for the training process are discussed.

The section will then discuss results from the proposed ANN model based on the daily average percentage error and the mean absolute percentage error for peak and hourly loads in all seasons. Results for multiple-day forecast are also discussed. To show the advantages of the proposed ANN model, comparisons are made with two other models.

2.4.1 Training and Test Data

Load and weather data are used for training and tests as discussed in the previous section. Daily load data are comprised of hourly loads, and daily weather data are comprised of the hourly temperatures of the day.

Load Data. The 1996 calendar year is used for this study starting with January 1 and peak load for the system is 2850 MW. We follow the RTS load data (see Appendix C) for weekly, daily, and hourly peak percentages.

As for the daily load, daily peak load occurs on Tuesday, lower loads occur during the weekend, and the lowest load occurs on Sunday. In determining hourly loads, we differentiate between the type of the day as weekdays and weekend. These hourly loads also differ for every season: spring and fall have the same pattern; winter and summer have their own patterns. Winter peak loads occur in the evening, and summer peaks occur in the afternoon when the temperature rises.

SHORT-TERM LOAD FORECASTING

In Table 2.1, the maximum load of the year occurs on week 51 or the third week of December, which is a winter peak. The second peak occurs on week 23 or the first week of June, as temperature starts to increase at the beginning of summer. The minimum load occurs on week 38 or the third week of September, which is in the fall season.

Figures 2.12 and 2.13 show the seasonal maximum and minimum loads, respectively. Comparing seasons, we learn that winter has the highest load, due to the cold weather and the use of electrical heaters. The lowest load of all seasons occurs during the fall season.

Table 2.1 Maximum and Minimum Loads

Season	Maximum Load		Minimum Load	
	Week	Load (MW)	Week	Load (MW)
Winter	51	2850	8	1026
Summer	23	2565	27	1006
Spring	16	2280	15	1001
Fall	43	2052	38	965

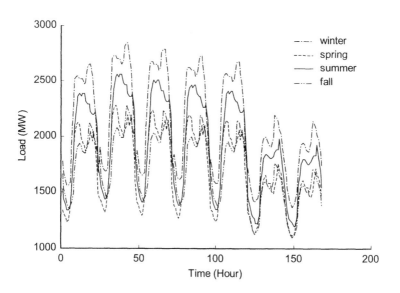

Figure 2.12 Seasonal Load Shape During Maximum Load

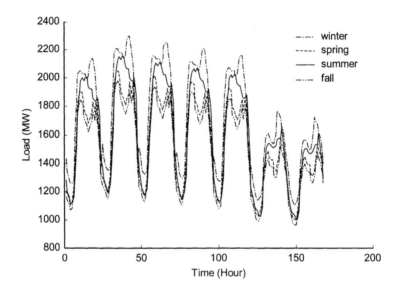

Figure 2.13 Seasonal Load Shape during Minimum Load

Weather Data. Based on the proposed ANN model the only necessary weather data are the hourly temperatures. For this purpose, temperature data at the O'Hare Airport during the year of 1996 shown in Figure 2.14 are used. These data are available from the National Climatic Data Center (NCDC).

There are two types of temperature data for the ANN model: hourly temperature data from previous day and hourly temperature data for the day of forecast. For example, to forecast the load for January 2, temperature data on January 1 represent the data from the previous day, and the temperature data on January 2 are used for the hourly temperature forecast. The maximum and minimum temperatures for all seasons during the year are shown in Table 2.2.

Training and Test Sets. All data are divided into three parts in Table 2.3 based on the season. The data for each season are divided again into two parts as training and test sets. Test sets will not be used for training; their purpose is only to examine errors produced by ANN after training.

SHORT-TERM LOAD FORECASTING 41

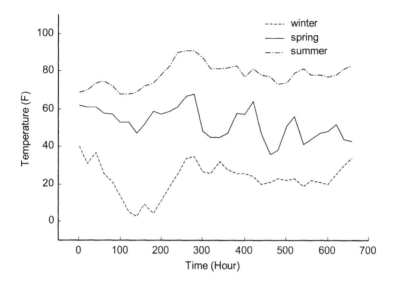

Figure 2.14 Typical Seasonal Temperature

Table 2.2 Maximum and Minimum Temperatures

Season	Max Temp		Min Temp	
	Week	Temp (F)	Week	Temp (F)
Winter	45	56	49	-0.5
Spring	16	59	9	13
Summer	28	93	18	43
Fall	33	87	38	39

Table 2.3 Training and Test Sets

Season	Training Sets		Test Sets	
	Weeks	No. of Input Vector	Weeks	No. of Input Vector
Winter	1-9 & 44-47 Jan, Feb, Nov	91	48-52 Dec	28
Spring/Fall	10-17 & 40-43 Mar, Apr, Oct	84	36-39 Sept	28
Summer	18-22 & 27-30 May, July	63	23-26 June	28

One month in each season will be used as a test set, based on the load in that particular month. For winter, December or week 48 to week 52 will be the test set, since the yearly maximum load occurs on week 51. For summer, the test set will be the month of June or week 23 to week 26, since the summer maximum load or the second highest load during a year occurs on week 23. For fall, the month of September, that is weeks 36 to 39, are chosen for the test set, since the yearly minimum load occurs on week 38.

Each seasonal model will have a different number of training vectors, which will result in different processing times for training and different errors produced by each model.

2.4.2 Stopping Criteria for Training Process

Stopping criteria for the training process are based on the error produced by ANN. To determine the error, the absolute percentage error (APE) and the mean absolute percentage error (MAPE) are used, which are defined as,

$$\text{APE} = \frac{|Load_{forecast} - Load_{actual}|}{Load_{actual}} \times 100\% \quad (2.4)$$

$$\text{MAPE} = \frac{1}{N_h} \sum_{N_h} \text{APE} \quad (2.5)$$

where N_h is the number of hours in the forecasting period.

Each model is trained with its training sets for a certain amount of epoch (iterations). After the maximum number of epochs is reached, the model is tested by the training set. Based on test results, APE and MAPE can be calculated. If the calculated MAPE is higher than 3%, another training must be done. This process continues until all MAPE from test results are below 3%. Table 2.4 shows the approximate time for training.

Table 2.4 Approximate Training Time for Each Model

Model	Approximate Training Time
Summer	1.0 hours
Spring/Fall	2.0 hours
Winter	1.5 hours

SHORT-TERM LOAD FORECASTING

According to Table 2.4, the required time for each model to reach the 3% MAPE is not the same. One reason is the size of the training vectors, which is easy to understand. Another reason may be the temperature difference in each model. A larger difference in temperature would make it harder for ANN to recognize the pattern.

2.4.3 ANN Models for Comparison

Another ANN model (referred to as the Alt1 model) is considered for comparison. The Alt1 model, depicted in Figure 2.15, is similar to the proposed architecture, except that it uses one ANN module for all seasons.

Alt1 has 73 neurons in the input layer, 90 neurons in the hidden layer, and 24 neurons in the output layer. The number of neurons used in Alt1 is the same as that in the proposed architecture. This is done intentionally for comparison. In using the same number of neurons in both architectures we can learn whether the partitioning method of the proposed architecture would have a better performance compared with the performance of the Alt1 model. The same method used for training the proposed architecture is applied to the Alt1 model.

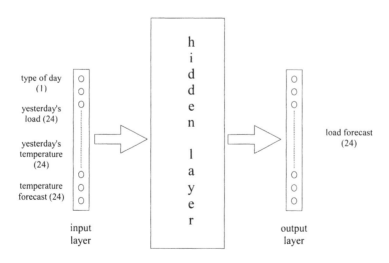

Figure 2.15 Alt1 Model

Another difference between Alt1 and the proposed architecture is the use of training data. In the proposed architecture, data are divided into three seasonal sets according to the partitioning method. Since Alt1 will only use one network, it is not necessary to divide the data. The three training sets used in the proposed architecture will be combined into one training set for Alt1. Alt1 will use the same test sets as that of the proposed architecture.

A third model is built based on [Kho97] (referred to as the Alt2 model shown in Figure 2.16). There are some similarities in input used between this model and the proposed architecture: both use previous day's load, previous day's temperature and temperature forecast. The only difference in input is that in Alt2, days of the week are represented with seven neurons instead of one neuron used in the proposed architecture.

Alt2 comprises two networks, Base Load Forecast (BLF) and Change Load Forecast (CLF). BLF is intended to forecast the regular load for the next 24 hours, while CLF predicts the change of load for 24 hours. The final forecast is obtained by a linear combination of BLF and CLF forecasts.

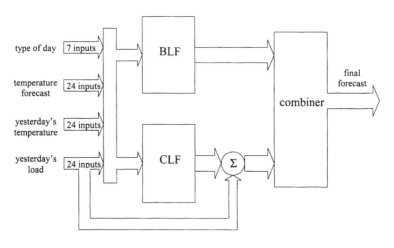

Figure 2.16 Alt2 model

SHORT-TERM LOAD FORECASTING

2.4.4 Performance of One-Day Forecast

This section will present results for one-day forecast. Each test set (June, September and December) has 28 input and output vectors. These vectors represent load and temperature data for the days of the week in that particular month. The testing process will follow the same procedure as that applied to the training process. Hourly load and temperature data from the previous day are given to input layer neurons. The next day temperature is given as forecast, and the next day's load is compared with the forecasted load to calculate APE. After the training process is done, ANN is validated with the test set. MAPE for each daily forecast is calculated and then compared with that of Alt1. Besides, the APE for the daily peak load and the maximum error are also calculated. Each seasonal model is validated with a test set from the same season.

Winter Result. Figure 2.17 shows the winter results for the proposed model. Table 2.5 compares the winter results of the proposed models, Alt1 and Alt2 models.

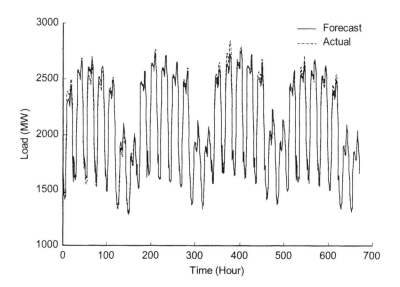

Figure 2.17 Winter Forecast and Actual Load for the Proposed Architecture

As Table 2.5 shows the winter model can predict the December load which is higher than the usual. On the third Tuesday of December, when the yearly peak load occurs, the maximum APE of the proposed model is 4.57%. This level of error could be expected since this amount of load is not in the training vector. Nevertheless, the proposed model can still make good predictions since the MAPE for this day is only 2.72%.

Table 2.5 Winter Model Performance

Week & Day		MAPE (%)			Peak Load APE (%)			Max APE (%)		
		Proposed	Alt1	Alt2	Proposed	Alt1	Alt2	Proposed	Alt1	Alt2
49	Mon	3.02	2.86	2.72	2.58	3.84	3.05	5.44	7.24	8.26
	Tue	1.03	2.16	1.43	0.59	1.86	0.35	2.92	6.10	5.36
	Wed	3.35	2.47	1.66	2.52	2.93	0.37	4.84	6.51	6.11
	Thu	2.14	1.92	2.02	1.46	0.49	0.31	3.41	5.63	7.48
	Fri	0.78	4.39	2.47	1.15	5.30	0.29	1.57	8.43	7.63
	Sat	2.46	8.01	3.29	1.55	7.92	1.40	5.23	27.69	8.44
	Sun	1.79	3.57	2.80	1.52	0.95	0.24	4.16	8.09	8.19
50	Mon	0.30	2.99	3.37	0.10	1.88	1.41	0.63	7.79	8.66
	Tue	0.72	2.17	2.90	1.51	0.42	0.27	1.78	5.68	9.28
	Wed	0.37	8.02	2.46	0.18	9.55	0.11	1.12	10.06	8.32
	Thu	0.33	3.39	2.01	0.18	3.25	0.21	0.97	7.99	8.16
	Fri	1.05	5.31	1.48	1.25	4.82	0.28	2.02	11.54	6.21
	Sat	0.50	5.14	2.43	0.22	4.84	1.04	1.69	12.24	6.24
	Sun	0.48	1.26	3.06	0.32	0.37	2.24	1.36	3.96	9.04
51	Mon	1.30	5.47	3.18	2.43	4.09	4.80	2.70	7.47	6.62
	Tue	2.72	1.64	2.15	4.47	2.55	1.59	4.57	5.03	7.27
	Wed	0.78	1.90	1.55	1.56	1.19	0.67	1.56	7.32	5.57
	Thu	0.58	2.93	2.13	1.01	2.86	0.74	1.59	7.70	8.03
	Fri	1.53	5.18	2.11	2.11	5.25	0.35	2.72	9.91	7.71
	Sat	1.36	3.99	4.07	1.92	4.55	0.83	2.86	10.52	8.63
	Sun	0.54	1.97	4.00	0.26	0.16	0.04	1.59	6.18	8.81
52	Mon	1.01	4.47	2.94	1.31	5.48	0.37	2.43	7.14	6.40
	Tue	1.79	1.61	2.18	2.48	0.46	0.30	2.58	5.42	6.90
	Wed	0.86	2.12	2.14	0.69	2.54	0.70	1.77	4.49	6.82
	Thu	1.24	1.43	2.21	0.68	1.11	0.21	2.61	4.31	7.04
	Fri	1.60	4.89	2.36	2.15	3.36	0.79	2.60	9.49	6.38
	Sat	0.41	2.78	2.85	0.12	1.24	2.65	1.59	7.00	7.45
	Sun	0.34	1.58	2.35	0.27	3.99	3.45	0.99	4.90	5.94
Average		1.23	3.42	2.51	1.30	3.12	1.04	2.47	8.06	7.39

The APE for daily peak load of the proposed model is within the range of 0.10% to 4.47 %, which indicates that the proposed model can forecast higher peak loads. Daily maximum APE of the proposed model

SHORT-TERM LOAD FORECASTING

falls in the range of 0.63% to 5.44%. The worst weather condition occurs on Sunday of the week 51 with a temperature of -0.5° F. The daily MAPE of the proposed model on that day is only 0.54%, which verifies the accuracy of the proposed model.

The results of Alt1 and Alt2 models are not as promising. The MAPE of Alt2 is between 1.43% and 4.07%, while for Alt1, it is between 1.26% and 8.02%. The maximum APE of Alt2 is between 5.36% and 9.28% and that of Alt1 is between 3.96% and 27.96%. On the coldest day of the year (i.e., Sunday of week 51), Alt1 has an MAPE of 1.97% and Alt2 has an MAPE of 4%.

Spring/Fall Results. Figure 2.18 shows the spring/fall result of the proposed model. Table 2.6 compares the spring/fall results of the proposed, Alt1 and Alt2 models.

The loads for September are used as a test case since the minimum load of the year occurs during the weekend of the third week of September. On Saturday and Sunday of the week, the spring/fall model could achieve 0.78% and 1.68% of MAPE, indicating that the spring/fall model could forecast the smallest load during the year. In the same two days, Alt1 has MAPEs of 1.78% and 1.95%, respectively, and Alt2 has MAPEs of 1.14% and 1.19%, respectively.

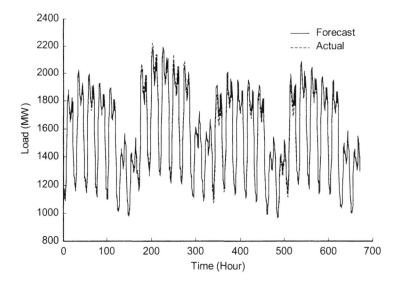

Figure 2.18 Spring/Fall Forecast and Actual Load for the Proposed Architecture

The MAPE of the proposed model is in the range of 0.51% to 3.30%. The MAPE of the Alt1 model is in the range of 1.51% and 2.92%, and for Alt2, it is between 0.65% and 5.88%. The APE for the daily peak load of the proposed model is within the range of 0.08% to 4.11%, and the highest occurs during the Monday of week 38. In comparison, Alt1 has a peak load APE between 1.03% and 5.11%, and for Alt2 it is between 0.07 and 9.77%.

The daily maximum APE of the proposed model is in the range of 1.77% to 5.57%, and the highest occurs during the Tuesday of week 38. The maximum APE of Alt1 is between 1.36% and 5.27%, and for Alt1, it is between 2.05% and 15.30%.

Table 2.6 Spring/Fall Model Performance

Week & Day		MAPE (%)			Peak Load APE (%)			Max APE (%)		
		Proposed	Alt1	Alt2	Proposed	Alt1	Alt2	Proposed	Alt1	Alt2
36	Mon	1.01	2.01	5.80	0.67	1.67	5.20	4.10	2.87	12.67
	Tue	0.58	1.58	3.15	0.30	1.30	3.62	4.30	1.98	6.92
	Wed	1.14	1.50	1.14	1.15	2.15	0.34	3.59	1.52	3.41
	Thu	0.60	1.60	0.98	0.08	1.08	1.86	2.55	2.76	2.05
	Fri	0.51	1.51	0.73	0.16	1.16	0.20	2.26	1.62	1.80
	Sat	1.01	2.01	1.23	0.95	1.42	1.89	1.82	3.35	2.45
	Sun	1.20	2.20	1.55	1.67	2.82	0.61	1.77	3.20	4.10
37	Mon	2.09	2.09	5.88	1.70	2.70	5.93	1.63	3.59	7.10
	Tue	2.02	2.48	1.82	1.91	2.91	0.63	2.78	2.98	3.35
	Wed	0.80	1.80	1.59	0.58	1.58	1.39	2.12	1.77	3.09
	Thu	2.15	2.60	1.67	1.90	2.90	1.28	1.97	3.12	3.56
	Fri	1.39	1.86	1.59	1.97	2.97	2.15	3.02	2.51	3.12
	Sat	0.78	1.78	2.23	0.80	1.42	1.82	2.69	1.88	5.06
	Sun	0.76	1.76	1.19	0.24	1.09	0.22	2.36	2.78	5.07
38	Mon	3.30	2.13	1.54	4.11	5.11	9.77	4.20	5.27	15.30
	Tue	1.74	2.74	0.82	1.35	2.35	0.45	5.57	2.90	2.40
	Wed	0.51	1.51	0.76	0.47	1.47	1.28	2.56	1.36	2.71
	Thu	1.92	2.92	0.95	2.77	3.77	1.34	2.59	3.22	2.63
	Fri	1.59	2.59	0.90	2.81	3.81	0.07	3.72	3.04	2.83
	Sat	0.78	1.78	1.14	0.84	1.03	0.24	3.86	2.44	4.43
	Sun	1.68	1.95	1.19	2.00	2.54	1.00	2.59	2.89	4.15
39	Mon	2.01	2.39	1.26	3.99	4.99	0.96	3.43	5.05	2.17
	Tue	0.90	1.90	0.67	1.19	2.19	0.17	3.58	1.95	1.81
	Wed	0.99	1.99	0.65	0.99	1.99	1.66	2.77	1.88	1.66
	Thu	0.76	1.76	0.73	0.63	1.63	1.01	3.61	1.74	2.51
	Fri	0.68	1.68	0.94	0.81	1.81	0.65	3.60	1.75	2.64
	Sat	0.64	1.64	1.87	0.42	2.07	1.13	2.59	2.10	4.46
	Sun	1.00	2.00	2.15	1.66	2.01	1.32	1.99	2.09	3.35
Average		1.34	1.99	1.65	1.36	2.28	1.72	2.97	2.63	4.17

SHORT-TERM LOAD FORECASTING

Summer Results. Figure 2.19 shows the summer result of the proposed model. Table 2.7 compares the spring/fall results of the proposed model, Alt1 model and the Alt2 model.

The test set is the hourly load from the month of June. This month is chosen because the second highest load during the year occurs on week 23 or the first week of June. During this period, the proposed model has good accuracy with a maximum 1.88% MAPE.

Alt1 has an MAPE in the range of 1.06% and 6.74%. Alt2 has an MAPE in the range of 1.31% and 4.73%. In comparison, the proposed model has an MAPE in the range of 1.27% and 2.43%, which is lower than the other two.

The highest APE for the daily peak load of the proposed model occurs on Friday of week 25, which is 2.87%. While Alt1 has an APE between 0.18% and 11.02% and Alt2 has an APE between 1.01% and 4.57%.

The maximum APE of the proposed model occurs on Friday of week 24, which is 3.36%. While the maximum APE of Alt1 is in the range of 2.05% to 8.51% and that of Alt2 is between 1.71% and 7.87%.

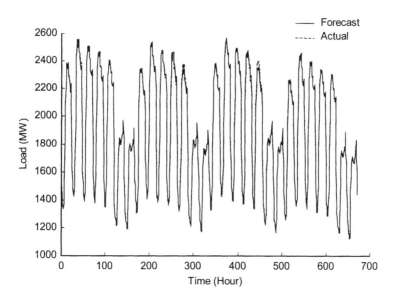

Figure 2.19 Summer Forecast and Actual Load for the Proposed Architecture

Table 2.7 Summer Model Performance

Week & Day		MAPE (%)			Peak Load APE (%)			Max APE (%)		
		Proposed	Alt1	Alt2	Proposed	Alt1	Alt2	Proposed	Alt1	Alt2
23	Mon	1.57	3.75	4.73	1.60	5.43	4.44	2.09	7.85	7.87
	Tue	1.88	2.22	1.36	2.19	4.68	1.09	2.48	7.28	2.01
	Wed	1.56	6.74	1.50	1.65	1.89	1.22	2.14	8.51	2.39
	Thu	1.39	3.99	1.43	1.37	9.98	1.12	2.60	5.79	1.99
	Fri	1.27	3.82	1.47	1.06	10.43	1.29	1.86	6.49	2.36
	Sat	1.59	5.15	2.05	1.43	9.87	1.01	2.46	7.23	3.78
	Sun	1.49	3.81	1.94	1.47	11.02	1.95	2.59	5.24	3.30
24	Mon	1.45	5.71	2.07	1.61	7.57	1.12	2.13	7.43	4.01
	Tue	1.44	1.66	1.35	1.17	4.56	1.28	2.90	3.39	2.21
	Wed	1.51	1.60	1.58	1.22	1.01	1.52	2.44	3.27	2.40
	Thu	2.36	4.96	1.32	2.58	3.52	1.52	3.13	6.96	2.10
	Fri	2.22	2.81	1.44	1.97	2.78	1.05	3.36	5.32	2.11
	Sat	1.58	1.08	1.47	1.19	4.71	1.72	2.19	2.05	2.13
	Sun	1.48	2.50	1.47	1.39	2.80	1.34	1.95	3.08	2.31
25	Mon	1.48	2.13	1.99	1.44	0.18	1.58	2.39	2.89	3.37
	Tue	1.47	1.06	1.77	1.67	2.96	1.61	2.03	3.68	2.53
	Wed	1.54	4.40	1.31	1.68	0.90	1.34	2.32	6.45	2.09
	Thu	1.98	1.70	1.38	2.39	3.00	1.56	3.15	2.48	2.12
	Fri	2.43	3.45	1.32	2.87	6.93	1.35	3.35	5.17	1.71
	Sat	1.73	4.42	1.63	1.11	7.06	1.62	2.40	5.05	2.64
	Sun	1.52	4.02	1.58	1.73	7.66	2.30	2.88	4.63	2.30
26	Mon	1.69	4.22	4.27	1.65	0.22	4.57	3.02	6.95	4.57
	Tue	1.44	3.15	1.48	1.39	3.83	1.78	1.99	6.29	2.45
	Wed	1.27	3.39	1.39	1.00	0.18	1.20	1.63	3.39	1.99
	Thu	1.29	5.87	1.75	1.85	1.94	2.30	2.06	3.84	3.22
	Fri	1.43	3.62	1.95	1.02	6.28	1.54	2.37	5.02	3.59
	Sat	1.40	3.61	2.22	1.11	6.63	2.54	2.46	4.05	3.41
	Sun	1.70	1.98	2.45	1.73	8.89	2.23	3.18	4.98	2.78
Average		1.61	3.46	1.84	1.59	4.89	1.76	2.48	5.17	2.85

The highest temperature for this month was $84°F$, which occurred on Wednesday of week 25. On this day, the MAPE of the proposed model was 1.54%, and the MAPEs of Alt1 and Alt2 were 4.4% and 1.31%, respectively. From this result, we learn that Alt1 is not capable of making an accurate forecast in extreme weather conditions. The best result is given by Alt2, while the proposed model shows a reasonably good performance.

Comments. Based on the test results, the proposed architecture shows a satisfactory performance for all seasons and outperforms both Alt1 and

Alt2 models. There are significant differences in peak load APE and maximum APE between the proposed model and the alternative models.

- The highest peak load APE of Alt1 is 11.02% on Sunday of week 23, and for the Alt2 model, the highest peak load APE occurs on Monday of week 38, which is 9.77%. These numbers are high compared with 4.47% by the proposed architecture on Tuesday of week 51.
- For the Alt1 model, the maximum APE can reach 27.69% on Saturday of week 49, and for the Alt2 model, the maximum APE occurs on Monday of week 38, which is 15.30%. These are very high compared with the maximum APE of the proposed architecture, which is only 5.57% on Tuesday of week 38.

2.4.5 Performance of Multiple-Day Forecast

The proposed model's capability in forecasting the load up to seven days is described in this section. As was mentioned earlier, a forecast for more than one day is possible by using a single day's output as input for the following day. In this case, hourly temperature data for all forecasting days has to be available.

Figure 2.20 shows 14-day forecast results for spring and Figure 2.21 shows the MAPE.

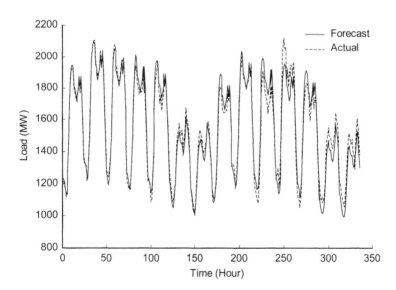

Figure 2.20 Spring Model's 14-Day Forecast

According to Figure 2.21, MAPE increases as we increase the number of forecasting days. The additional error stems from the cumulative daily load forecasting error. According to Figure 2.21, it is reasonable to forecast up to seven days as the MAPE of the first seven days can still be tolerated. However, the MAPE beyond seven days will be quite high.

Table 2.8 shows the MAPE analysis for the 7-day forecast in three seasons. From Table 2.8, for forecasts up to seven days, the proposed architecture still performs very well. MAPE during this test is between 0.83% and 4.01%. The maximum MAPE for this test is slightly higher than the 3% standard used for the one-day forecast. However, this amount of MAPE can still be tolerated.

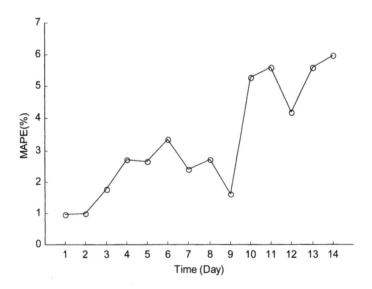

Figure 2.21 MAPE for 14-Day Forecast in Spring

Table 2.8 MAPE for 7-Day Forecast

Season	Monday	Tuesday	Wednesday	Thursday	Friday	Saturday	Sunday
Winter	0.83	1.58	1.97	1.10	1.67	2.57	4.01
Spring	2.04	1.46	2.54	2.05	2.57	2.45	3.03
Summer	0.97	0.98	1.77	2.72	2.64	3.35	2.40

SHORT-TERM LOAD FORECASTING

2.5 SENSITIVITY ANALYSIS

The main purpose of sensitivity analysis is to study the impact of various factors, especially the weather factor on the proposed architecture for load forecasting. Here, load data from week 10 to week 17 and from week 40 to week 43 are selected for analyses.

2.5.1 Possible Models

Depending on input factors, there are five possible models as listed in Table 2.9. In this table, time information is considered in all the models. Model 1 considers load. Model 2 considers load and temperature. Model 3 considers load, temperature, and humidity. Model 4 considers load, temperature, humidity, and wind. Model 5 considers all these factors and some implicitly, when sufficient data are not available, or when considering all the factors in the ANN would require an excessive amount of training time. The corresponding numbers of layers in different models are shown in Table 2.10.

Table 2.9 Factors Considered for Different Models

Factors	Model 1	Model 2	Model 3	Model 4	Model 5
Type of day	*	*	*	*	*
Load of Previous day	*	*	*	*	*
Temperature of Previous day		*	*	*	*
Temperature of Forecast day		*	*	*	*
Humidity of Previous day			*	*	* (Implicitly)
Humidity of Forecast day			*	*	* (Implicitly)
Wind of Previous day				*	* (Implicitly)
Wind of Forecast day				*	* (Implicitly)

Table 2.10 Typical Number of Layers for Different Models

Layer	Model 1	Model 2	Model 3	Model 4	Model 5
Input	25	73	121	169	73
Hidden	50	90	150	200	90
Output	24	24	24	24	24

2.5.2 Sensitivity to Input Factors

The $\varepsilon-\delta$ method is employed in the sensitivity analyses of models 2 and 4. The idea behind the $\varepsilon-\delta$ method is that if we perturb the ANN input by a small amount (ε), we can analyze the sensitivity of ANN, which represents the input–output mapping, to the input by checking the variation of the output (δ).

Table 2.11 shows the results of Model 2 and Table 2.12 shows the results of model 4, in which MAPE is used to measure the forecasting accuracy. For instance, 0.623% in Table 2.11 corresponds to MAPE when the previous day's load is preturbed by 1%. From Table 2.11, we learn that:

- The larger the variation of an input factor, the larger is the variation of MAPE.
- Historical loads have a larger impact on MAPE than historical and forecasting temperatures.
- If the input perturbation were limited to 5%, the change in MAPE would be relatively small and limited to 2.76%. This highest MAPE occurs due to the perturbation of all factors.
- The impact of the historical load is very similar to that of all factors. This point reveals that the load forecast is dominated by the historical load.

Table 2.11 MAPEs of Model 2

	Magnitude of Variation					
	0	1%	2%	3%	4%	5%
Load of previous day	0.374	0.623	1.129	1.670	2.213	2.754
Temperature of previous day	0.374	0.526	0.799	1.104	1.427	1.758
Temperature of forecast day	0.374	0.511	0.763	1.045	1.343	1.654
All the above	0.374	0.625	1.131	1.672	2.216	2.760

In model 4, all weather information is considered as an input to ANN. Based on Table 2.12, we learn that:

- The larger the variation of a factor, the larger is the MAPE.
- Historical load will have the largest impact on forecasting. Temperature will have a larger impact than humidity and wind. The impact of humidity is similar to that of wind.

SHORT-TERM LOAD FORECASTING

- If the variation is limited to 5%, the change in MAPE will be relatively small and limited to 2.214%, which occurs at variation of history load.

- The variation of all factors has a smaller impact than the variation of only history load. This means that a variation of some weather factors may cancel out other's impact.

Table 2.12 MAPEs of Model 4

	Magnitude of Variation					
	0	1%	2%	3%	4%	5%
Load of previous day	0.298	0.499	0.896	1.331	1.772	2.214
Temperature of previous day	0.298	0.358	0.509	0.705	0.920	1.142
Temperature of forecast day	0.298	0.340	0.450	0.599	0.766	0.945
Humidity of previous day	0.298	0.305	0.325	0.356	0.396	0.443
Humidity of forecast day	0.298	0.308	0.337	0.379	0.429	0.485
Wind of previous day	0.298	0.302	0.313	0.331	0.353	0.379
Wind of forecast day	0.298	0.307	0.332	0.367	0.410	0.458
All the above	0.298	0.460	0.797	1.179	1.568	1.959

Comparison of Models 2 and 4. Comparing Tables 2.11 and 2.12, we learn that:

- Variation of factors in model 4 will have a smaller impact on forecasting accuracy. This means that a more realistic model for the load forecast, with more factors included, would provide better performance.

- As long as the variation of a factor is small, the accuracy of the load forecast will not be compromised much. This result supports the notion of using similar-day historical information for forecasting.

2.5.3 Inclusion of Temperature Implicitly

Using model 4, we learned that the inclusion of more factors could improve the forecasting performance. However, it will also take more time to train ANN, and this time could increases exponentially. We also note that the temperature is the dominant weather factor. So we chose to include all weather information in which temperature is considered as a function of humidity and wind in training the ANN. That is, we use an effective temperature to replace the original temperature to represent the impact of humidity and wind inclusively. The effective temperature [Kho98] is defined as

$$Te = \begin{cases} T + \lambda \dfrac{H(T-75)}{100}, & T > 75 \\ T, & 75 \geq T \geq 65 \\ T - \rho \dfrac{W(65-T)}{100}, & T < 65 \end{cases} \quad (2.6)$$

where H is relative humidity, W is wind speed, λ and ρ are effective coefficients that include the effects of humidity and wind. For our example, λ is 0.5 and ρ is 1. Table 2.13 shows the results of model 5.

Table 2.13 MAPEs of Model 5

	Magnitude of Variation					
	0	1%	2%	3%	4%	5%
Load of previous day	0.363	0.596	0.998	1.449	1.908	2.371
Temperature of previous day	0.363	0.439	0.513	0.616	0.734	0.860
Temperature of forecast day	0.363	0.448	0.537	0.654	0.791	0.925
All the above	0.363	0.587	0.969	1.406	1.849	2.295

Comparing Tables 2.13, 2.11, and 2.12, we learn that:

- Model 5 has a better performance than model 2. It takes the same amount of time to train models 2 and 5.
- The performance of model 4 is better than that of Model 5. However, it takes much less to train model 5.
- By replacing temperature with the effective temperature, we consider the impact of humidity and wind implicitly. The key is selecting λ and ρ. The model in (2.6) is one of the models that can be altered depending on the power system.

Chapter 3

Electricity Price Forecasting

3.1 INTRODUCTION

With the introduction of restructuring into the electric power industry, the price of electricity has become the focus of all activities in the power market. Based on its specific application, price forecasting can be categorized into short-term (few days), mid-term (few months) and long-term (few years). In this chapter, we are mainly interested in short-term price forecasting (STPF) of electricity in the restructured power market.

Price forecasting has long been at the center of intense studies in other commodity markets like stocks and agriculture [Mal94, Yoo91, Ati97, Bab92, Sny92]. In recent years, electricity has also been traded as a commodity in various markets. However, electricity has distinct characteristics from other commodities. For example, electricity cannot be stored economically, and transmission congestion may prevent a free exchange among control areas. Thus, electricity price movements can exhibit a major volatility, and the application of forecasting methods prevailed in other commodity markets, can pose a large error in forecasting the price of electricity.

Power engineers have been familiar with load forecasting for some time. With the introduction of restructuring in the electric power industry, more literature has been devoted to price forecasting [Bas99, Kor98, Wan97, Szk99]. Among the proposed methods, some are presumed to be too complex to implement [Bas99] (e.g., simulation method) or too simple to yield sufficient accuracy [Kor98] (e.g., modular time series analysis employing heuristic logic). Among the proposed methods, the ANN method provides a simple and powerful tool for forecasting in practical systems [Wan97, Szk99]. However, the forecasting accuracy is still an issue which could be due to the limited number of physical factors considered in price forecasting.

It is apparent that many physical factors could impact the electricity price in which some factors are more dominant than others. So it is imperative to select the factors that affect the price forecasting intensely in short-term applications. We perform sensitivity studies based on the simulation to explore physical factors that are dominating the price forecasting process.

As for the measure of forecasting accuracy, the most widely used index is the mean absolute percentage error (MAPE). However, when used in its present form for electricity price forecasting, MAPE could create errors because of the perceived behavior of electricity price. For instance, the price of electricity can rise to tens of or even hundreds of times its normal value during some periods, or drop to zero and even to negative values in other periods. Accordingly, MAPE could reach the extreme values in those cases. So, a more reasonable measure for MAPE is discussed in this chapter for considering the special behavior of electricity price.

Price spikes are distinctive aspects of electricity which could impact the forecasting accuracy. In practice, it is difficult to forecast accurate price spikes which can occur when the load level in a system approaches its generating capacity limit. So it is useful to study the probability of price spikes under different load levels and, further, the probability distribution of prices under different load levels.

In this chapter, we will introduce a comprehensive framework for price forecasting, denoted as *ForePrice*. It has four functional modules: price simulation, price forecasting by ANN, performance analysis, and volatility analysis as summarized below:

Price Simulation Module. *ForePrice* can create a detailed price curve in simulating an actual system dispatch constrained with the system operating requirements. A sensitivity analysis can identify potential price drivers such as line limits, line outages, generator outages, load patterns, and bidding patterns.

Price Forecasting Module. Taking the price simulation results, *ForePrice* selects the most influential price drivers and establishes a relationship between these price drivers and the electricity price via ANN. *ForePrice* uses an adaptive scheme to adjust the parameters of ANN with the latest available data. *ForePrice* employs different data pre-processing techniques to improve the qualities of available raw data which are fed into ANN. *ForePrice* automatically decides how much historical data are necessary to achieve the most accurate forecasting. According to price forecasts provided by *ForePrice*, one can perform price-based unit commitment to optimize the generation resources for achieving the maximum profit.

Performance Analysis Module. This module defines a reasonable error analysis index to evaluate the forecasting performance. It compares the price forecasting results based on ANN and alternative techniques. The alternative techniques are based on linear and non-linear interpolations as well as brute force methods for forecasting the electricity price. The results are presented in this chapter for practical data.

Volatility Analysis Module. In restructured power market, a single-point price forecast that refers to forecasting the expected value may not be sufficient for system analyses. Volatility analysis, the most distinct feature of *ForePrice*, provides more insight on price forecasting and extends the application of price forecasting. In the volatility analysis module, the probability of a price spike is analyzed based on different load levels and different price forecast levels. In addition, the volatility analysis module analyzes the probability distribution of an electricity price using both the statistical method and the ANN method. The statistical method studies the probability distribution of price at different load levels and time periods based on historical data analyses. The ANN method studies the probability distribution of price by considering the probability distribution of price drivers such as load. In utilizing the distribution of price and its spikes, engineers and marketers can perform generation asset valuations, risk management, and option valuations as discussed in this book.

The proposed framework has the following features:

- An actual dispatch simulation with system operating requirements and constraints to provide detailed insights on price movement.

- Potential price driver identification, such as line limits, line outages, generator outages, load patterns, and bidding patterns.

- Probability analysis of electricity price spikes to provide more insight on price forecasting.

- Probability analysis of distribution of electricity price at different load levels and time periods with applications to generation asset valuation, risk management, and option valuation.

- ANN analyses to establish the non-trivial relationships between price drivers and electricity price.

- Adaptive updating of ANN's parameters to reflect the latest available data.
- Enhanced forecasting accuracy due to improved quality of raw data with different data pre-processing techniques.
- Automatic determination of quantity of necessary historical data for achieving the best forecasting accuracy.
- A more reasonable error analysis index for evaluating the forecasting performance.

3.2 ISSUES OF ELECTRICITY PRICING AND FORECASTING

3.2.1 Electricity Price Basics

In a power market, the price of electricity is the most important signal to all market participants and the most basic pricing concept is market-clearing price (MCP). Generally, when there is no transmission congestion, MCP is the only price for the entire system. However, when there is congestion, the zonal market clearing price (ZMCP) or the locational marginal price (LMP) could be employed. ZMCP may be different for various zones, but it is the same within a zone. LMP can be different for different buses.

MCP Calculation. After receiving bids, ISO aggregates the supply bids into a supply curve (*S*) and aggregates the demand bids into a demand curve (*D*). In Figure 3.1, the intersection of (*S*) and (*D*) is the MCP.

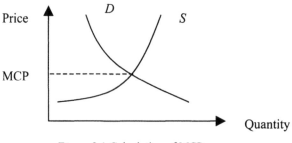

Figure 3.1 Calculation of MCP

ELECTRICITY PRICE FORECASTING

ZMCP Calculation. If at a given period, the ISO detects congestion along any transmission paths, it will adjust its zonal schedules at the two ends of each path to relieve the congestion. Accordingly, the MCPs in the two regions could be different which are denoted as zonal MCP (or ZMCP). Using ZMCP, we calculate the congestion charge (or usage charge) for each congested transmission path across that path. See Chapter 10 for more information on congestion charges.

LMP Calculation. LMP is the cost of supplying the next MW of load at a specific location, after considering the generation marginal cost, cost of transmission congestion, and losses [Pjm01]. That is, LMP is the sum of generation marginal cost, transmission congestion cost, and cost of marginal losses, although the cost of losses is usually ignored. When there is no congestion, LMP is the same as MCP. When there is congestion, the optimal power flow (OPF) solution considers transmission line constraints in order to balance supply and demand at each bus. The marginal cost of each bus is the LMP.

3.2.2 Electricity Price Volatility

The most distinct property of electricity is its volatility. Volatility is the measure of change in the price of electricity over a given period of time. It is often expressed as a percentage and computed as the annualized standard deviation of percentage change in the daily price (other prices such as weekly or monthly prices can also be used). Compared with load, the price of electricity in a restructured power market is much more volatile. Figure 3.2 displays load and price curves for the PJM power market from 1/1/99 to 1/14/99 [Pjm01]. From these curves, we learn that:

- The load curve is relatively homogeneous and its variations are cyclic.
- The price curve is non-homogeneous and its variations show a little cyclic property.

Although electricity price is very volatile, it is not regarded as random. Hence, it is possible to identify certain patterns and rules pertaining to market volatility. For example, transmission congestion usually incurs a price spike which is not sustained as electricity price would revert to a more reasonable level (this is known as mean reversion in statistics). It is conceivable to use historical prices to forecast electricity prices. Accordingly, we use a training scheme to capture perceived patterns for forecasting electricity prices.

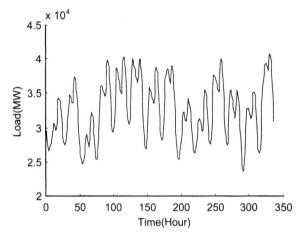

Figure 3.2a Load Curve of PJM Power Market from 1/1/99 to 1/14/99

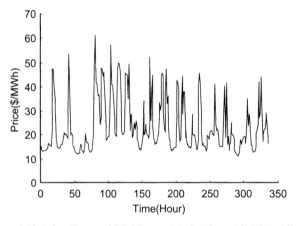

Figure 3.2b Price Curve of PJM Power Market from 1/1/99 to 1/14/99

The fundamental reason for electricity price volatility is that the supply and demand must be matched on a second-by-second basis. Other reasons follow [Cal00a]:

- Volatility in fuel price
- Load uncertainty
- Fluctuations in hydroelectricity production

ELECTRICITY PRICE FORECASTING

- Generation uncertainty (outages)
- Transmission congestion
- Behavior of market participant (based on anticipated price)
- Market manipulation (market power, counterparty risk)

Because of the special properties of electricity, the price of electricity is far more volatile than that of other relatively volatile commodities. The annualized volatility of WTI oil future contracts is reported as 28%; it is 54% for NYMEX Henry Hub natural gas future contracts, while 62% for COB and 57% for Palo Verde are normally noted for electricity future contracts. In electricity spot markets, annualized volatility at COB is about 180%, 300% for MAIN, and 450% for the Power Pool of Alberta [Ene98].

Because of the significant volatility, it is difficult to make an accurate forecast for the spot market of electricity. This is evidenced by the fact that the existing price forecasting accuracy is far lower than that of load forecasting. As of the year 2001, the least error reported for price forecasting was about 10% as compared to a 3% for load forecasting. However, price forecasting accuracy is not as stringent as that of load forecasting.

3.2.3 Categorization of Price Forecasting

We categorize the price forecasting process by duration of time, point of forecasting, and type of customers.

Duration of Time. Basically, there are two kinds of price forecasting based on the duration of time: short-term price forecasting and long-term price forecasting. Short-term forecasting is mainly used to determine GENCOs' bidding strategies in the spot market or to set up bilateral transactions. It is envisioned that more accuracy in forecasting would reduce the risk of under- or over-estimating revenues from generation sales and provide better risk management. Long-term forecasting is usually used for planning, such as determining the future sites of generators; accurate forecasting of electricity prices would enable power marketers and companies to make sound business decisions in a volatile environment [Bas99].

Point of forecasting. There are MCP, ZMCP, and LMP forecasts for the entire system, a specific zone, and a specific bus, respectively.

Type of Customers. The ISO and GENCOs/marketers are two main market participants. They have different goals for price forecasting. In this sense, this category includes ISO price forecasting and GENCO price forecasting.

- Price forecasting for the ISO is the same as determining MCP. That is the case in the National Energy Market (NEM) in Australia and the Power Pool in Albert in Canada. Price forecasting for the ISO is not a true forecasting process because, once the ISO receives the participants' bids, it can calculate the MCP numerically.

- For a GENCO, price forecasting means predicting MCP, ZMCP, or LMP before submitting bids. A GENCO would know very little about other GENCOs and would only have access to the publicly available information, including forecasted load and historical data such as loads and MCP. Because of the limited information, the accuracy of a GENCO price forecasting is not high. However, an accurate estimation of price helps a GENCO determine its bidding strategy or set up bilateral contracts more precisely. A bid closer to MCP would result in a higher income for a GENCO. Besides, a GENCO could exert market power if it could predict the price more accurately, as a bidder who has a generating unit with marginal cost close to the expected MCP could benefit from withholding that generating capacity.

In this chapter, we mainly discuss short-term price forecasting by GENCOs.

3.2.4 Factors Considered in Price Forecasting

The factors considered in electricity price forecasting can be summed as follows:

- Time: Hour of the day, day of the week, month, year, and special days.
- Reserve: Historical and forecasted reserve.
- Price: Historical price.
- Load: Historical and forecasted loads. Load fluctuations could impact price. On the other hand, price fluctuations could impact load values. Thus, load forecasting and price forecasting can be combined into a single forecasting model.

- Fuel price: It is possible to approximate the impact of a gas price on MCP. In California, a 10% increase in the PG&E's gas price caused about 5% increase in MCP [Kle97].

In the following we elaborate on the four modules mentioned earlier in this chapter.

3.3 ELECTRICITY PRICE SIMULATION MODULE

By simulating the electricity price, we intend to study the impact of the physical characteristics of the system, market participants' bidding strategies, and load distribution on prices.

In making the simulation, we mimic an actual dispatch using system operating requirements and constraints. The simulation can yield many insights on price curve following intensive input data. Two commercially known packages for simulation are MAPS and UPLAN-E. The simulation must include a detailed model of the power systems and a procedure for pricing. Then, based on the model and the procedure, the simulation method sets up mathematical models and solve them for price forecasting.

The following issues must be addressed in the simulation of electricity price [Bas99]:

- Transmission model
- Unit commitment
- Transmission constrained dispatch
- Transmission security dispatch
- Chronological simulation
- Large-scale system simulation capability

The required data are sampled as follows:

Generating Unit Data. Include heat rates at different generating points, generating capacity, maintenance schedule, forced outage rates, environmental factors, emission data, variable operation and maintenance (O&M) cost, fixed O&M cost, minimum down/up time, fuel price, fuel constraints, ramping capability, quick start capacity, and unit ownership.

Transmission Data. Include ac load flow data, generator siting, line ratings, interface limits, voltage and stability limits, and regulating transformers.

Transaction Data. Include transferring capability, routing, and wheeling charges.

Hydro Data. Include hydro unit capacity, unit ownership, minimum rating, available flow of water, dispatch strategy, sequential dam data, maximum pumping rates, and storage tank limits.

Others. Include load forecasts, load curves, non-conforming loads, spinning reserve requirement, inflation pattern, and emergency costs.

3.3.1 A Sample of Simulation Strategies

A power system is comprised of three parts: generation, transmission, and distribution (load). To simulate a power system, all the three parts should be included.

The main generation factors impacting the price are generator availability (e.g., generator outages and generator sites) and generator bidding strategy. The information may not be readily available but can be obtained by reviewing the historical data. For instance, historical data often point out that bidding strategies remain unchanged for days of the week, hours of the day or both. Generator outage is another factor that we should consider in the generation simulation. If the location of a generator within a zone does not impact the price, we may adopt a "zonal generator capacity outage factor" to represent generator outages within a zone.

On the transmission side, forecasting the value of MCP does not require any information on the transmission system. However, for ZMCP and LMP forecasts, we require the transmission system information. The main transmission information factors are network configuration and line limit. Line outages are viewed as changing a network configuration. A power system is divided into zones in which prices are the same or similar within a zone.

We use the load pattern information to describe loads. The pattern could include load distribution factors, load values, and bidding strategies for loads. Usually, we have abundant information on load values. However, detailed information on load distribution factors and bidding strategies may not be available. Similar to generation, bidding strategies for loads may be inferred from historical data. We may use the zonal load to simplify the representation of load distribution.

After simulating a practical power system based on generation, transmission and load, we can simulate prices. Generally, price simulation is an OPF problem, with the objective of minimizing the cost of generation purchases.

3.3.2 Simulation Example

In the following, we simulate the price for the 118-bus system. Table 3.1 shows important inter-zonal lines studied in this example. The power system is depicted in Figure 3.3 with system data given in Appendix D.5. We assume that the limits for other lines in the system will not be violated in our simulation.

The simulation results will show the correlation between prices and price drivers, including line limits, line outages, generator outages, load patterns, and bidding patterns. These sensitivity analyses can provide some insight into how the simulated prices will be impacted as a system undergoes various outages and changes. The results of such sensitivity studies would help marketers and GENCO operators decide on how to operate a system to maximize revenues. In the following, some of the sensitivity results for the example system are discussed.

Table 3.1 Important Inter-zonal Lines of 118-Bus System

Line Index	From Bus	To Bus	R (p.u.)	X (p.u.)	Line Limit (p.u.)
120	75	77	0.0601	0.1999	0.3
128	77	82	0.0298	0.0853	1
148	80	96	0.0356	0.182	0.5
158	98	100	0.0397	0.179	1.5
159	99	100	0.018	0.0813	1

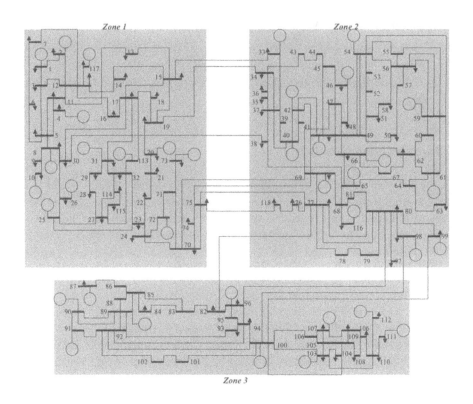

Figure 3.3 One-Line Diagram of 118-Bus System

Figure 3.4 shows the simulation results for prices of bus 69 and bus 75 as the load is changed for various generator outages. In this example, we consider generator 26 and another outaged generator. In Figure 3.4, the behavior of price (vs. load) of bus 69 is homogeneous, while that of bus 75 is non-homogeneous. Now consider more closely the price behavior of bus 75. The uppermost curve represents the base case, where no generator is outaged. In the base case, the price always increases with an increase in system load. The adjacent curves represent cases with generator outages. In those cases, the price first increases with an increase in the system load but then falls with further increase of system load.

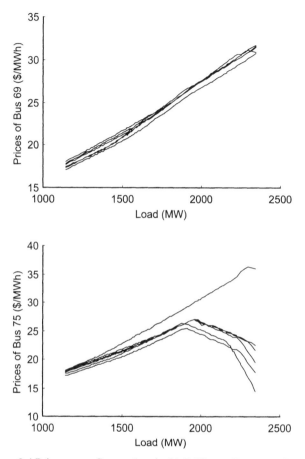

Figure 3.4 Price versus System Load with Different Generator Outages

3.4 PRICE FORECASTING MODULE BASED ON ANN

The classical methods for forecasting include regression and state-space methods. The more modern methods include expert systems, evolutionary programming, fuzzy systems, ANN, and various combinations of these tools. ANN has gained more attention among the existing tools because of its clear model, easy implementation, and good performance. We apply ANN to price forecasting in this section and use MATLAB for training the ANN. ANN provides a very powerful tool for analyzing factors that could influence electricity prices.

In using ANN in price forecasting, we identify parameters that would fit historical data, and with the resulting models predict future electricity prices based on actual inputs. The ANN method is comparatively easy to implement. However, it fails to capture temporal variations (e.g., congestion and contingency). Our ANN input layer includes: time factor (day of the week and hour of the day), load factor (system load and bus load), and line factor (line status, line limit); the output layer represents individual bus prices; the number of neurons in the hidden layer is equal to the average number of neurons in input and output layers.

For a large-scale system, it may be difficult to include all bus and line information since the ANN scale could grow much and the training time could increase significantly. So we must reduce the ANN scale while maintaining a reasonable accuracy by (1) decreasing the number of input neurons, or (2) decreasing the number of input vectors, or (3) both.

To decrease the number of input neurons, we can use zonal loads instead of bus loads. We can also disregard the time factor since input load information would include the information on time. To decrease the number of input vectors, we could divide them into several groups. Suppose that we train the ANN for the first group. If the second group has similarities with the first group (i.e., they have similar ANN outputs), we disregard the second group.

3.4.1 ANN Factors in Price Forecasting

We learned in Section 3.3 that the impact of line limits, line outages, and generator outages on price are homogeneous. In other words, if we have enough historical data, we can employ an ANN to find the relationships between these factors and price. Also, the impact of the load pattern and bidding pattern on price seemed non-homogeneous. Using OPF, we should be able to draw a relationship between price and load pattern using the load distribution.

An index is needed to supply the bidding pattern for training the ANN. One option is to suppose that bidding patterns are dependent on the period of study. If bidding patterns depend on the day of the week, we add a 'day of the week' factor to the input data to represent the pattern. For hours of the day, an 'hour of day' factor is added. For both days of the week and hours of the day, the two factors are added simultaneously. In other words, we introduce the time factor input to the ANN for representing the bidding strategy.

ELECTRICITY PRICE FORECASTING

The larger the system, the smaller is the impact of a generator outage on market prices. For a generator outage representation, we use an outaged zonal generator capacity. However, as we see in sensitivity analyses, the location of generator is also very important. So we use 'adaptive learning' to modify the trained ANN for generator outages.

3.4.1.1 Impact of Transmission Congestion

Congestion can introduce differential bus prices (zones). So predicting the severity of congestion is an important factor in price forecasting. Congestion occurs when a transmission line flow exceeds its limit. So line flow and line limit information together could reveal line flow congestion and its severity. Thus, to find the relationship between congestion and price, we would calculate the effects of line flow and line limit on price.

There are two ways for representing this relationship using ANN. First, we may take line limits and line flows as direct inputs to ANN, as shown in Figure 3.5. The problem may escalate if we have a large number of lines. Hence, we may opt to consider major (e.g., inter-zonal) lines only. Another option would be to define a congestion index that includes line flow and line limit information and is able to convey physical meaning to system behavior. One example is as follows:

$$I_C = \sum_i f(L_i - F_i) \qquad (3.1)$$

where, I_C is the congestion index, L_i is the flow limit of line i and F_i is the flow of line i.

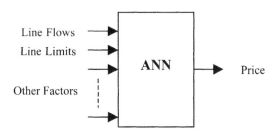

Figure 3.5 ANN Model for Considering Congestion Explicitly

The *f* function is illustrated in Figure 3.6a.

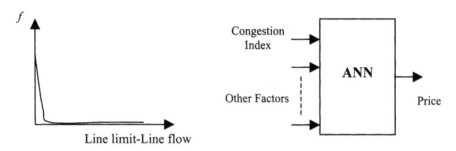

Figure 3.6a Congestion Index   *Figure* 3.6b ANN Model for Congestion Index

Figure 3.6a shows that when a line flow is close to the line limit, the possibility of congestion is high; when the line flow is much less than the line limit, the congestion possibility would be smaller. This index value may be used as an input to ANN as depicted in Figure 3.6b. The difference between the two options depicted in Figures 3.5 and 3.6b is that the latter would only have one input with respect to congestion.

3.4.2 118-Bus System Price Forecasting with ANN

Figure 3.7 depicts the bus price profile for the 118-bus system. Note that many bus prices could represent similar behavior as we consider network alterations (generator line outages, load fluctuations, etc.). So we basically focus our attention on zones or buses with non-homogeneous price fluctuations. For instance, Figure 3.4 showed that the price at bus 75 has non-homogeneous behavior. Hence, for illustration purposes, we primarily study the price at bus 75.

The process for forecasting bus prices based on ANN is as follows:
1. Run the simulation module to determine bus prices (e.g., cases discussed in Section 3.3 for the 118-bus system).
2. Use the sensitivity results based on simulation for training the ANN.
3. Use the trained ANN for forecasting bus prices.

ELECTRICITY PRICE FORECASTING

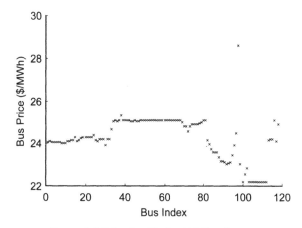

Figure 3.7 Price Profile for 118-Bus System

In the following ANN cases, we follow this procedure to forecast bus prices and study the sensitivity of bus price forecasts for the 118-bus system. We refer to the simulation results as training cases since they are used for training the ANN.

A. Price versus Line Limit. To study the relationship between bus prices and line limits, we consider two cases. In Case A, limits of line 158 are considered to be 0.6 and 0.8 in the training set and 0.7 in the testing set. Case B is similar to Case A except that more samples are included in the training set. In Case B, limits of line 158 include 0.6, 0.65, 0.75, and 0.8. The results are shown in Figures 3.8a and 3.8b for Cases A and B respectively.

Since the difference between simulation prices for line limits of 0.8 and 0.6 is large (Case A), we need to introduce additional ANN training cases (between 0.8 and 0.6) in order to forecast bus prices more accurately (Case B). By adding training sets (i.e., 0.8, 0.75, 0.65, and 0.6 in Case B), we improve the price forecasting accuracy significantly.

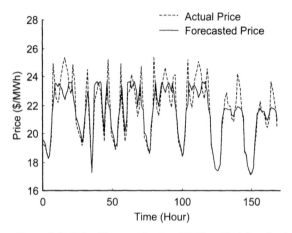

Figure 3.8a Price Forecast for Bus 75 (Few Training Sets)

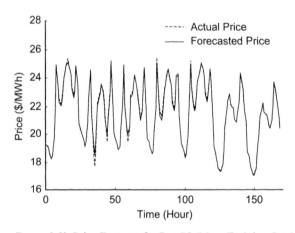

Figure 3.8b Price Forecast for Bus 75 (More Training Sets)

B. Price versus Load Pattern. To study the relationship between price forecasts and load patterns, we consider the four cases presented in Figures 3.9a to 3.9d. The ANN results reveal that the more training data we include, the more accurate the price forecasts become. The definition of MAPE is given in Section 3.5.2. Accordingly, we conclude that:

- If no bus loads are considered in ANN (Case *A*), the MAPE is 1.582%.

ELECTRICITY PRICE FORECASTING

- If bus loads greater than 1.0 (about 10) are included in ANN (Case *B*), the MAPE is 1.025%.

- If all non-zero bus loads (about 81) are included in ANN (Case *C*), the MAPE is 0.903%.

- If the three zonal loads are included in ANN (Case *D*), the MAPE is 1.211%.

For data availability and data accuracy in real applications (i.e., individual bus loads are not easily specified and the bus load information may not be accurate), Case *D* including zonal loads would be a good choice.

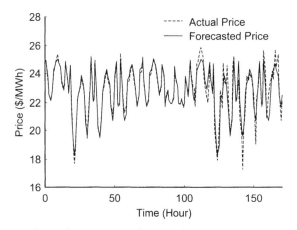

Figure 3.9a Price versus Load Pattern for Bus 75 (No Bus Loads)

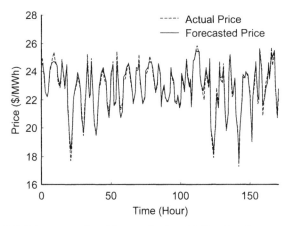

Figure 3.9b Price versus Load Pattern for Bus 75 (Bus Loads Larger than 1.0)

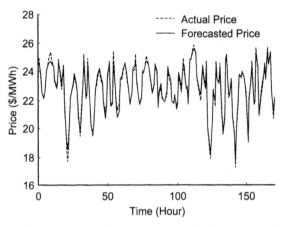

Figure 3.9c Price versus Load Pattern for Bus 75 (All Non-zero Bus Loads)

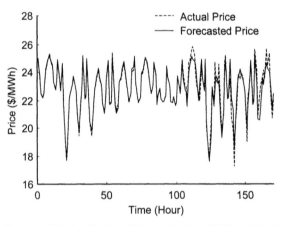

Figure 3.9d Price Vs. Load Pattern for Bus 75 (Zonal Loads)

C. Price versus Outages. It would be impractical to include all line and generator outages in price forecasting. The solution is that for different outage patterns, to employ different ANNs and select a relevant ANN for forecasting when the related outage information is made available. In practice, one needs to train the ANN for the most frequent and the most important outage cases.

3.5 PERFORMANCE EVALUATION OF PRICE FORECASTING

To evaluate the performance of an ANN module, we compare its forecasts with those of alternative methods. The alternative methods are given as follows.

3.5.1 Alternative Methods

3.5.1.1 Alternative Method 1 (AM1)

Let i be the index of day and t be the index of hour. Then,

$$Price(i,t) = \frac{Load(i,t)}{Load(i-1,t)} \times Price(i-1,t) \quad (3.2)$$

3.5.1.2 Alternative Method 2 (AM2)

We would match the load profile of the forecasting day with historical load profiles of previous days to find N similar days with matching load profiles. Then,

$$Price(i,t) = \frac{1}{N} \sum_{j=1}^{N} k \times Price(j,t) \quad (3.3)$$

Let L be the forecasted load (for one day) and HL be the historical load (for one day). The basic idea of AM2 is to find similar days in the sense of load profile and use prices of these similar days to forecast price. Suppose that we find a linear relationship to exist between L and HL, say $HL = k \times L + b$. Note that when $b = 0$, we say that L and HL are similar (presumably, k would be between 0.5 and 2). So we use the difference between HL and $k \times L$ to define similarity.

$$S' = \sqrt{\sum_{i=1}^{24}(HL_i - k \times L_i)^2} \quad (3.4)$$

To normalize, we have

$$S = \sqrt{\sum_{i=1}^{24}\left(\frac{HL_i - k \times L_i}{k \times L_i}\right)^2} \quad (3.5)$$

Substitute $HL_i = k \times L_i + b$ into (3.5):

$$S = \sqrt{\sum_{i=1}^{24}\left(\frac{b}{k \times L_i}\right)^2} = \frac{b}{k}\sqrt{\sum_{i=1}^{24}\frac{1}{L_i^2}} \qquad (3.6)$$

Since $\sqrt{\sum_{i=1}^{24}\frac{1}{L_i^2}}$ is constant (forecasted load is given), we use $\frac{b}{k}$ as a similarity index.

3.5.1.3 Alternative Method 3 (AM3)

Using the historical data, we establish a price versus load curve as shown in Figure 3.10. As in Figure 3.10, if we could express the analytical relationship for any of the given curves as $Price = f(Load)$, then we would calculate the corresponding price for a given load value.

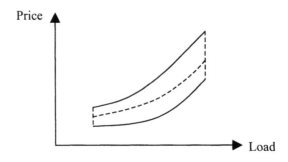

Figure 3.10 Alternative Method 3

3.5.2 Alternative MAPE Definition

3.5.2.1 Traditional MAPE

Let V_a be the actual value and V_f the forecast value. Then, Percentage Error (PE) is defined as

$$PE = (V_f - V_a)/V_a \times 100\% \qquad (3.7a)$$

and the absolute percentage error (APE) is

$$APE = |PE| \tag{3.7b}$$

Then, the mean absolute percentage Error (MAPE) is given as

$$MAPE = \frac{1}{N}\sum_{i=1}^{N} APE_i \tag{3.7c}$$

MAPE is widely used to evaluate the performance of load forecasting. However, in price forecasting, MAPE is not a reasonable criterion as it may lead to an inaccurate representation.

3.5.2.2 Problem with the Traditional MAPE

If the actual value is large and the forecasted value is small, then APE will be close to 100%. If the actual value is small, APE could be very large even if the difference between actual and forecasted values is small. For instance, when the actual value is zero, APE could reach infinity if the forecast is not zero. So, there is a problem with using APE for price forecasting. It should be noted that this problem does not arise in load forecasting, since actual load values are rather large, while price could be very small, or even zero.

3.5.2.3 A Proposed Alternative Definition of MAPE

One proposed alternative is as follow. First we define the average value for a variable V_a:

$$\bar{V} = \frac{1}{N}\sum_{i=1}^{N} V_a \tag{3.8a}$$

Then, we redefine PE, APE and MAPE as follows:

$$PE = (V_f - V_a)/\bar{V} * 100\% \tag{3.8b}$$

$$APE = |PE| \tag{3.8c}$$

$$MAPE = \frac{1}{N}\sum_{i=1}^{N} APE_i \tag{3.8d}$$

The point here is that we would use the average value as the basis to avoid the problem caused by very small or zero prices. In the following discussion, the proposed MAPE definition will be applied.

3.5.2.4 Example for the Proposed MAPE Definition

In this section, we apply the new definition of MAPE to practical data and compare it with the traditional MAPE definition. Two examples are presented.

Figure 3.11 shows price forecast from 05/29 to 06/04 where only half of the forecasts are acceptable. So, a reasonable index to measure the quality of the forecast should be close to 50%.

According to its traditional definition, the MAPE will be infinite since the actual prices during some time periods are zero (e.g., 6:00 AM in 05/29 which is the hour 6 in Figure 3.11) or close to zero (e.g., 6:00 AM in 06/04 which is the hour 150 in Figure 3.11). However, according to its new definition, the MAPE will be about 46.68%, which is reasonable since that is close to 50%.

Figure 3.12 shows price forecast from 03/01 to 03/07 with 9.23% and 8.05% for the traditional and the new MAPEs, respectively. The MAPEs are close to each other. The main differences come in hour 101 (47% vs. 19%), hour 126 (70% vs. 24%) and hour 127 (56% vs. 23%). At these hours, the actual prices are very low (less than 5 $/MWh) and the forecasted prices are not at all accurate.

Clearly, the new MAPE definition is a more reasonable measure to use in electricity price forecasting even in an extreme situations (the price is very low or even zero). In comparison, the traditional MAPE definition fails to provide a reasonable index by which to measure the quality of the forecast.

ELECTRICITY PRICE FORECASTING

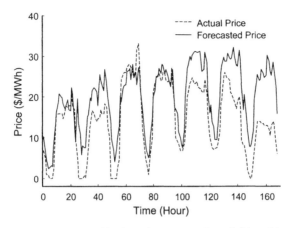

Figure 3.11 Application of New MAPE Definition (1)

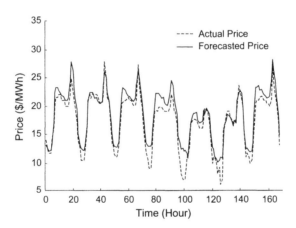

Figure 3.12 Application of New MAPE Definition (2)

3.6 PRACTICAL CASE STUDIES

In this section, we use ANN for price forecasting and performance evaluation based on practical data. We study the impact of data pre-processing, quantity of training vectors, quantity of impacting factors, and adaptive forecasting on price forecasting. We also compare the ANN results with those of alternative methods.

We use the data for the California power market in our study, including system loads and unconstrained MCP, from 1/1/99 to 9/30/99.

The curves are given in Figures 3.13a and 3.13b. Other data used in this study can be obtained from [Uce01].

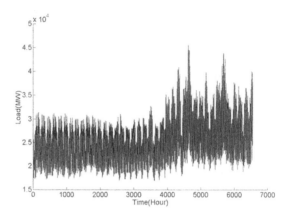

Figure 3.13a Load Curve of California Power Market from 1/1/99 to 9/30/99

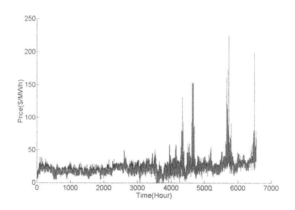

Figure 3.13b Price Curve of California Power Market from 1/1/99 to 9/30/99

3.6.1 Impact of Data Pre-processing

Now, we turn to study the impact of data pre-processing on ANN forecasting and compare the effects of two pre-processing methods. Figure 3.13b shows the actual prices from 7/1 to 8/4 with spikes at 7/1, 7/12, 7/13, 7/14, and 7/15. The training period is from 7/1 to 7/28, and the testing is

ELECTRICITY PRICE FORECASTING

for a period from 7/29 to 8/4. Without any pre-processing, the training MAPE is 40.67% and the testing MAPE is 15.89%.

Two data pre-processing methods for eliminating price spikes are considered: limiting price spikes and excluding price spikes. In limiting price spikes, we have two options:

First, we can set an upper limit (UL) on price. In other words, in pre-processing, if the price is higher than UL, it will be set to UL. For example, if the price is higher than 50 $/MWh, we set it to 50 $/MWh, Accordingly, the training and testing performances are both improved (i.e., training MAPE will be 7.66% and testing MAPE will be 13.82%).

Second, we can set an upper limit of price, UL, and apply the following pre-processing scheme for handling the actual prices that are higher than UL:

$$P' = \begin{cases} P & \text{if } P \leq UL \\ UL + UL \log \dfrac{P}{UL} & \text{if } P > UL \end{cases} \quad (3.9)$$

Figure 3.14 illustrates this idea, where P is the original price and P' is the processed value. If a price is higher than UL, its processed value is UL plus UL times the logarithm (base 10) of the ratio of price to UL.

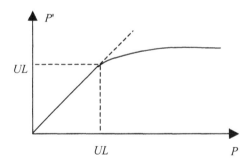

Figure 3.14 Illustration of Data Pre-processing

We also introduce a post-processing scheme for recovering the original price after curtailing its spikes. The post-processing scheme for a forecasted price is as follows:

$$k = \begin{cases} k' & \text{if } k' \leq UL \\ UL10^{\frac{k'-UL}{UL}} & \text{if } k' > UL \end{cases} \quad (3.10)$$

where k' is the forecasted price and k is the modified forecasted price (i.e., in post-processing). Accordingly, both the training and testing performances will be improved with MAPEs of 8.23% and 14.05%, respectively.

On the other hand, if we exclude days with price spikes from the training data, training and testing performances will both be improved more remarkably (i.e., in training and testing MAPEs will be 5.35% and 11.43%, respectively). The improvement in training MAPE is because of the exclusion of price spikes. However, price spikes are indicative of abnormality in the system, so we do not intend to delete them from the training process. Consequently, we stay with the option of limiting the magnitude of spikes rather than eliminating them entirely.

3.6.2 Impact of Quantity of Training Vectors

In this section, we study the impact of a number of training vectors on forecasting performance. In Table 3.2, the training period is from 2/1 to 4/4. The testing period is fixed from 3/29 to 4/4 (1 week). The training period can vary from one week to eight weeks as indicated in column 1 of Table 3.2 where case numbers correspond to the number of weeks of training. In Case 1, the training period is from 3/22 to 3/28 (1 week). In Case 2, the training period is from 3/15 to 3/28 (2 weeks). In Case 8, the training period is from 2/1 to 3/28 (8 weeks). The training periods of Cases 3, 4, 5, 6, and 7 are defined similarly.

Since the ANN weights are initialized randomly, every time we train and test the ANN, we can get a different result. To decrease the randomness of error, we repeat the "training and testing" procedure five times for each case with the average, minimum and maximum results shown in Table 3.2. Figure 3.15 compares the different cases.

According to Figure 3.15, the testing MAPEs would first decrease with an increase in training vectors (from Case 1 to Case 4), then remain flat (from Case 4 to Case 6), and finally increase as the number of training

ELECTRICITY PRICE FORECASTING

vectors increases (from Case 6 to Case 8). There are a number of reasons for this occurrence.

Table 3.2 Impact of Quantity of Training Vectors on Forecasting Performance

Case No	Training Vectors	Testing Vectors	Testing MAPE (%)		
			Average	Minimum	Maximum
1	3/22 – 3/28 (7)	3/29 – 4/4 (7)	12.77	12.59	12.96
2	3/15 – 3/28 (14)		12.21	11.68	12.56
3	3/8 – 3/28 (21)		11.88	11.64	12.13
4	3/1 – 3/28 (28)		11.19	11.11	11.25
5	2/22 – 3/28 (35)		11.21	11.13	11.45
6	2/15 – 3/28 (42)		11.26	11.14	11.48
7	2/8 – 3/28 (49)		11.63	11.57	11.70
8	2/1 – 3/28 (56)		11.80	11.40	12.09

Note: In "training vectors" and "testing vectors" (7) means "7 vectors"

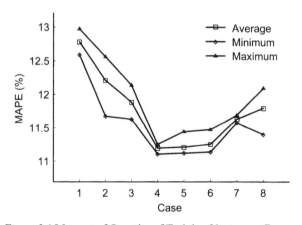

Figure 3.15 Impact of Quantity of Training Vectors on Forecasting

First, in introducing more training vectors, we present a more diverse set of training samples that encourage more general input-output mapping. Thus, the forecasting performance, as measured by the testing MAPE, improves. However, as we increase further the number of training vectors, the diversity of training samples no longer increases, so any additional training does not improve the forecasting results. Thus, the forecasting performance remains flat. Now, by even further increasing the number of training vectors (Cases 6 to 8), ANN could become over-trained. In other words, ANN has to adjust its weights to accommodate the input-output mapping of a large number of training vectors that may not be similar to the testing data to a large extent. Thus, the forecasting performance can get worse with further increasing the number of training vectors.

Based on the preceding analysis, the training quantity could depend on the similarity of training vectors. In our study, Cases 4 through 6 represent reasonable compromises. However, since Case 4 would require less training time than Cases 5 and 6, it is the best choice. It is also a good forecast (i.e., a smaller testing MAPE) made with a less effort (i.e., less training time). This is not to say that Case 4 is the best for all systems. For other systems, we would first start with a test similar to that described in this section and find the best choice accordingly.

3.6.3 Impact of Quantity of Input Factors

In this section, we study the forecasting process for MCP and ZMCP. The computation of ZMCP is more complicated than MCP, since ZMCP is related to the system congestion. As described before, it is not easy to consider the effect of congestion because very little public information on congestion is available. However, other factors such as the system's reserve may indirectly provide the congestion information. So we will use such information in order to improve the forecasting accuracy of ZMCP. The ZMCP in our study is that of Zone "NP-15," one of the 24 zones of the California market in 1999.

Table 3.3 presents three models with the factors we will consider in studying the impact of input vectors.

Table 3.3 Factors Considered in Different Model Types

Factors	Type 1 (T1M)	Type 2 (T2M)	Type 3 (T3M)
Time	√	√	√
Historical MCP	√	√	√
Historical Load		√	√
Forecasted Load		√	√
Historical Reserve			√
Forecasted Reserve			√

Note: "Historical" information refers to the "previous day" information.

ELECTRICITY PRICE FORECASTING

The study period is from 3/1 to 4/4. The training period is from 3/1 to 3/28 (four weeks). The testing period is from 3/29 to 4/4 (one week). Again, we repeat the training and testing procedures five times for each model type and present the average MAPE results. MCP and ZMCP results are shown in Tables 3.4 and 3.5, respectively. In Figures 3.16a and 3.16b we compare the testing MAPEs for the three different models.

Table 3.4 Forecasting Performance of Different Models– MCP Case

Type	Network Structure	Testing MAPE (%)		
		Average	Minimum	Maximum
T1M	25-40-24	12.81	12.44	13.20
T2M	73-100-24	11.19	11.11	11.25
T3M	121-150-24	11.75	11.56	12.11

Table 3.5 Forecasting Performance of Different Models–ZMCP Case

Type	Network Structure	Testing MAPE (%)		
		Average	Minimum	Maximum
T1M	25-40-24	12.75	12.31	13.16
T2M	73-100-24	11.61	11.37	11.94
T3M	121-150-24	10.88	10.56	11.12

Note: In the "Network Structure" column, 25-40-24 means 25 input neurons, 40 hidden neurons and 24 output neurons.

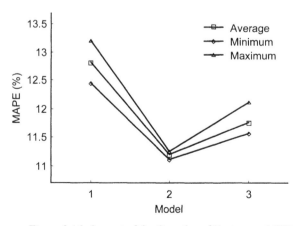

Figure 3.16a Impact of the Quantity of Factors on MCP

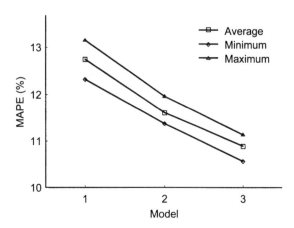

Figure 3.16b Impact of the Quantity of Factors on ZMCP

Comments.

As Figure 3.16a shows, for MCP, if we only consider the historical price as input to ANN (i.e., T1M), we obtain the worst forecasting performance. If we consider additional load information (historical and forecasted load) as input to ANN (i.e., T2M), we obtain a better forecasting performance than that of T1M. However, if we further consider the reserve information (historical and forecasted reserve) as input (i.e., T3M), the forecasting performance does not improve; it might even get worse compared with that of T2M.

According to Figure 3.16b, the more factors we consider for ZMCP, the better the forecasting quality will be. T3M includes the most factors and obtains the best forecast. In Figure 3.16a, the price forecasted for MCP is closely related to the historical price and load information, and the reserve information does not affect MCP significantly. This is what we may expect since MCP is determined simply by matching supply and demand bids without considering the power system's structure and operating constraints. The ZMCP price forecasting in Figure 3.16b is affected by historical price, load, and reserve information. Here, the reserve information is an indicator of system congestion because it affects the zonal price.

If a factor does not have an impact on price forecasting, such as the reserve information in T3M for the MCP case, its inclusion in forecasting may rather aggravate the results. This is because the factor could interfere with the ANN training and make it more difficult to map the price onto the impacting factors. If we fail to consider a factor that does impact price forecasting, such as the reserve information in T2M for ZMCP, the forecasting performance will be affected adversely.

In general, it is inappropriate to claim that the more factors we consider in training, the better the forecasting performance will be. However, it is possible to claim that we can improve the forecasting results by only considering those factors that impact the results. In other words, what is important is not the number of factors, but the number of factors that would impact the forecasting results.

3.6.4 Impact of Adaptive Forecasting

In the performance evaluation of ANN, we can either use the fixed training weights or update the weights frequently based on the ANN results. We refer to the latter method as an adaptive forecasting method.

In studying the profile of a price, we could expect the adaptive modification of network weights to yield a better forecast. In Table 3.6, a Type 2 model (T2M) is employed, and results are shown for comparing non-adaptive and adaptive methods.

Table 3.6 Comparison of Non-adaptive and Adaptive Forecasting

Case No.	Training Vectors	Testing Vectors	Testing MAPE (%)	
			Non-adaptive	Adaptive
1	2/1 – 2/28 (28)	3/1 – 3/7 (7)	14.04	8.71
2	5/1 – 5/28 (28)	5/29 – 6/4 (7)	52.94	25.81
3	7/1 – 7/28 (28)	7/29 – 8/4 (7)	12.53	12.59
4	8/1 – 8/28 (28)	8/29 – 9/4 (7)	11.59	10.23

Note: In "training vectors" and "testing vectors" (28) means 28 vectors.

From the table, we learn that in most cases adaptive forecasting provides better accuracy. This is because adaptive forecasting takes the latest information into consideration. Case 2 deserves more attention where there are zero prices in 5/29, 5/30, and 5/31, and non-adaptive forecasting would not identify this information. In comparison, adaptive forecasting can identify this information and modify network weights accordingly.

3.6.5 Comparison of ANN Method with Alternative Methods

In this section, we compare the ANN method with alternative methods described in Section 3.5.1 and present the results in Table 3.7. In this table, the new MAPE definition is used to compare the different methods.

Table 3.7 Comparisons of Different Forecasting Methods

Method	Strategy	MAPE (%)
ANN	Non-adaptive	8.25
	Adaptive	6.57
AM1	Using current day data	7.87
	Using previous day data	9.89
AM2	Similar error = 0.05	11.35
	Similar error = 0.1	11.12
AM3	1st order curve fitting	11.99
	2nd order curve fitting	12.12
	3rd order curve fitting	12.06

From the results of Sections 3.6.3 and 3.6.4, we consider the Type 2 Model (i.e., previous day MCP, previous day load, and forecast load to forecast MCP). Here four weeks' history data are used for training and the data preprocessing technique is used.

In AM1, "using current day data" means using the data of day i to forecast the price of day $i+1$, while "using previous day data" means using the data of day i-1 to forecast the price of day $i+1$. The former is an ideal situation since in practice it is impossible to obtain current day data when forecasting the next day price. However, the latter is the normal situation in practice.

In AM2, the following strategy is employed to determine the so-called similar error. Suppose that we only consider the load information to forecast the price (the idea can be easily extended to consider more information). L is the forecasted load and HL is the history load. We could derive the relationship between L and HL as $HL = k \times L + b$ and define b/k as the similar error. When the similar error is less than a specified value, we say that L is similar to HL. Consequently, a history price corresponding to HL is selected to compute the price forecast.

ELECTRICITY PRICE FORECASTING

In AM3, by a "first-order curve fitting" we mean using the first order curve to fit the mapping between price and load. "Second order curve fitting" and "third order curve fitting" can be similarly defined.

In Table 3.7, the ANN method would provide better results than alternative methods with appropriate training strategies (data pre-processing, Type 2 Model, four weeks' data training) and appropriate forecasting strategy (adaptive forecasting).

3.7 PRICE VOLATILITY ANALYSIS MODULE

In price forecasting, it would be premature to rely solely on the hourly price forecast. The volatility of the electricity price must also be analyzed. In this section, we take an additional look at price spikes, which relate to the distinct volatility of electricity prices, and study their probability distribution.

3.7.1 Price Spikes Analysis

It is believed that spikes cannot be forecasted accurately. However, the probability of spikes can be derived from historical data. In this section, we determine the probability distribution spikes based on

- Different load levels
- Different price levels.

These topics are discussed next.

3.7.1.1 Price Spikes at Different Load Levels

Using historical data, we can find the relationship between price spikes and loads. Figure 3.17 illustrates the probability of spikes as a function of load. As is obvious, the higher the load, the larger is the probability of spikes, and at different load levels, the probability distributions of price are different. Figure 3.17 gives three load levels: low (I), medium (II) and high (III). Figure 3.18 shows the probability distributions of price corresponding to these three load levels.

In the following we consider a practical case in order to understand the construction of Figures 3.17 and 3.18. Consider Figure 3.19. The figure gives an example of the relationship between price and load in the California power market from 7/1/99 to 8/4/99.

Suppose that we define prices higher than 50 $/MWh in Figure 3.19 as spikes. Then, we have in Figure 3.20 the probability of spikes at the different load levels, where the minimum load is 0, the maximum load is 47500 MW, and the step is 2500 MW. Figure 3.20 verifies our expectation that the higher is the load, the higher the probability of price spikes.

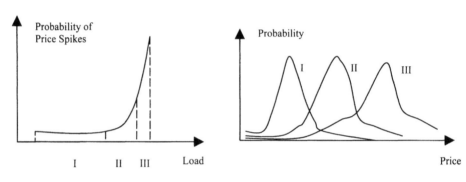

Figure 3.17 Probability of Price Spikes

Figure 3.18 Probability Distributions of Prices at Different Load Levels

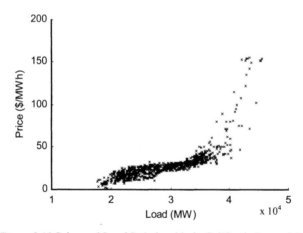

Figure 3.19 Price and Load Relationship in California Power Market

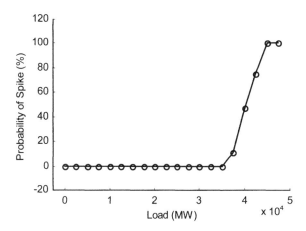

Figure 3.20 Probability of Spikes at Different Load Levels

Let us define three load levels as follows: low load level (less than 35,000 MW), medium load level (between 35,000 MW and 40,000 MW), and high load level (higher than 40,000 MW). Then, using Figure 3.19 and considering prices higher than 50 $/MWh as spikes, we calculate the probability of spikes at the three load levels as shown in Table 3.8.

Table 3.8 Probabilities of Spikes at Different Load Levels

Load levels (MW)	≥40000	[35000, 40000]	<35000
Probability of price spikes	86.20%	21.90%	0

We use Figure 3.19 to construct Figure 3.21, which shows the probability distribution of price at three load levels. Comparing the three curves in Figure 3.21, we come to the conclusion that electricity prices are more volatile at the high load level than in other levels, since they span from very low (20 $/MWh) to very high (180 $/MWh) and the distribution is relatively flat.

Using Figure 3.21, we determine the probability distribution of forecasted price. Suppose the forecasted load is 42,000 MW, which is at the high load level, and the forecasted price is 120 $/MWh. The high load level has an expected price of 105 $/MWh. Accordingly, the two distributions are similar and the latter has a 15 $/MWh disposition compared to the former (i.e., Using Appendix B.1, $\delta = k - k^* = 120 - 105 = 15$). Figure 3.22 depicts the probability distributions of price at the high load level and the forecasted price.

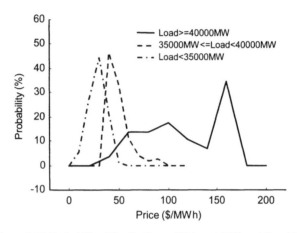

Figure 3.21 Probability Distributions of Price at Different Load Levels

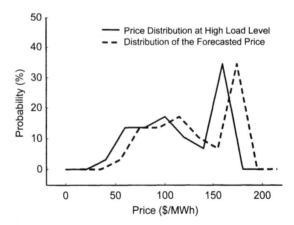

Figure 3.22 Probability Distributions of the High Load Level and the Forecasted Price

The distribution in Figure 3.21 could be used for risk analysis to study the impact of price forecast inaccuracy. However, one of the disadvantages of using the high load level curve in Figure 3.21 is that the distribution is not so smooth around 150 $/MWh. One option that we may

ELECTRICITY PRICE FORECASTING

consider is to utilize a longer period for price data. Figure 3.23 shows the results with price data extending from 1/1/99 through 9/30/99. The high load level curve in Figure 3.23 is more realistic than that in Figure 3.21 while the low and medium load levels results are similar to those in Figure 3.21.

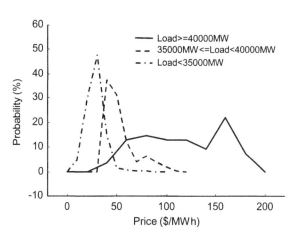

Figure 3.23 Probability Distributions of Price at Different Load Levels using More Samples

3.7.1.2 Price Spikes at Different Price Forecast Levels – Approach 1

In the ANN application to price forecasting, we could set an upper limit (UL) for price. In other words, if a price is larger than UL, the price is set to UL at the pre-processing stage. We refer to this approach for price forecasting as approach 1 and will discuss approach 2 in the next section. Correspondingly, we expect that if the forecast is close to UL, the probability of spikes will increase as depicted in Figure 3.24. In labeling this figure, (1) refers to approach 1. At every price forecast level, we encounter a different probability distribution of prices. In Figure 3.24, we identify three price forecast levels: low (I), medium (II), and high (III); the probability distributions corresponding to these levels are presented in Figure 3.25.

Again, we consider an example from California in constructing the curves for a practical case. Figure 3.26 shows the actual and forecasted price curves of the California power market from 7/1/99 to 8/4/99. Figure 3.27 shows the relationship between actual and forecasted prices.

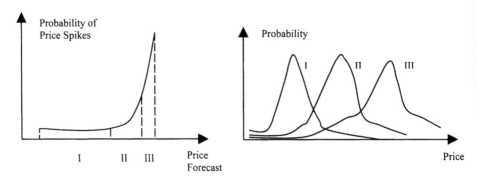

Figure 3.24 Probability of Spikes as a Function of Price Forecast (1)

Figure 3.25 Probability Distributions of Price at Different Price Forecast Levels (1)

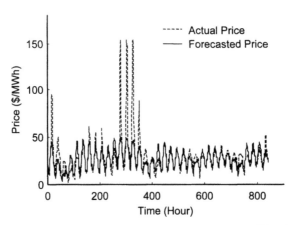

Figure 3.26 Actual and Forecasted Price Curves of California (1)

ELECTRICITY PRICE FORECASTING

Let us now consider prices higher than 50 $/MWh as spikes. Figure 3.28 shows the probability of spikes at different price forecast levels. The minimum price is 0, the maximum price is 50 $/MWh, and the step is 5 $/MWh.

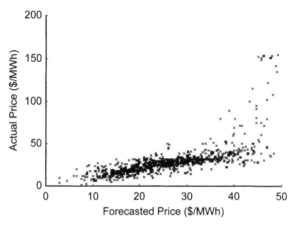

Figure 3.27 Relationship between Actual and Forecasted Price of California (1)

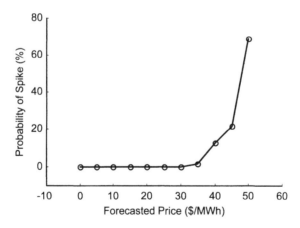

Figure 3.28 Probability of Spikes at Different Price Forecast Levels (1)

Likewise, we define three price forecast levels as follows: low price forecast level (less than 35 $/MWh), medium price forecast level (between 35 $/MWh and 45 $/MWh), and high price forecast level (higher than 45 $/MWh). Table 3.9 shows the probability of spikes at the three price forecast levels according to Figure 3.27. Also, Figure 3.29 shows probability distributions of price at the three price forecast levels.

Table 3.9 Probability of Spikes at Different Price Forecast Levels (1)

Price forecast range ($/MWh)	≥45	[35, 45)	<35
Probability of price spikes	68.97%	16.08%	0.30%

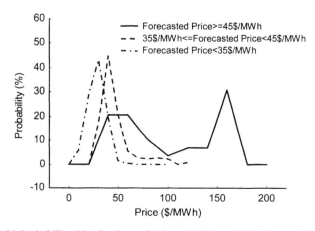

Figure 3.29 Probability Distributions of Price at Different Price Forecast Levels (1)

Using Figure 3.29, we determine the probability distribution of the forecasted price. Take the high price forecast level as an example. Since we do not differentiate among prices that are higher than the upper limit, we cannot directly recover the actual price from the forecasted price. One option is to base the process on historical data. Consider the following case. Assume that the forecasted price is k, which is at the high price forecast level. If a forecasted price and its corresponding actual price represent a price pair, the price pairs at the high price forecast level will be between $k + \varepsilon$ and $k - \varepsilon$. The mapped forecasted price is the expected value of the price pairs. The selection of ε depends on the number of price pairs

ELECTRICITY PRICE FORECASTING

at the high price forecast level. If the number is large, ε can be small. If the number is small, ε should be large. If ε is large enough, all price pairs at the high forecast price level will be included.

Suppose that the forecasted price is 47 $/MWh, which is at the high price forecast level. The expected value of the price at the high price forecast level is 90 $/MWh. Consider ε as 1 $/MWh. By calculation, the mapped forecasted price is 87 $/MWh. The probability distribution of price at the high price forecast level and the probability distribution of the forecasted price are similar since the latter has a -3 $/MWh disposition (according to the Appendix B.1, $\delta = k - k^* = 87 - 90 = -3$) compared to the former. Figure 3.30 illustrates the two probability distributions (the expected value of the distribution of the forecasted price is 87 $/MWh).

The distribution of high price forecast level in Figure 3.29 has similar disadvantage to that of the high load level in Figure 3.21. Alternatively, Figure 3.31 shows the results for price data extending from 1/1/99 through 9/30/99. The high level curve in Figure 3.31 is more realistic than that in Figure 3.29 while there are no significant changes in the low and medium level curves.

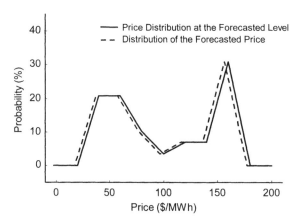

Figure 3.30 Probability Distributions at the Price Forecast Level and the Forecasted Price (1)

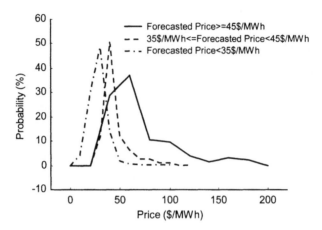

Figure 3.31 Probability Distributions of Price at Different Price Forecast Levels using More Samples (1)

3.7.1.3 Price Spikes at Different Price Forecast Levels – Approach 2

We recognize that the purpose of pre-processing is to limit prices. However, we consider a different approach here to differentiate between prices levels over UL. This concept is different from that of the previous approach, where all prices over UL were set to be UL (See Section 3.6.1). That is, if a price is higher than UL, its processed value would be UL plus UL times the logarithm (10 as the base) of the ratio between the original value and UL. We refer to this approach as approach 2.

We expect that the probability of spikes will increase as depicted in Figure 3.32, if the forecast is close to or larger than UL. At different price forecast levels, the probability distributions of price differ. Therefore, in Figure 3.32, we consider four price forecast levels: low (I), medium (II), high (III) and extra-high (IV). In labeling this figure, (2) refers to approach 2. Figure 3.33 shows the probability distributions of the prices corresponding to these four price forecast levels.

In relating this result to a practical situation, we depict in Figure 3.34a (before post-processing) and Figure 3.34b (after post-processing) the California power market from 7/1/99 to 8/4/99. Figure 3.35 shows the relationship between actual and forecasted prices.

ELECTRICITY PRICE FORECASTING

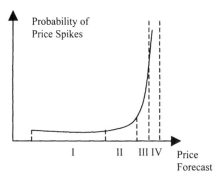

Figure 3.32 Probability of Spikes as a Function of Price Forecast (2)

Figure 3.33 Probability Distributions of Price at Different Price Forecast Levels (2)

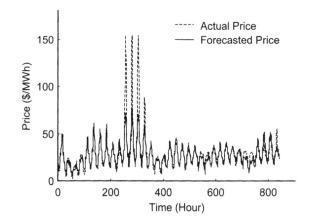

Figure 3.34a Actual and Forecasted Price Curves (Before Post-processing)

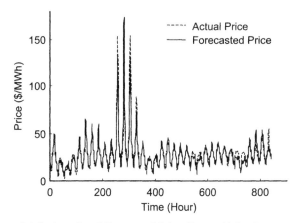

Figure 3.34b Actual and Forecasted Price Curves (After Post-processing)

If we define prices higher than 50 $/MWh as spikes, Figure 3.36 shows the probability of spikes at different price forecast levels. The minimum price is 0, the maximum price is 80 $/MWh and the step is 10 $/MWh.

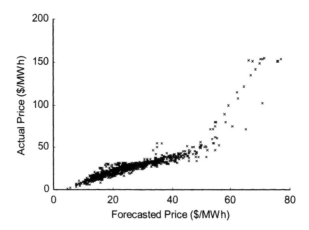

Figure 3.35 Relationship between Actual and Forecasted Prices of California (2)

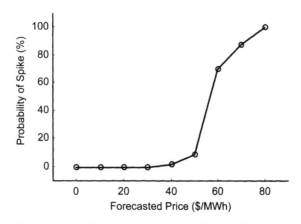

Figure 3.36 Probability of Spikes at Different Price Forecast Levels (2)

In Figure 3.35, we define four price forecast levels as follows: low price forecast level (less than 35 $/MWh), medium price forecast level (between 35 $/MWh and 45 $/MWh), high price forecast level (between 45 $/MWh and 50 $/MWh), and extra-high price forecast level (higher than 50

ELECTRICITY PRICE FORECASTING

$/MWh). Since the spikes are larger than 50 $/MWh, we determine the probability of spike at each level of Figure 3.35 as shown in Table 3.10. In addition, we use the same information to show probability distributions of price at the four price forecast levels in Figure 3.37.

We use Figure 3.37 to determine the probability distribution of price and its spikes. The forecasted price is given as 60 $/MWh, which is at the extra-high price forecast level. The expected value of price at the high price forecast level is 94 $/MWh. According to equation (3.13), the post-processed price forecast is 79 $/MWh. The probability distribution of price at the extra-high price level and the probability distribution of the forecasted price are similar and the latter has a -15 $/MWh disposition (according to the Appendix B.1, $\delta = k - k^* = 79 - 94 = -15$) compared to the former. Figure 3.38 illustrates the two distributions (the expected value of the distribution of the forecasted price is 79 $/MWh).

Table 3.10 Probability of Spikes at Different Price Forecast Levels (2)

Price forecast range ($/MWh)	≥50	[45,50)	[35, 45)	<35
Probability of price spikes	89.47%	23.08%	1.92%	0.14%

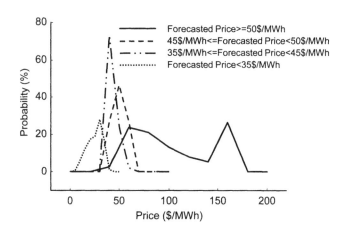

Figure 3.37 Probability Distributions of Prices at Different Price Forecast Levels (2)

The distribution of extra-high price forecast level in Figure 3.37 has similar disadvantage to that of the high price forecast level in Figure 3.29. Figure 3.39 shows the results for price data extending from 1/1/99 through

9/30/99. The extra-high level curve in Figure 3.39 is more realistic than that in Figure 3.37 while there are no significant changes in other level curves.

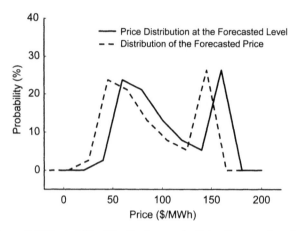

Figure 3.38 Probability Distributions at the Price Forecast Level and the Forecasted Price (2)

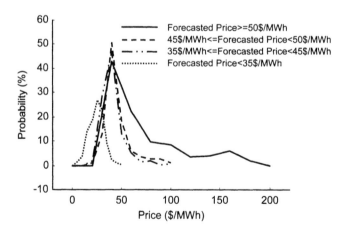

Figure 3.39 Probability Distributions of Prices at Different Price Forecast Levels using More Samples (2)

ELECTRICITY PRICE FORECASTING

3.7.2 Probability Distribution of Electricity Price

In this section, we discuss the calculation of the probability distribution of electricity price. Two methods are available: statistical method and ANN method.

3.7.2.1 Statistical Method

By this method, we obtain a price distribution in analyzing historical data for price forecasts. The analysis may be based on load levels, price levels or price difference (actual price - forecasted price) levels, as discussed below.

Statistical Method Based on Load Levels. In this method, the probability distribution of price is based on different load levels. This method was discussed earlier when we analyzed price spikes at different load levels.

Statistical Method Based on Price Forecast Levels. In this method, the probability distribution of price is based on different price forecast levels. This method was discussed earlier when we analyzed price spikes at different price levels.

Statistical Method Based on Price Difference. In this method, we use historical forecasts to obtain a price difference (actual price - forecasted price) distribution and then the probability distribution of electricity price. Figure 3.40a shows a historical market price forecast and the actual market price. Accordingly, we derive the price difference distribution as shown in Figure 3.40b.

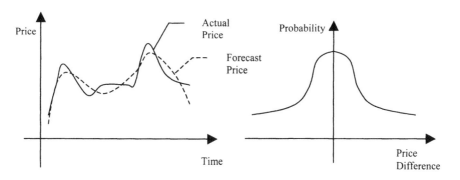

Figure 3.40a Historical Forecast   *Figure* 3.40b Price Difference Distribution

To obtain a price difference distribution, two methods are considered and discussed as follows:

- Do not differentiate between individual hours. We assume that the price difference distributions at various hours are the same and that all are based on historical forecasts.

- Differentiate between individual hours in analyzing the price difference distribution. We use historical forecasts at a given hour to derive a price difference distribution at that hour.

Compare the forecast in Figure 3.34b with its price difference shown in Figure 3.41. Correspondingly, we construct the price difference distribution in Figure 3.42, if we do not differentiate between individual hours.

If we assume that the forecasted price at hour 10 is 40.5 $/MWh and the expected price difference is about 0.5 $/MWh, Figure 3.43 illustrates the probability distribution of price difference and the probability distribution of the forecasted price (expected value of the latter distribution is 40.5 $/MWh). The two distributions are similar, and the latter has a 40 $/MWh disposition (according to the Appendix B.1, $\delta = k - k^* = 40.5 - 0.5 = 40$) compared to the former.

Figure 3.44 shows the price difference distribution at hour 10 when individual hours are differentiated. The expected price difference is about 0.2 $/MWh. Suppose that the forecasted price at hour 10 is 40.5 $/MWh.

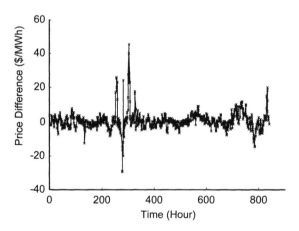

Figure 3.41 Price Difference Analysis

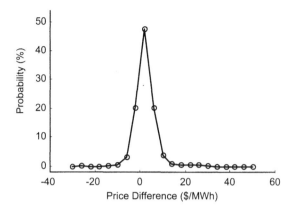

Figure 3.42 Price Difference Distribution for All Hours

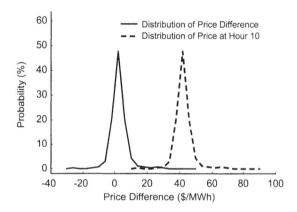

Figure 3.43 Distribution Analysis of Price Difference and Price at Hour 10

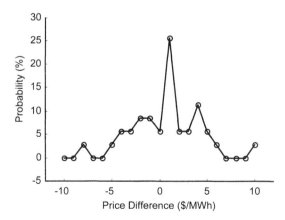

Figure 3.44 Price Difference Distribution at Hour 10

Figure 3.45 illustrates the probability distribution of price difference and the probability distribution of the forecasted price (expected value of the latter distribution is 40.5 $/MWh). The two distributions are similar and the latter has a 40.3 $/MWh disposition (according to the Appendix B.1, $\delta = k - k^* = 40.5 - 0.2 = 40.3$) compared to the former.

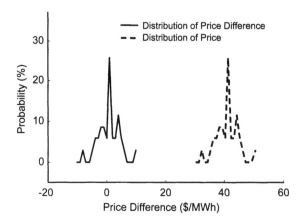

Figure 3.45 Distribution Analysis of Price Difference and Price at Hour 10

3.7.2.2 ANN Method

In [Deb01a], the authors indicate that:

- For price forecasting, a single-point forecast is not enough. Here, a single-point forecast refers to using expected values of input factors to forecast the expected value of output.

- A probability distribution of input factors should be considered in order to calculate a probability distribution of output.

- The forecasted price and its volatility are intrinsically connected.

We propose an indirect method for calculating the probability distribution of electricity price based on these observations. In [Deb01], the mapping from impact factors to price forecast is implemented by OPF. Here, we adopt the ANN method described in Section 3.4, and Figure 3.46a shows the mapping from load forecast to price forecast. Once we know the probability distribution of load forecast, we can calculate the

ELECTRICITY PRICE FORECASTING

probability distribution of price forecast using ANN and a Monte Carlo simulation as illustrated in Figure 3.46b.

Let us consider an example. First, as in Section 3.4, for the study system, we use ANN to find the relationship between loads for 24 hours (and other factors) and prices for 24 hours. Then, with this ANN, for a specific load sample (24 hours of loads), we get a corresponding price sample (24 hours of prices). Now, if we vary the load samples (keeping other factors unchanged in this case), we can get various price samples. For this example, we could vary the loads by uniformly multiplying all 24-hour loads with a factor that follows a normal distribution (the mean is 1 and the standard deviation is 0.1 as shown in Figure 3.47). Each factor corresponds to a load sample, and thus a price sample. Each try amounts to one Monte Carlo simulation.

Repeating this scheme, we could make 1000 Monte Carlo simulations. Then we could make a probability analysis for prices at each hour. Figure 3.48 shows the probability distribution of forecasted prices at hours 5, 10, 15, and 20. For convenience and clarity, we normalize those prices by their corresponding expected values. The actual probability distribution of prices is determined by mapping from loads to prices as is represented by ANN. As Figure 3.48 shows, although the loads follow a normal distribution, the prices do not necessarily, for example, at hour 5. Prices at hours 10, 15, and 20 do have normal distributions, but their standard deviations differ.

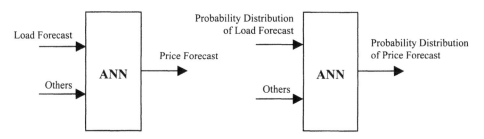

Figure 3.46a Price Forecasting by ANN

Figure 3.46b Probability Distribution of Price Forecast

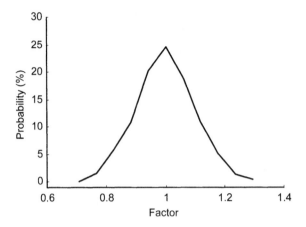

Figure 3.47 Probability Distribution of the Factor

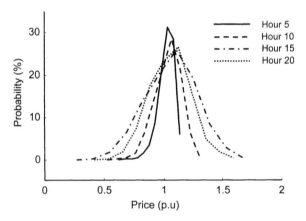

Figure 3.48 Probability Distribution of Price

3.7.2.3 Comparison of Statistical Method and ANN Method

In comparing the statistical method with the ANN method, it is difficult to say which method is better, since both methods have advantages and disadvantages. In practice, the method used depends on the specific case.

The statistical method is a direct method. It is a way to directly analyze the properties of historical price data, such as the relationship between price and load or relationship between actual price and forecasted price. The assumption is that the probability distribution of price (or price difference) forms a pattern that repeats over time. So the applicability of this method depends very much on repeatable intervals. There are two advantages to this method. First, it is easy to implement because only simple statistical methods are needed. Second, the way to implement is very flexible. It could be based on load level, price forecast level, or price difference.

The ANN method is an indirect method. Through ANN, one studies the distribution of prices by analyzing the distribution of price drivers (load in the example). The basic assumption is that the ANN will correctly map the relationship between the input and the output, even for cases not included in training samples. So, its accuracy depends on how accurately one models the probability distribution of the price driver and how accurately the ANN represents the mapping from price driver to price. The beauty of the ANN method is that it can use many samples that are created by the simulation of the practical system. However, compared to the statistical method, the ANN method may be more complicated in implementation.

3.8 APPLICATIONS OF PRICE FORECASTING

So far in this chapter, we have discussed point forecast of electricity prices and probability distribution of electricity prices. In this section, we look at the applications of these tools to power systems market operations.

3.8.1 Application of Point Price Forecast to Making Generation Schedule

Point forecast refers to forecasting expected values. In Section 3.4, using neural network, we found the relationship between the impacting factors and electricity prices. From the impacting factors, some of which have forecasted values such as load, we can forecast the electricity prices.

We have only one price for each period, which is the expected price. This kind of forecast is called a point forecast. The point forecast is used to produce generation schedules. In Chapter 4 of this book, we will discuss the PBUC, for which forecasted market prices are used to optimize generation resources and profits. Different point price forecasts will lead to different commitment schedules and generation schedules.

3.8.2 Application of Probability Distribution of Price to Asset Valuation and Risk Analysis

In electricity markets, it is not enough to forecast a single (expected) value of the market price. The probability distribution of the prices needs to be simulated for generation asset valuations. The probability distribution of the prices is indispensable in evaluating the risk associated with the deviation of actual prices from a particular forecast. These topics will be discussed in Chapter 7 of this book.

3.8.3 Application of Probability Distribution of Price to Options Valuation

The values of options are the expected premiums a buyer or seller pays to ensure that he can conduct his trade at the average or forecasted price [Deb01]. From the probability distribution of electricity prices, the value of the call option is calculated as the difference between the average of the prices above the mean and the mean price of the entire distribution, and the value of the put option is calculated as the difference between the average of the prices less than the mean and the mean of the entire distribution.

Mathematically, we would say that the probability distribution of electricity price is $f(P)$. The expected value is

$$E(P) = \int_{0}^{+\infty} P * f(P) dP \qquad (3.11)$$

The values of the call option and put option are

$$VC = \int_{E(P)}^{+\infty} P * f(P) dP - E(P) \qquad (3.12a)$$

$$VP = \int_{0}^{E(P)} P * f(P) dP - E(P) \qquad (3.12b)$$

where $E(P)$ denotes expected value of price, VC denotes value of the call option, and VP denotes value of the put option.

3.8.4 Application of Conditional Probability Distribution of Price on Load to Forward Price Forecasting

In this chapter, we mainly discuss short-term price forecasting. A conditional probability distribution of prices on load allows us to make long-term price forecasts with reasonable accuracy.

ELECTRICITY PRICE FORECASTING

If we know the probability distribution of the load and the conditional probability distribution of prices on that load, we can calculate the expected value of price as follows [Cam01]:

$$E(P) = \int_0^{+\infty}\int_0^{+\infty} P*f(P|L)*f(L)dPdL = \int_0^{+\infty} E(P|L)dL \quad (3.13)$$

where L denotes load, P denotes price, $f(L)$ is the probability distribution of load, $f(P|L)$ is the conditional probability distribution of price on load, $E(P|L)$ is the expected price when the load is L, and $E(P)$ denotes expected value of price.

Equation (3.13) can be used to calculate the forward price in the long term, since statistically both the probability distribution of a load and the conditional probability distribution of price on that load are easy to obtain with relatively high accuracy. We should be cautious, however, if we want to use this formula to forecast price in the short term, since the accuracy may be compromised. That is because the probability distribution of a load and the conditional probability distribution of price on that load cannot be obtained accurately for short-term studies. The following is a simple example for calculating the forward price in the long term.

Suppose that we calculate the forward price for the third month from now. There are three load levels as in Figure 3.35. The probability distribution of load is shown in Table 3.11, which has a high accuracy statistically. Also shown in the table are the expected values of prices at different load levels, which are calculated from Figure 3.35.

Table 3.11 Load Level Distribution and Expected Price

Load range (MW)	≥40000	[35000, 40000)	<35000
Probability	3%	25%	72%
Expected prices ($/MWh)	105	43	24

So, the forward price (*FP*) can be calculated as follows.

$FP = 24 \times 72\% + 43 \times 25\% + 105 \times 3\% = 31.18$ $/MWh

Chapter 4

Price-Based Unit Commitment

4.1 INTRODUCTION

In the regulated power industry, unit commitment (UC) refers to optimizing generation resources to satisfy load demand at least cost. Since the related objective would be to minimize the operational cost, unit commitment is commonly referred to as cost-based unit commitment. If maintaining security is emphasized in the UC solution, the new UC is referred to as security-constrained unit commitment (SCUC). Three elements are included in the SCUC paradigm: supplying load, maximizing security, and minimizing cost. Satisfying the load is a hard constraint and an obligation for SCUC. Maximizing security is often satisfied by maintaining sufficient spinning reserve at less congested regions that could easily be accessed by loads. Cost minimization is realized by committing less expensive units while satisfying the corresponding constraints and dispatching the committed units economically. SCUC is the subject of Chapter 8 in this book. In several of the restructured markets, the ISO would use SCUC to plan the day-ahead schedule. However, the SCUC discussed in Chapter 8 is different from that of the regulated power industry since the ISO would have no control over generation bids submitted by GENCOs.

In comparison, the UC used by individual GENCOS refers to optimizing generation resources in order to maximize the GENCO's profit. This UC has a different objective than that of SCUC and is referred to as price-based unit commitment[1] (PBUC) to emphasize the importance of the price signal. In PBUC, satisfying load is no longer an obligation and the objective would be to maximize the profit, and security would be

---

[1] PBUC is occasionally referred to as profit-based unit commitment.

unbundled from energy and priced as an ancillary service. In this new paradigm, the signal that would enforce a unit's ON/OFF status would be the price, including the fuel purchase price, energy sale price, ancillary service sale price, and so on.

In comparing SCUC with PBUC, it is wrong to assume that maximizing the profit is essentially the same as minimizing the cost. Profit is defined as the revenue minus cost. That is, profit not only depends on cost but on revenue as well. If the incremental revenue is larger than the incremental cost, we may generate more energy for attaining more profit. On the contrary, if the incremental revenue is smaller than the incremental cost, it may be less attractive to sell energy. In an extreme case, if the objective in the new paradigm is to minimize the cost, a GENCO might not opt to generate because it would have no incentive to serve a load at zero cost.

Compared to cost-based unit commitment, the distinct feature of price-based unit commitment is that all market information is reflected in the market price. Some examples are as follows: Although the system load is not a hard constraint in PBUC, load forecasting would be required for market price forecasting. Likewise, security would not be a consideration in formulating PBUC, but the ISO's criteria for maintaining security would impact the market price. It may be difficult for certain participants in an energy market to foresee other market participants' bidding strategies. Of course, gaming techniques could be utilized by participants to incorporate strategies in their market price forecasting. On the issue of transmission congestion, such constraints could be included explicitly in SCUC. However, it is impractical to assume transmission network information in PBUC since that information is unavailable to GENCOs. Since transmission congestion would incur price differences among different regions, transmission congestion will be incorporated through locational marginal prices in PBUC.

In the restructured power industry, there may exist price discrepancies between electricity and other commodities, such as fuel and emission allowance. Price discrepancies mean opportunities for arbitrage. PBUC can be used to discover those opportunities, which will be discussed in Chapter 5.

In a competitive energy market, the valuation of generation assets is an important issue. An efficient PBUC could maximize the value of generation assets. In generation asset valuation, price forecasting is indispensable. No matter how good a price forecast is, discrepancies will always occur in committing enough generation, which could lead to

PRICE–BASED UNIT COMMITMENT

additional risk. The application of PBUC to generation asset valuation and risk analysis will be treated in Chapter 7.

In this chapter, the PBUC formulation and its solution methodology are discussed. Some examples are presented to illustrate the formulation and the solution methodology.

4.2 PBUC FORMULATION

The objective of PBUC is to maximize the profit (i.e., revenue minus cost) subject to all prevailing constraints. For unit i at time t, the profit is given as[2]

$$F(i,t) = \begin{cases} \rho_{gm}(t)[P(i,t) - P_0(i,t)] + \rho_{rm}(t)R(i,t) + \rho_{nm}(t)N(i,t) \\ -C_i(P(i,t) + R(i,t) + N(i,t)) - S(i,t) + f_i(P_0(i,t)) \end{cases} I(i,t)$$
$$+ \begin{cases} \rho_{nm}(t)N(i,t) - \rho_{gm}(t)B(i,t) \\ -C_i(N(i,t)) + f_i(P_0(i,t)) \end{cases}(1 - I(i,t))$$
(4.1)

We let P represent energy, R represent spinning reserve and N represent non-spinning reserve. The first part of (4.1) represents the profit when the unit is ON. Profit is defined as the revenue from the sales of energy and ancillary services minus production costs. The profit from bilateral contracts would also be included, though it is assumed to be constant. The second part of (4.1) represents the profit when the unit is OFF. Here, profit represents revenue from the non-spinning reserve sales minus production costs and the cost of any energy purchases. Similarly, profit from bilateral contracts would also be included.

In the scheduling horizon, the profit for all scheduled units is given as

$$F = \sum_i \sum_t F(i,t) \qquad (4.2)$$

The PBUC problem is formulated as
maximize $\sum_i \sum_t F(i,t)$

subject to system and unit constraints, as discussed below.

---

[2] The list of symbols is given in Appendix A.

4.2.1 System Constraints

a) System Energy and Reserve Limits

$$P^{min}(t) \leq \sum_i P(i,t)I(i,t) \leq P^{max}(t) \tag{4.3a}$$

$$R^{min}(t) \leq \sum_i R(i,t)I(i,t) \leq R^{max}(t) \tag{4.3b}$$

$$N^{min}(t) \leq \sum_i N(i,t) \leq N^{max}(t) \tag{4.3c}$$

These constraints represent a GENCO's special requirements. For example, a GENCO may have minimum and maximum generation requirements in order to play the game in the energy market. Because of reliability requirements, a GENCO may pose lower and upper limits on its spinning and no-spinning reserves. These constraints can be relaxed otherwise.

b) System Fuel Constraints (For a "FT" type of fuel)

$$F^{min}(FT) \leq \sum_{i \in FT} \sum_t C_{fi}\big(P(i,t)I(i,t) + R(i,t)I(i,t) + N(i,t)\big) + S_f(i,t) \leq F^{max}(FT)$$

$$\tag{4.4}$$

c) System Emission Constraint

$$\sum_i \sum_t C_{ei}\big(P(i,t)I(i,t) + R(i,t)I(i,t) + N(i,t)\big) + S_e(i,t) \leq E^{max} \tag{4.5}$$

4.2.2 Unit Constraints

a) Unit Generation Limits

$$P(i,t)I(i,t) + B(i,t) \geq P_0(i,t) \tag{4.6a}$$

$$0 \leq B(i,t) \leq P_0(i,t) \tag{4.6b}$$

$$P_{gmin}(i) \leq P(i,t)I(i,t) + R(i,t)I(i,t) + N(i,t) \leq P_{gmax}(i) \tag{4.6c}$$

$$R(i,t)I(i,t) \leq r_s(i,t)I(i,t) \tag{4.6d}$$

where,

$$r_s(i,t) = \min\{10 \times MSR(i), P_{gmax}(i,t) - P(i,t)\}$$

$$N(i,t) \leq n_o(i,t) \tag{4.6e}$$

PRICE-BASED UNIT COMMITMENT

where, $n_o(i,t) = \begin{cases} q(i), & \text{if unit is OFF} \\ n_s(i,t), & \text{if unit is ON} \end{cases}$

and $n_s(i,t) = \min\{10 \times MSR(i), P_{gmax}(i) - P(i,t) - R(i,t)\}$

Note that MSR is the maximum sustained ramp rate (MW/min). The spinning reserve is the unloaded synchronized generation that can ramp up in 10 minutes. Non-spinning reserve is the unsynchronized generating capacity that can ramp up in 10 minutes.

b) Unit Minimum ON/OFF Durations

$$[X^{on}(i,t) - T^{on}(i)] \times [I(i,t-1) - I(i,t)] \geq 0 \tag{4.7a}$$

$$[X^{off}(i,t-1) - T^{off}(i,t)] \times [I(i,t-1) - I(i,t)] \geq 0 \tag{4.7b}$$

c) Unit Ramping Constraints

$$P(i,t) - P(i,t-1) \leq UR(i) \quad \text{as unit } i \text{ ramps up} \tag{4.8a}$$

$$P(i,t-1) - P(i,t) \leq DR(i) \quad \text{as unit } i \text{ ramps down} \tag{4.8b}$$

See Section 8.2 for more general ramping constraints modeling.

d) Unit Fuel Constraints

$$F^{min}(i) \leq \sum_t C_{fi}\big(P(i,t)I(i,t) + R(i,t)I(i,t) + N(i,t)\big) + S_f(i,t) \leq F^{max}(i) \tag{4.9}$$

4.3 PBUC SOLUTION

Lagrangian relaxation is used to solve PBUC. The basic idea is to relax coupling constraints (i.e., coupling either units, time periods, or both) into the objective function by using Lagrangian multipliers. The relaxed problem is then decomposed into subproblems for each unit. The dynamic programming process is used to search the optimal commitment for each unit. Lagrangian multipliers are then updated based on violations of coupling constraints.

We begin our solution methodology with a simple case that ignores emission or fuel constraints. We will show that emission and fuel constraints can be included without much modification to this methodology.

4.3.1 Solution without Emission or Fuel Constraints

The original maximization objective is equivalent to the minimization a revised objective function. Here, we specify the objective as

$$\text{minimize} \sum_i \sum_t -F(i,t) \tag{4.10}$$

Using Lagrangian multipliers to relax system constraints (i.e., energy and reserve), we write the Lagrangian function as

$$L = \sum_i \sum_t -F(i,t) \\ + \sum_t \lambda_g(t) \sum_i P(i,t) I(i,t) + \sum_t \lambda_r(t) \sum_i R(i,t) I(i,t) + \sum_t \lambda_n(t) \sum_i N(i,t) \tag{4.11}$$

4.3.1.1 Single-Unit Dynamic Programming

The Lagrangian term for one unit at a single period is given as follows:

When the unit is ON ($I(i,t) = 1$),

$$\begin{aligned} L(i,t) = & -\rho_{gm}(t)[P(i,t) - P_0(i,t)] + \lambda_g(t) P(i,t) \\ & - \rho_{rm}(t) R(i,t) + \lambda_r(t) R(i,t) \\ & - \rho_{nm}(t) N(i,t) + \lambda_n(t) N(i,t) \\ & + C_i(P(i,t) + R(i,t) + N(i,t)) + S(i,t) - f_i(P_0(i,t)) \end{aligned} \tag{4.12}$$

When unit is OFF ($I(i,t) = 0$),

$$\begin{aligned} L(i,t) = & -\rho_{nm}(t) N(i,t) + \lambda_n(t) N(i,t) \\ & + \rho_{gm}(t) P_0(i,t) + C_i(N(i,t)) - f_i(P_0(i,t)) \end{aligned} \tag{4.13}$$

The separable single-unit problem is formulated as

$$\text{minimize } L(i) = \sum_t L(i,t) I(i,t) \tag{4.14}$$

Dynamic programming is used to solve this problem. The key is to determine P, R, and N since they are coupled through the cost function.

PRICE-BASED UNIT COMMITMENT

4.3.1.1.1 Optimality Condition When the Unit is ON. When the unit is ON, the derivatives of the Lagrangian function with respect to P, R, and N are

$$\frac{\partial L}{\partial P(i,t)} = -\rho_{gm}(t) + \lambda_g(t) + \frac{\partial C_i}{\partial P(i,t)} \tag{4.15a}$$

$$\frac{\partial L}{\partial R(i,t)} = -\rho_{rm}(t) + \lambda_r(t) + \frac{\partial C_i}{\partial R(i,t)} \tag{4.15b}$$

$$\frac{\partial L}{\partial N(i,t)} = -\rho_{nm}(t) + \lambda_n(t) + \frac{\partial C_i}{\partial N(i,t)} \tag{4.15c}$$

Interestingly, λ_g, λ_r, and λ_n act as price signals for their relaxed constraints. Since

$$\begin{aligned} C_i &= C_i(P(i,t) + R(i,t) + N(i,t)) \\ &= a(i) + b(i)[P(i,t) + R(i,t) + N(i,t)] + c(i)[P(i,t) + R(i,t) + N(i,t)]^2 \end{aligned} \tag{4.16}$$

we have

$$\begin{aligned} \frac{\partial C_i}{\partial P(i,t)} &= \frac{\partial C_i}{\partial R(i,t)} = \frac{\partial C_i}{\partial N(i,t)} \\ &= b(i) + 2c(i)[P(i,t) + R(i,t) + N(i,t)] \\ &\equiv \lambda(i,t) \end{aligned} \tag{4.17}$$

So, when the unit is ON, the optimality condition is

$$-\rho_{gm}(t) + \lambda_g(t) + \lambda(i,t)$$
$$= \frac{\partial L}{\partial P(i,t)} \begin{cases} = 0, & \text{when } P_{min}(i,t) < P(i,t) < P_{max}(i,t) \\ < 0, & \text{when } P(i,t) = P_{max}(i,t) \\ > 0, & \text{when } P(i,t) = P_{min}(i,t) \end{cases} \tag{4.18a}$$

$$-\rho_{rm}(t) + \lambda_r(t) + \lambda(i,t)$$
$$= \frac{\partial L}{\partial R(i,t)} \begin{cases} = 0, & \text{when } R_{min}(i,t) < R(i,t) < R_{max}(i,t) \\ < 0, & \text{when } R(i,t) = R_{max}(i,t) \\ > 0, & \text{when } R(i,t) = R_{min}(i,t) \end{cases} \tag{4.18b}$$

$$\frac{\partial L}{\partial N(i,t)} = -p_{nm}(t) + \lambda_n(t) + \lambda(i,t) \begin{cases} = 0, & \text{when } N_{min}(i,t) < N(i,t) < N_{max}(i,t) \\ < 0, & \text{when } N(i,t) = N_{max}(i,t) \\ > 0, & \text{when } N(i,t) = N_{min}(i,t) \end{cases} \quad (4.18c)$$

Since $\lambda(i,t)$ is a function P, R, and N, we could presumably solve the following equations for optimality conditions (i.e., three variables and three equations) to compute P, R, and N. However, the coefficient matrix in this case would be singular, and we would not be able to calculate P, R, and N.

$$\begin{bmatrix} 1 & 1 & 1 \\ 1 & 1 & 1 \\ 1 & 1 & 1 \end{bmatrix} \begin{bmatrix} P(i,t) \\ R(i,t) \\ N(i,t) \end{bmatrix} = \begin{bmatrix} \dfrac{p_{gm}(t) - \lambda_g(t) - b(i)}{2c(i)} \\ \dfrac{p_{rm}(t) - \lambda_r(t) - b(i)}{2c(i)} \\ \dfrac{p_{nm}(t) - \lambda_n(t) - b(i)}{2c(i)} \end{bmatrix}$$

Instead, we propose the following approach. According to the formulation of $L(i,t)$, equations for P, R, and N would be identical except for their price coefficients and Lagrangian multipliers. In order to minimize $L(i,t)$, we should compare $p_{gm}(t) - \lambda_g(t)$, $p_{rm}(t) - \lambda_r(t)$ and $p_{nm}(t) - \lambda_n(t)$, and apply the optimality conditions to compute P, R, and N sequentially.

If we suppose that $p_{gm}(t) - \lambda_g(t) < p_{rm}(t) - \lambda_r(t) < p_{nm}(t) - \lambda_n(t)$ (other cases can be handled similarly), the following procedure would solve the optimization problem:

1. Determine $N_{min}(i,t)$ and $N_{max}(i,t)$, and based on optimality conditions for N (see 4.18c), compute $N(i,t)$ as follows subject to $N_{min}(i,t)$ and $N_{max}(i,t)$:

$$N(i,t) = \frac{p_{nm}(t) - \lambda_n(t) - b(i)}{2c(i)} - P_{gmin}(i)$$

$$N_{min}(i,t) \le N(i,t) \le N_{max}(i,t)$$

PRICE-BASED UNIT COMMITMENT

2. Determine $R_{min}(i,t)$ and $R_{max}(i,t)$, and based on optimality conditions for R (see 4.18b), compute $R(i,t)$ as follows subject to $R_{min}(i,t)$ and $R_{max}(i,t)$:

$$R(i,t) = \frac{\rho_{rm}(t) - \lambda_r(t) - b(i)}{2c(i)} - P_{gmin}(i) - N(i,t)$$

$$R_{min}(i,t) \le R(i,t) \le R_{max}(i,t)$$

3. Determine $P_{min}(i,t)$ and $P_{max}(i,t)$, and based on optimality conditions for P (see 4.18a), compute $P(i,t)$ as follows subject to $P_{min}(i,t)$ and $P_{max}(i,t)$:

$$P(i,t) = \frac{\rho_{gm}(t) - \lambda_g(t) - b(i)}{2c(i)} - R(i,t) - N(i,t)$$

$$P_{min}(i,t) \le P(i,t) \le P_{max}(i,t)$$

4.3.1.1.2 Optimality Condition When the Unit is OFF. When the unit is OFF, the derivative of Lagrangian function with respect to N is given as

$$\frac{\partial L}{\partial N(i,t)} = -\rho_{nm}(t) + \lambda_n(t) + \frac{\partial C_i}{\partial N(i,t)} \quad (4.19)$$

Since

$$\begin{aligned} C_i &= C_i(N(i,t)) \\ &= a(i) + b(i)[N(i,t)] + c(i)[N(i,t)]^2 \end{aligned} \quad (4.20)$$

we would have

$$\frac{\partial C_i}{\partial N(i,t)} = b(i) + 2c(i)[N(i,t)] \equiv \lambda(i,t) \quad (4.21)$$

So, when the unit is OFF, the optimality condition is

$$-\rho_{nm}(t) + \lambda_n(t) + \lambda(i,t)$$

$$= \frac{\partial L}{\partial N(i,t)} \begin{cases} = 0, & \text{when } N_{min}(i,t) < N(i,t) < N_{max}(i,t) \\ < 0, & \text{when } N(i,t) = N_{max}(i,t) \\ > 0, & \text{when } N(i,t) = N_{min}(i,t) \end{cases} \quad (4.22)$$

Correspondingly, we compute N as follows. Determine $N_{min}(i,t)$ and $N_{max}(i,t)$, and based on optimality conditions for N (see 4.22), compute $N(i,t)$ as follows subject to $N_{min}(i,t)$ and $N_{max}(i,t)$. In this case,

$$N(i,t) = \frac{\rho_{nm}(t) - \lambda_n(t) - b(i)}{2c(i)}$$

$$N_{min}(i,t) \leq N(i,t) \leq N_{max}(i,t)$$

4.3.1.2 Multipliers Update

From the optimality conditions (4.18) and (4.22):

- P is proportional to $\rho_{gm}(t) - \lambda_g(t)$ and a larger $\lambda_g(t)$ corresponds to a smaller P (and vice versa).
- R is proportional to $\rho_{rm}(t) - \lambda_r(t)$ and a larger $\lambda_r(t)$ corresponds to a smaller R (and vice versa).
- N is proportional to $\rho_{nm}(t) - \lambda_n(t)$ and a larger $\lambda_n(t)$ corresponds to a smaller N (and vice versus).

Considering energy and reserve constraints (4.3), we use the following rule for updating multipliers based on the subgradient method. See Section 4.4.2.1 for computing the step component.

$$\lambda_g(t) = \lambda_g(t) + step \times \left(\sum_i P(i,t)I(i,t) - P^{max}(t) \right)$$

$$\text{if } \sum_i P(i,t)I(i,t) > P^{max}(t) \tag{4.23a}$$

$$\lambda_g(t) = \lambda_g(t) + step \times \left(\sum_i P(i,t)I(i,t) - P^{min}(t) \right)$$

$$\text{if } \sum_i P(i,t)I(i,t) < P^{min}(t) \tag{4.23b}$$

$$\lambda_r(t) = \lambda_r(t) + step \times \left(\sum_i R(i,t)I(i,t) - R^{max}(t) \right)$$

PRICE-BASED UNIT COMMITMENT

$$\text{if } \sum_i R(i,t)I(i,t) > R^{max}(t) \tag{4.23c}$$

$$\lambda_r(t) = \lambda_r(t) + step \times \left(\sum_i R(i,t)I(i,t) - R^{min}(t) \right)$$

$$\text{if } \sum_i R(i,t)I(i,t) < R^{min}(t) \tag{4.23d}$$

$$\lambda_n(t) = \lambda_n(t) + step \times \left(\sum_i N(i,t) - N^{max}(t) \right)$$

$$\text{if } \sum_i N(i,t)I(i,t) > N^{max}(t) \tag{4.23e}$$

$$\lambda_n(t) = \lambda_n(t) + step \times \left(\sum_i N(i,t) - N^{min}(t) \right)$$

$$\text{if } \sum_i N(i,t)I(i,t) < N^{min}(t) \tag{4.23f}$$

4.3.1.3 Economic Dispatch

Once the unit commitment status is determined, an economic dispatch problem is formulated and solved to ensure the feasibility of the original unit commitment solution. The economic dispatch problem at time t is given as

$$\text{minimize } \sum_i -F(i,t) \tag{4.24}$$

subject to energy, reserve, and unit generation limits. Here unlike (4.1), the value of $I(i,t)$ is already determined and is no longer a variable. Therefore, quadratic or linear programming can be applied to solve this problem. However, the process is time-consuming for a large-scale system.

Analyzing $F(i,t)$, we learn that the only item that distinguishes P, R, and N from one another is their market price. So we convert the original problem into three subproblems for P, R, and N and solve the subproblems sequentially, starting with the most expensive (in the sense of market price).

In the following, we consider $\rho_{gm}(t) < \rho_{rm}(t) < \rho_{nm}(t)$. Other cases can be considered similarly.

4.3.1.3.1 Economic Dispatch for Non-spinning Reserve (N). Since $I(i,t)$ is given in the unit commitment section, the formulation is as follows:

minimize

$$\sum_i \{-\rho_{nm}(t)N(i,t) + C_i(P_{gmin}(i) + N(i,t))\} \cdot I(i,t)$$
$$+ \sum_i \{-\rho_{nm}(t)N(i,t) + C_i(N(i,t))\} \cdot (1 - I(i,t)) \qquad (4.25a)$$

subject to

$$0 \le N(i,t) \le G_{max}(i,t) - P_{gmin}(i) \quad \text{(Unit is ON)} \qquad (4.25b)$$

$$0 \le N(i,t) \le G0_{max}(i,t) \qquad \text{(Unit is OFF)} \qquad (4.25c)$$

$$N^{min}(t) \le \sum_i N(i,t) \le N^{max}(t) \qquad (4.25d)$$

In (4.25b), $G_{max}(i,t)$ is the maximum capability of unit i at time t when unit i is ON. In some ramping cases, $G_{max}(i,t)$ may be different from the maximum physical capacity of the unit. In (4.25c), $G0_{max}(i,t)$ is the maximum capability of unit i at time t when unit i is OFF, which is the quick start capability of the unit, if available, and otherwise it is zero. The solution method is given as follows.

1. Determine $N_{min}(i,t)$ and $N_{max}(i,t)$, and using $\rho_{nm}(t)$, find $N(i,t)$ subject to $N_{min}(i,t)$ and $N_{max}(i,t)$.

$$N(i,t) = \frac{\rho_{nm}(t) - b(i)}{2c(i)} - P_{gmin}(i) \quad \text{(unit is ON)}$$

$$N(i,t) = \frac{\rho_{nm}(t) - b(i)}{2c(i)} \quad \text{(unit is OFF)}$$

$$N_{min}(i,t) \le N(i,t) \le N_{max}(i,t)$$

2. If $N^{min}(t) \le \sum_i N(i,t) \le N^{max}(t)$, end.

PRICE-BASED UNIT COMMITMENT

3. If $\sum_i N(i,t) > N^{max}(t)$, let $\sum_i N(i,t) = N^{max}(t)$ and compute $N(i,t)$ again.

4. If $\sum_i N(i,t) < N^{min}(t)$, let $\sum_i N(i,t) = N^{min}(t)$ and compute $N(i,t)$ again.

The size of the decomposed problem is one-third that of the original problem. Note that classical economic dispatch methods, such as lambda iteration method, first-order gradient method and second-order gradient method can be used to compute N. Also, quadratic or linear programming may be used in this approach.

4.3.1.3.2 Economic Dispatch for Spinning Reserve (R). In the following formulation, $I(i,t)$ is given in the unit commitment section:

minimize

$$\sum_i \{-\rho_{rm}(t)R(i,t) + C_i(P_{gmin}(i) + N(i,t) + R(i,t))\} \cdot I(i,t) \quad (4.26a)$$

subject to (for all ON units)

$$0 \le R(i,t) \le G_{max}(i,t) - P_{gmin}(i) - N(i,t) \quad (4.26b)$$

$$R^{min}(t) \le \sum_i R(i,t) \le R^{max}(t) \quad (4.26c)$$

The solution method is given as follows.

1. Determine $R_{min}(i,t)$ and $R_{max}(i,t)$, and using $\rho_{rm}(t)$, find $R(i,t)$ subject to $R_{min}(i,t)$ and $R_{max}(i,t)$.

$$R(i,t) = \frac{\rho_{rm}(t) - b(i)}{2c(i)} - P_{gmin}(i) - N(i,t), \quad R_{min}(i,t) \le R(i,t) \le R_{max}(i,t)$$

2. If $R^{min}(t) \le \sum_i R(i,t) \le R^{max}(t)$, end.

3. If $\sum_i R(i,t) > R^{max}(t)$, let $\sum_i R(i,t) = R^{max}(t)$ and compute $R(i,t)$ again.

4. If $\sum_i R(i,t) < R^{min}(t)$, let $\sum_i R(i,t) = R^{min}(t)$ and compute $R(i,t)$ again.

4.3.1.3.3 Economic Dispatch for Energy (P). In the following formulation, $I(i,t)$ is given in the unit commitment section.

minimize

$$\sum_i \{-\rho_{gm}(t)[P(i,t)-P_0(i,t)] + C_i(N(i,t)+R(i,t)+P(i,t))\} \cdot I(i,t) \quad (4.27a)$$

subject to (for all ON units)

$$P_{gmin}(i) \le P(i,t) \le G_{max}(i,t) - N(i,t) - R(i,t) \quad (4.27b)$$

$$P^{min}(t) \le \sum_i P(i,t) \le P^{max}(t) \quad (4.27c)$$

The solution method is given as follows.

1. Determine $P_{min}(i,t)$ and $P_{max}(i,t)$, and using $\rho_{gm}(t)$, find $P(i,t)$ subject to $P_{min}(i,t)$ and $P_{max}(i,t)$.

$$P(i,t) = \frac{\rho_{gm}(t) - b(i)}{2c(i)} - R(i,t) - N(i,t), \quad P_{min}(i,t) \le P(i,t) \le P_{max}(i,t)$$

2. If $P^{min}(t) \le \sum_i P(i,t) \le P^{max}(t)$, end.

3. If $\sum_i P(i,t) > P^{max}(t)$, let $\sum_i P(i,t) = P^{max}(t)$ and compute $P(i,t)$ again.

4. If $\sum_i P(i,t) < P^{min}(t)$, let $\sum_i P(i,t) = P^{min}(t)$ and compute $P(i,t)$ again.

4.3.1.4 Convergence Criterion

The convergence criterion may be defined as the relative duality gap to be less than a pre-specified threshold. The relative duality gap is defined as follows.

Suppose that the solution from unit commitment is *SU* and the solution from economic dispatch is *SE*. Substituting *SU* into the Lagrangian function, we would get the Lagrangian value, denoted as *LU*. Substituting *SE* into the Lagrangian function we would get the Lagrangian value, denoted as *LE*. The relative duality gap (*RDG*) is defined as

PRICE-BASED UNIT COMMITMENT

$$RDG = \frac{LU - LE}{LU} \tag{4.28}$$

The flowchart of PBUC is shown in Figure 4.1.

4.3.2 Solution with Emission and Fuel Constraints

With emission and fuel constraints, the Lagrangian function is given as

$$\begin{aligned}
L = &\sum_i \sum_t -F(i,t) \\
&+ \sum_t \lambda_g(t) \sum_i P(i,t)I(i,t) + \sum_t \lambda_r(t) \sum_i R(i,t)I(i,t) + \sum_t \lambda_n(t) \sum_i N(i,t) \\
&+ \lambda_e \sum_i \sum_t C_{ei}\{P(i,t)I(i,t) + R(i,t)I(i,t) + N(i,t)\} \\
&+ \sum_{FT} \lambda_f(FT) \sum_{i \in FT} \sum_t C_{fi}\{P(i,t)I(i,t) + R(i,t)I(i,t) + N(i,t)\} \\
&+ \sum_i \lambda_{fu}(i) \sum_t C_{fi}\{P(i,t)I(i,t) + R(i,t)I(i,t) + N(i,t)\}
\end{aligned} \tag{4.29}$$

Lagrangian term for one unit at one time period is given as follows:
When the unit is ON ($I(i,t) = 1$),

$$\begin{aligned}
L(i,t) = &-\rho_{gm}(t)[P(i,t) - P_0(i,t)] + \lambda_g(t)P(i,t) \\
&- \rho_{rm}(t)R(i,t) + \lambda_r(t)R(i,t) \\
&- \rho_{nm}(t)N(i,t) + \lambda_n(t)N(i,t) \\
&+ C_i(P(i,t) + R(i,t) + N(i,t)) + S(i,t) - f_i(P_0(i,t)) \\
&+ \lambda_e C_{ei}(P(i,t) + R(i,t) + N(i,t)) \\
&+ (\lambda_f(FT) + \lambda_{fu}(i))C_{fi}(P(i,t) + R(i,t) + N(i,t))
\end{aligned} \tag{4.30}$$

When the unit is OFF ($I(i,t) = 0$),

$$\begin{aligned}
L(i,t) = &-\rho_{nm}(t)N(i,t) + \lambda_n(t)N(i,t) \\
&+ \rho_{gm}(t)P_0(i,t) + C_i(N(i,t)) - f_i(P_0(i,t)) \\
&+ \lambda_e C_{ei}(N(i,t)) + (\lambda_f(FT) + \lambda_{fu}(i))C_{fi}(N(i,t))
\end{aligned} \tag{4.31}$$

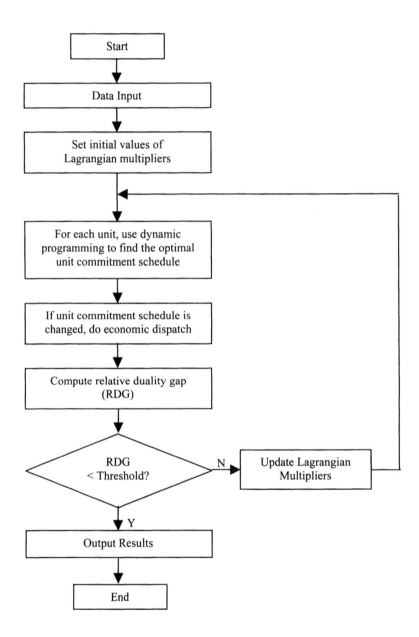

Figure 4.1 PBUC Flowchart

4.3.2.1 Optimality Condition

When the unit is ON, derivatives of the Lagrangian function with respect to P, R, and N are

$$\frac{\partial L}{\partial P(i,t)} = -\rho_{gm}(t) + \lambda_g(t) + \frac{\partial C_i}{\partial P(i,t)} + \lambda_e \frac{\partial C_{ei}}{\partial P(i,t)} + (\lambda_f(FT) + \lambda_{fu}(i))\frac{\partial C_{fi}}{\partial P(i,t)} \quad (4.32a)$$

$$\frac{\partial L}{\partial R(i,t)} = -\rho_{rm}(t) + \lambda_r(t) + \frac{\partial C_i}{\partial R(i,t)} + \lambda_e \frac{\partial C_{ei}}{\partial R(i,t)} + (\lambda_f(FT) + \lambda_{fu}(i))\frac{\partial C_{fi}}{\partial R(i,t)} \quad (4.32b)$$

$$\frac{\partial L}{\partial N(i,t)} = -\rho_{nm}(t) + \lambda_n(t) + \frac{\partial C_i}{\partial N(i,t)} + \lambda_e \frac{\partial C_{ei}}{\partial N(i,t)} + (\lambda_f(FT) + \lambda_{fu}(i))\frac{\partial C_{fi}}{\partial N(i,t)} \quad (4.32c)$$

Since

$$\begin{aligned} C_i &= C_i(P(i,t) + R(i,t) + N(i,t)) \\ &= a(i) + b(i)[P(i,t) + R(i,t) + N(i,t)] + c(i)[P(i,t) + R(i,t) + N(i,t)]^2 \end{aligned} \quad (4.33a)$$

$$\begin{aligned} C_{ei} &= C_{ei}(P(i,t) + R(i,t) + N(i,t)) \\ &= a_e(i) + b_e(i)[P(i,t) + R(i,t) + N(i,t)] + c_e(i)[P(i,t) + R(i,t) + N(i,t)]^2 \end{aligned} \quad (4.33b)$$

$$\begin{aligned} C_{fi} &= C_{fi}(P(i,t) + R(i,t) + N(i,t)) \\ &= a_f(i) + b_f(i)[P(i,t) + R(i,t) + N(i,t)] + c_f(i)[P(i,t) + R(i,t) + N(i,t)]^2 \end{aligned} \quad (4.33c)$$

We have

$$\begin{aligned} &\frac{\partial C_i}{\partial P(i,t)} + \lambda_e \frac{\partial C_{ei}}{\partial P(i,t)} + (\lambda_f(FT) + \lambda_{fu}(i))\frac{\partial C_{fi}}{\partial P(i,t)} \\ &= \frac{\partial C_i}{\partial R(i,t)} + \lambda_e \frac{\partial C_{ei}}{\partial R(i,t)} + (\lambda_f(FT) + \lambda_{fu}(i))\frac{\partial C_{fi}}{\partial R(i,t)} \\ &= \frac{\partial C_i}{\partial N(i,t)} + \lambda_e \frac{\partial C_{ei}}{\partial N(i,t)} + (\lambda_f(FT) + \lambda_{fu}(i))\frac{\partial C_{fi}}{\partial N(i,t)} \\ &= [b(i) + \lambda_e b_e(i) + (\lambda_f(FT) + \lambda_{fu}(i))b_f(i)] \\ &\quad + 2[c(i) + \lambda_e c_e(i) + (\lambda_f(FT) + \lambda_{fu}(i))c_f(i)] \cdot [P(i,t) + R(i,t) + N(i,t)] \\ &\equiv \lambda(i,t) \end{aligned} \quad (4.34)$$

If we define

$$b'(i) = b(i) + \lambda_e b_e(i) + (\lambda_f(FT) + \lambda_{fu}(i))b_f(i) \tag{4.35a}$$

$$c'(i) = c(i) + \lambda_e c_e(i) + (\lambda_f(FT) + \lambda_{fu}(i))c_f(i) \tag{4.35b}$$

the only difference between including and not including emission and fuel constraints in the formulation is represented by b and b', c and c'.

The optimality condition when the unit is OFF is given as

$$\frac{\partial L}{\partial N(i,t)} = -\rho_{nm}(t) + \lambda_n(t) + \frac{\partial C_i}{\partial N(i,t)} + \lambda_e \frac{\partial C_{ei}}{\partial N(i,t)} + (\lambda_f(FT) + \lambda_{fu}(i))\frac{\partial C_{fi}}{\partial N(i,t)} \tag{4.36}$$

Since

$$C_i = C_i(N(i,t)) = a(i) + b(i)[N(i,t)] + c(i)[N(i,t)]^2 \tag{4.37a}$$

$$C_{ei} = C_{ei}(N(i,t)) = a_e(i) + b_e(i)[N(i,t)] + c_e(i)[N(i,t)]^2 \tag{4.37b}$$

$$C_{fi} = C_{fi}(N(i,t)) = a_f(i) + b_f(i)[N(i,t)] + c_f(i)[N(i,t)]^2 \tag{4.37c}$$

We have

$$\begin{aligned}&\frac{\partial C_i}{\partial N(i,t)} + \lambda_e \frac{\partial C_{ei}}{\partial N(i,t)} + (\lambda_f(FT) + \lambda_{fu}(i))\frac{\partial C_{fi}}{\partial N(i,t)} \\ &= [b(i) + \lambda_e b_e(i) + (\lambda_f(FT) + \lambda_{fu}(i))b_f(i)] \\ &\quad + 2[c(i) + \lambda_e c_e(i) + (\lambda_f(FT) + \lambda_{fu}(i))c_f(i)] \cdot [N(i,t)] \\ &\equiv \lambda(i,t)\end{aligned} \tag{4.38}$$

If we define

$$b'(i) = b(i) + \lambda_e b_e(i) + (\lambda_f(FT) + \lambda_{fu}(i))b_f(i) \tag{4.39a}$$

$$c'(i) = c(i) + \lambda_e c_e(i) + (\lambda_f(FT) + \lambda_{fu}(i))c_f(i) \tag{4.39b}$$

the only difference in including emission and fuel constraints is represented by b and b', c and c'.

4.3.2.2 Multipliers Update for Emission and Fuel Constraints

For convenience, we define

$$E^{\Sigma} = \sum_i \sum_t C_{ei}\big(P(i,t)I(i,t) + R(i,t)I(i,t) + N(i,t)\big) + S_e(i,t) \quad (4.40a)$$

$$F^{\Sigma}(FT) = \sum_{i \in FT} \sum_t C_{fi}\big(P(i,t)I(i,t) + R(i,t)I(i,t) + N(i,t)\big) + S_f(i,t) \quad (4.40b)$$

$$F^{\Sigma}(i) = \sum_t C_{fi}\big(P(i,t)I(i,t) + R(i,t)I(i,t) + N(i,t)\big) + S_f(i,t) \quad (4.40c)$$

Consider an example for the multiplier updating process of emission constraint. Since $(P + R + N)$ is proportional to λ_e and a larger λ_e corresponds to a smaller $(P + R + N)$ (and vice versa), and emission is a quadratic function of $(P + R + N)$ and a larger $(P + R + N)$ corresponds to a larger emission (and vice versa), we use the following multipliers updating rule (i.e., the subgradient method) for emission multipliers:

$$\lambda_e = \lambda_e + step \times \big(E^{\Sigma} - E^{max}\big), \text{ if } E^{\Sigma} > E^{max} \quad (4.41a)$$

Similarly, for system and unit fuel constraints, we implement the following updating rule for multipliers (i.e., subgradient method).

$$\lambda_f(FT) = \lambda_f(FT) + step \times \big(F^{\Sigma}(FT) - F^{max}(FT)\big), \text{ if } F^{\Sigma}(FT) > F^{max}(FT) \quad (4.41b)$$

$$\lambda_f(FT) = \lambda_f(FT) + step \times \big(F^{\Sigma}(FT) - F^{min}(FT)\big), \text{ if } F^{\Sigma}(FT) < F^{min}(FT) \quad (4.41c)$$

$$\lambda_{fu}(i) = \lambda_{fu}(i) + step \times \big(F^{\Sigma}(i) - F^{max}(i)\big), \text{ if } F^{\Sigma}(i) > F^{max}(i) \quad (4.41d)$$

$$\lambda_{fu}(i) = \lambda_{fu}(i) + step \times \big(F^{\Sigma}(i) - F^{min}(i)\big), \text{ if } F^{\Sigma}(i) < F^{min}(i) \quad (4.41e)$$

See Section 4.4.2.2 for computing the step components.

4.3.2.3 Economic Dispatch

Given λ_e, $\lambda_f(FT)$, and $\lambda_{fu}(i)$, we perform an economic dispatch without considering emission and fuel constraints as in Section 4.3.1.3 but replacing b with b' and c with c'. Then we check emission and fuel constraints. If no constraint is violated, the economic dispatch is completed. Otherwise, we modify λ_e, $\lambda_f(FT)$, and $\lambda_{fu}(i)$, and perform another economic dispatch until all constraints are satisfied.

4.4 DISCUSSION ON SOLUTION METHODOLOGY

Additional discussions are presented here on the solution methodology.

4.4.1 Energy Purchase

Based on $P(i,t)I(i,t) + B(i,t) \geq P_o(i,t)$ and $0 \leq B(i,t) \leq P_o(i,t)$, and assuming that energy sales and purchases are not made simultaneously for the same unit, we will have the following:

When the unit is ON,

$$B(i,t) = P_0(i,t) - P(i,t), \text{ if } P(i,t) < P_0(i,t) \tag{4.42a}$$

$$B(i,t) = 0, \text{ if } P(i,t) > P_0(i,t) \tag{4.42b}$$

When the unit is OFF,

$$B(i,t) = P_0(i,t) \tag{4.42c}$$

(4.42) indicates that the energy purchase variable (B) depends on unit status (I) and generation (P).

4.4.2 Derivation of Steps in Update of Multipliers

4.4.2.1 Step Calculation for P, R, and N

Take P as an example. Let $\lambda_g^{(1)}$ correspond to $\left(\sum P\right)^{(1)}$ and $\lambda_g^{(2)}$ correspond to $\left(\sum P\right)^{(2)}$, where the superscript refers to the iteration number. Then, for $\lambda_g = \lambda_g^{(0)} + step * \left(\left(\sum P\right) - \left(\sum P\right)^{(0)}\right)$, the steps can be developed as $step = \dfrac{\lambda_g^{(2)} - \lambda_g^{(1)}}{\left(\sum P\right)^{(2)} - \left(\sum P\right)^{(1)}}$. At the limit,

$$step = \frac{d\lambda_g}{d\left(\sum P\right)} = \frac{1}{\dfrac{d\left(\sum P\right)}{d\lambda_g}} = \frac{1}{\sum \dfrac{dP}{d\lambda_g}} = \frac{1}{\sum \dfrac{1}{\dfrac{d\lambda_g}{dP}}}$$

PRICE–BASED UNIT COMMITMENT

Since

$$C_i(P) = a(i) + b(i)P + c(i)P^2$$

$$\lambda_i(P) = \frac{dC_i(P)}{dP} = b(i) + 2c(i)P$$

$$\frac{d\lambda_i(P)}{dP} = 2c(i)$$

Then, $\quad step = \dfrac{1}{\sum_i \dfrac{1}{2c(i)}}$ \hfill (4.43)

We adopt a similar approach for R and N.

4.4.2.2 Step Calculation for Emission and Fuel Constraints

Take emission constraint as an example. Let $\lambda_e^{(1)}$ correspond to $\left(\sum E\right)^{(1)}$ and $\lambda_e^{(2)}$ correspond to $(E)^{(2)}$.

Then, in $\lambda_e = \lambda_e^{(0)} + step \times \left(\left(\sum E\right) - \left(\sum E\right)^{(0)}\right)$, the step can be developed as $step = \dfrac{\lambda_e^{(2)} - \lambda_e^{(1)}}{\left(\sum E\right)^{(2)} - \left(\sum E\right)^{(1)}}$. At the limit we have

$$step = \frac{d\lambda_e}{d\left(\sum E\right)} = \frac{1}{\dfrac{d\left(\sum E\right)}{d\lambda_e}} = \frac{1}{\sum \dfrac{dE}{d\lambda_e}}$$

Let $X(i,t) = P(i,t)I(i,t) + R(i,t)I(i,t) + N(i,t)$. Since

$$C_{ei}(X) = a_e(i) + b_e(i)X + c_e(i)X^2$$

$$\lambda_{ei}(X) = \frac{dC_{ei}(X)}{dX} = b_e(i) + 2c_e(i)X$$

$$\frac{d\lambda_{ei}(X)}{dX} = 2c_e(i)$$

$$\frac{dC_{ei}(X)}{d\lambda_{ei}(X)} = \frac{\frac{dC_{ei}(X)}{dX}}{\frac{d\lambda_{ei}(X)}{dX}} = \frac{\lambda_{ei}(X)}{\frac{d\lambda_{ei}(X)}{dX}} = \frac{b_e(i) + 2c_e(i)X}{2c_e(i)} = \frac{b_e(i)}{2c_e(i)} + X$$

So

$$step = \frac{1}{\sum_i \sum_t \left(\frac{b_e(i)}{2c_e(i)} + X(i,t) \right)} \quad (4.44)$$

Similarly, for system fuel constraint,

$$step = \frac{1}{\sum_{i \in FT} \sum_t \left(\frac{b_f(i)}{2c_f(i)} + X(i,t) \right)} \quad (4.45)$$

For unit fuel constraint,

$$step = \frac{1}{\sum_t \left(\frac{b_f(i)}{2c_f(i)} + X(i,t) \right)} \quad (4.46)$$

Comparing (4.44) with (4.43), we learn that in (4.43) the step is irrelevant to P (or X). However, in (4.44) the step is relevant to X (i.e., the current operating point). The reason is that P is a linear function of λ while emission (or fuel) is a quadratic function of λ. If we develop a linear relationship between emission and λ, we could derive a formula similar to (4.43). Since

$$\lambda_{ei}(X) = \frac{dC_{ei}(X)}{dX} = b_e(i) + 2c_e(i)X$$

We have

$$X = \frac{\lambda_{ei}(X) - b_e(i)}{2c_e(i)}$$

Substituting into $C_{ei}(X) = a_e(i) + b_e(i)X + c_e(i)X^2$, we have

PRICE-BASED UNIT COMMITMENT

$$C_{ei}(\lambda_{ei}) = a_e(i) + b_e(i)\left(\frac{\lambda_{ei} - b_e(i)}{2c_e(i)}\right) + c_e(i)\left(\frac{\lambda_{ei} - b_e(i)}{2c_e(i)}\right)^2$$

$$= \frac{\lambda_{ei}^2}{4c_e(i)} + \left(a_e(i) - \frac{b_e(i)^2}{4c_e(i)}\right)$$

$$\frac{dC_{ei}(\lambda_{ei})}{d(\lambda_{ei}^2)} = \frac{1}{4c_e(i)}$$

which shows that the emission is a linear function of λ^2. This suggests another kind of multiplier update,

$$\lambda_e^2 = \lambda_e^{(0)2} + step \times \left(\left(\sum E\right) - \left(\sum E\right)^{(0)}\right)$$

$$step = \frac{d(\lambda_e^2)}{d\left(\sum E\right)} = \frac{1}{\frac{d\left(\sum E\right)}{d(\lambda_e^2)}} = \frac{1}{\sum \frac{dE}{d(\lambda_e^2)}}$$

So

$$step = \frac{1}{\sum_i \sum_t \left(\frac{1}{4c_e(i)}\right)} \quad (4.47)$$

and

$$\lambda_e = \sqrt{\lambda_e^2 + step \times \left(E^\Sigma - E^{max}\right)}, \text{ if } E^\Sigma > E^{max} \quad (4.48)$$

Lagrangian multipliers for system fuel constraints and unit fuel constraints would be updated similarly.

4.4.3 Optimality Condition

Here, we analyze the optimal conditions in (4.18) to see when the equal sign holds. Suppose that $\lambda_g(t)$, $\lambda_r(t)$, and $\lambda_n(t)$ are all zero, which indicates that no system constraint is violated. Further, suppose that $\rho_{gm}(t) < \rho_{rm}(t) < \rho_{nm}(t)$.

If $\dfrac{\partial L}{\partial N(i,t)} = 0$, then $N_{min}(i,t) < N(i,t) < N_{max}(i,t)$. Accordingly, we derive the following conclusions:

- (4.18c) shows that $\lambda(i,t) = \rho_{nm}(t)$

- (4.18b) shows that $\dfrac{\partial L}{\partial R(i,t)} = \lambda(i,t) - \rho_{rm}(t) = \rho_{nm}(t) - \rho_{rm}(t) > 0$,

  which means $R(i,t) = R_{min}(i,t)$

- (4.18a) shows that $\dfrac{\partial L}{\partial P(i,t)} = \lambda(i,t) - \rho_{gm}(t) = \rho_{nm}(t) - \rho_{gm}(t) > 0$,

  which means that $P(i,t) = P_{min}(i,t)$

Other cases can be derived similarly. Results are given as follows:

If $\dfrac{\partial L}{\partial N(i,t)} > 0$, which means $N(i,t) = N_{min}(i,t)$,

then $\dfrac{\partial L}{\partial R(i,t)} > 0$, which means $R(i,t) = R_{min}(i,t)$

and $\dfrac{\partial L}{\partial P(i,t)} > 0$, which means $P(i,t) = P_{min}(i,t)$.

If $\dfrac{\partial L}{\partial N(i,t)} < 0$, which means $N(i,t) = N_{max}(i,t)$,

then $\dfrac{\partial L}{\partial R(i,t)} = 0$, which means $R_{min}(i,t) < R(i,t) < R_{max}(i,t)$

and $\dfrac{\partial L}{\partial P(i,t)} > 0$, which means $P(i,t) = P_{min}(i,t)$,

or $\dfrac{\partial L}{\partial R(i,t)} > 0$, which means $R(i,t) = R_{min}(i,t)$,

$\dfrac{\partial L}{\partial P(i,t)} > 0$, which means $P(i,t) = P_{min}(i,t)$,

or $\dfrac{\partial L}{\partial R(i,t)} < 0$, which means $R(i,t) = R_{max}(i,t)$,

PRICE-BASED UNIT COMMITMENT

$$\frac{\partial L}{\partial P(i,t)} = 0, \text{ which means } P_{min}(i,t) < P(i,t) < P_{max}(i,t),$$

or $\quad \dfrac{\partial L}{\partial P(i,t)} > 0$, which means $P(i,t) = P_{min}(i,t)$,

or $\quad \dfrac{\partial L}{\partial P(i,t)} < 0$, which means $P(i,t) = P_{max}(i,t)$.

One interesting conclusion is that $\dfrac{\partial L}{\partial P(i,t)} = 0$, $\dfrac{\partial L}{\partial R(i,t)} = 0$ and $\dfrac{\partial L}{\partial N(i,t)} = 0$ cannot be satisfied simultaneously. Another is that two conditions cannot be satisfied simultaneously. These observations point out that at least two of P, R, and N are at their limits.

4.5 ADDITIONAL FEATURES OF PBUC

4.5.1 Different Prices among Buses

If there is no transmission congestion, the bus prices will be the same. Otherwise, the bus prices will differ. The PBUC formulation could simulate a transmission congestion with different bus prices whereby the market price forecasts would include the conditions causing the congestion. Because bus prices differ in the market, the following change could be made in calculating the profit. However, the solution methodology will not be affected.

$$F(i,t) = \begin{Bmatrix} \rho_{gm}(i,t)[P(i,t) - P_0(i,t)] + \rho_{rm}(i,t)R(i,t) + \rho_{nm}(i,t)N(i,t) \\ -C_i(P(i,t) + R(i,t) + N(i,t)) - S(i,t) + f_i(P_0(i,t)) \end{Bmatrix} I(i,t)$$

$$+ \begin{Bmatrix} \rho_{nm}(i,t)N(i,t) - \rho_{gm}(i,t)B(i,t) \\ -C_i(N(i,t)) + f_i(P_0(i,t)) \end{Bmatrix} (1 - I(i,t))$$

(4.49)

4.5.2 Variable Fuel Price as a Function of Fuel Consumption

A difficult problem to solve is when the fuel price is not constant and would also depend on the total MBtu consumption. Then the cost function coefficients *a, b,* and *c* are not constant either, and the original formulation methodology cannot be used directly. Consider the case illustrated in Figure 4.2.

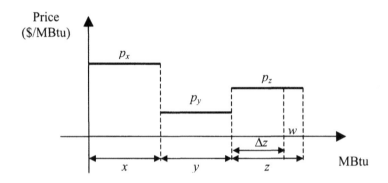

Figure 4.2 Fuel Purchase Price Curve (Illustration)

In the figure, there are three segments with different prices; p_x, p_y, and p_z in segments *x, y* and *z*, respectively. Suppose that a GENCO is interested in buying *w* MBtu (i.e., $w = x + y + \Delta z$), which is in segment *z*. The actual cost is given as

$$C = x \cdot p_x + y \cdot p_y + \Delta z \cdot p_z \qquad (4.50a)$$

Consider that if all MBtu were purchased at p_z, the cost would be

$$C_1 = (x + y + \Delta z) \cdot p_z \qquad (4.50b)$$

The difference between these two costs, called additional cost, is

$$\begin{aligned}\Delta C &= C - C_1 \\ &= (x \cdot p_x + y \cdot p_y + \Delta z \cdot p_z) - (x + y + \Delta z) \cdot p_z \\ &= (x \cdot p_x + y \cdot p_y) - (x + y) \cdot p_z\end{aligned} \qquad (4.50c)$$

Further analyzing the cost difference, we learn that the first part is the actual cost of purchasing $(x + y)$ MBtu and the second part is the cost of purchasing $(x + y)$ MBtu at p_z. Since *x, y,* p_x, p_y, and p_z are constant, the difference between the two costs would also be constant. Hence, we could

PRICE-BASED UNIT COMMITMENT

use p_z as the average price plus a constant for calculating the total cost. Since the average price is constant, cost function coefficients a, b and c would also be constant. So, we could use the original formulation methodology to solve the problem. Although the analysis is provided for a three-segment example, it can be extended to any number of segments.

4.5.3 Application of Lagrangian Augmentation

Since emission and fuel constraints are quadratic terms, it would be unnecessary to augment these constraints. Only energy and reserve constraints, which are linear, are augmented here to improve the convergence performance. Let L be the original Lagrangian function (without energy and reserve constraints):

$$\begin{aligned}L = &\sum_i \sum_t -F(i,t) \\ &+ \lambda_e \sum_i \sum_t C_{ei}(P(i,t)I(i,t) + R(i,t)I(i,t) + N(i,t)) \\ &+ \sum_{FT} \lambda_f(FT) \sum_{i \in FT} \sum_t C_{fi}(P(i,t)I(i,t) + R(i,t)I(i,t) + N(i,t)) \\ &+ \sum_i \lambda_{fu}(i) \sum_t C_{fi}(P(i,t)I(i,t) + R(i,t)I(i,t) + N(i,t))\end{aligned} \quad (4.51)$$

Let La be the augmented Lagrangian function.

$$La = L + \sum_t Lg(t) + \sum_t Lr(t) + \sum_t Ln(t) \quad (4.52)$$

where $Lg(t)$ is the augmentation term for energy constraints (4.3a), $Lr(t)$ is the augmentation term for spinning reserve constraints (4.3b), and $Ln(t)$ is the augmentation term for non-spinning reserve constraints (4.3c).

In the following, we list the formation of augmentation terms. See Appendix B.2 for formulating the augmentation.

$$Lg(t) = \begin{cases} \lambda_g(t) \cdot \left(\sum_i P(i,t)I(i,t) - P^{max}(t)\right) + \frac{1}{2}c \cdot \left(\sum_i P(i,t)I(i,t) - P^{max}(t)\right)^2 \\ \quad \text{if } \sum_i P(i,t)I(i,t) > P^{max}(t) - \frac{\lambda_g(t)}{c} \\ \frac{-\lambda_g(t)^2}{2c}, \quad \text{if } P^{min}(t) + \frac{\lambda_g(t)}{c} \le \sum_i P(i,t)I(i,t) \le P^{max}(t) - \frac{\lambda_g(t)}{c} \\ \lambda_g(t) \cdot \left(P^{min}(t) - \sum_i P(i,t)I(i,t)\right) + \frac{1}{2}c \cdot \left(P^{min}(t) - \sum_i P(i,t)I(i,t)\right)^2 \\ \quad \text{if } \sum_i P(i,t)I(i,t) < P^{min}(t) + \frac{\lambda_g(t)}{c} \end{cases}$$ (4.53a)

$$Lr(t) = \begin{cases} \lambda_r(t) \cdot \left(\sum_i R(i,t)I(i,t) - R^{max}(t)\right) + \frac{1}{2}c \cdot \left(\sum_i R(i,t)I(i,t) - R^{max}(t)\right)^2 \\ \quad \text{if } \sum_i R(i,t)I(i,t) > R^{max}(t) - \frac{\lambda_r(t)}{c} \\ \frac{-\lambda_r(t)^2}{2c} \quad \text{if } R^{min}(t) + \frac{\lambda_r(t)}{c} \le \sum_i R(i,t)I(i,t) \le R^{max}(t) - \frac{\lambda_r(t)}{c} \\ \lambda_r(t) \cdot \left(R^{min}(t) - \sum_i R(i,t)I(i,t)\right) + \frac{1}{2}c \cdot \left(R^{min}(t) - \sum_i R(i,t)I(i,t)\right)^2 \\ \quad \text{if } \sum_i R(i,t)I(i,t) < R^{min}(t) + \frac{\lambda_r(t)}{c} \end{cases}$$ (4.53b)

$$Ln(t) = \begin{cases} \lambda_n(t) \cdot \left(\sum_i N(i,t) - N^{max}(t)\right) + \frac{1}{2}c \cdot \left(\sum_i N(i,t) - N^{max}(t)\right)^2 \\ \quad \text{if } \sum_i N(i,t) > N^{max}(t) - \frac{\lambda_n(t)}{c} \\ \frac{-\lambda_n(t)^2}{2c} \quad \text{if } N^{min}(t) + \frac{\lambda_n(t)}{c} \le \sum_i N(i,t) \le N^{max}(t) - \frac{\lambda_n(t)}{c} \\ \lambda_n(t) \cdot \left(N^{min}(t) - \sum_i N(i,t)\right) + \frac{1}{2}c \cdot \left(N^{min}(t) - \sum_i N(i,t)\right)^2 \\ \quad \text{if } \sum_i N(i,t) < N^{min}(t) + \frac{\lambda_n(t)}{c} \end{cases}$$ (4.53c)

PRICE-BASED UNIT COMMITMENT

The quadratic terms in the augmentation function are coupled across units and thus are non-separable. However, the auxiliary principle [Coh84] states that every function that is not separable from a given decomposition of the decision variable space can be replaced by a linear approximation around the current operating point. Moreover, in order to improve the convergence property, we may impose convex quadratic terms of the decision variables (P, R, or N) to limit them from deviating from the previous solution. These quadratic terms are separable, convex, and differentiable with respect to the decision variables. It should be emphasized that the added terms do not affect optimality, since they vanish at the optimal solution.

In the following, we list the linearized formulations of $Lg(t)$, $Lr(t)$, and $Ln(t)$. To ensure convergence, as was proved in [Coh84] we consider $1/\varepsilon \geq 2c$.

$$Lg(t) = \begin{cases} \lambda_g(t) \cdot \left(\sum_i P(i,t) I(i,t) - P^{max}(t) \right) \\ \quad + c \cdot \left[\sum_i P(i,t)^{(k)} I(i,t)^{(k)} - P^{max}(t) \right] \cdot \left[\sum_i P(i,t) I(i,t) - P^{max}(t) \right] \\ \quad + \dfrac{1}{2\varepsilon} \sum_i \left[P(i,t)^{(k)} I(i,t)^{(k)} - P(i,t) I(i,t) \right]^2 \\ \qquad \text{if } \sum_i P(i,t) I(i,t) > P^{max}(t) - \dfrac{\lambda_g(t)}{c} \\[6pt] \dfrac{-\lambda_g(t)^2}{2c} \quad \text{if } P^{min}(t) + \dfrac{\lambda_g(t)}{c} \leq \sum_i P(i,t) I(i,t) \leq P^{max}(t) - \dfrac{\lambda_g(t)}{c} \\[6pt] \lambda_g(t) \cdot \left(P^{min}(t) - \sum_i P(i,t) I(i,t) \right) \\ \quad + c \cdot \left[\sum_i P(i,t)^{(k)} I(i,t)^{(k)} - P^{min}(t) \right] \cdot \left[\sum_i P(i,t) I(i,t) - P^{min}(t) \right] \\ \quad + \dfrac{1}{2\varepsilon} \sum_i \left[P(i,t)^{(k)} I(i,t)^{(k)} - P(i,t) I(i,t) \right]^2 \\ \qquad \text{if } \sum_i P(i,t) I(i,t) < P^{min}(t) + \dfrac{\lambda_g(t)}{c} \end{cases} \quad (4.54a)$$

$$Lr(t) = \begin{cases} \lambda_r(t) \cdot \left(\sum_i R(i,t)I(i,t) - R^{max}(t) \right) \\ + c \cdot \left[\sum_i R(i,t)^{(k)} I(i,t)^{(k)} - R^{max}(t) \right] \cdot \left[\sum_i R(i,t)I(i,t) - R^{max}(t) \right] \\ + \frac{1}{2\varepsilon} \sum_i \left[R(i,t)^{(k)} I(i,t)^{(k)} - R(i,t)I(i,t) \right]^2 \\ \qquad \text{if } \sum_i R(i,t)I(i,t) > R^{max}(t) - \frac{\lambda_r(t)}{c} \\ \frac{-\lambda_r(t)^2}{2c} \quad \text{if } R^{min}(t) + \frac{\lambda_r(t)}{c} \leq \sum_i R(i,t)I(i,t) \leq R^{max}(t) - \frac{\lambda_r(t)}{c} \\ \lambda_r(t) \cdot \left(R^{min}(t) - \sum_i R(i,t)I(i,t) \right) \\ + c \cdot \left[\sum_i R(i,t)^{(k)} I(i,t)^{(k)} - R^{min}(t) \right] \cdot \left[\sum_i R(i,t)I(i,t) - R^{min}(t) \right] \\ + \frac{1}{2\varepsilon} \sum_i \left[R(i,t)^{(k)} I(i,t)^{(k)} - R(i,t)I(i,t) \right]^2 \\ \qquad \text{if } \sum_i R(i,t)I(i,t) < R^{min}(t) + \frac{\lambda_r(t)}{c} \end{cases}$$

(4.54b)

$$Ln(t) = \begin{cases} \lambda_n(t) \cdot \left(\sum_i N(i,t) - N^{max}(t) \right) \\ + c \cdot \left[\sum_i N(i,t)^{(k)} - N^{max}(t) \right] \cdot \left[\sum_i N(i,t) - N^{max}(t) \right] \\ + \frac{1}{2\varepsilon} \sum_i \left[N(i,t)^{(k)} - N(i,t) \right]^2 \\ \qquad \text{if } \sum_i N(i,t) > N^{max}(t) - \frac{\lambda_n(t)}{c} \\ \frac{-\lambda_n(t)^2}{2c} \quad \text{if } N^{min}(t) + \frac{\lambda_n(t)}{c} \leq \sum_i N(i,t) \leq N^{max}(t) - \frac{\lambda_n(t)}{c} \\ \lambda_n(t) \cdot \left(N^{min}(t) - \sum_i N(i,t) \right) \\ + c \cdot \left[\sum_i N(i,t)^{(k)} - N^{min}(t) \right] \cdot \left[\sum_i N(i,t) - N^{min}(t) \right] \\ + \frac{1}{2\varepsilon} \sum_i \left[N(i,t)^{(k)} - N(i,t) \right]^2 \\ \qquad \text{if } \sum_i N(i,t) < N^{min}(t) + \frac{\lambda_n(t)}{c} \end{cases}$$

(4.54c)

4.5.4 Bidding Strategy Based on PBUC

The PBUC solution provides a generation schedule. We consider the bidding strategy to be a function of the generation schedule.

Suppose that P_0 represents bilateral contracts. The bidding quantity for the spot market is ΔP, which is given by

$$\Delta P = P - P_0 \tag{4.55a}$$

We let the cost function be a quadratic polynomial

$$C(P) = a + bP + cP^2 \tag{4.55b}$$

Thus, the marginal cost is given by

$$\begin{aligned}\frac{dC(P)}{dP} &= b + 2c\,P = b + 2c(P_0 + \Delta P) \\ &= b + 2c\,P_0 + 2c\Delta P\end{aligned} \tag{4.55c}$$

The incremental cost is given by

$$\begin{aligned}\frac{\Delta C(P)}{\Delta P} &= \frac{C(P) - C(P_0)}{\Delta P} \\ &= \frac{\left(a + bP + cP^2\right) - \left(a + bP_0 + cP_0^2\right)}{\Delta P} \\ &= \frac{b(P - P_0) + c(P^2 - P_0^2)}{\Delta P} \\ &= \frac{b(P - P_0) + c(P + P_0)(P - P_0)}{\Delta P} \\ &= \frac{b(P - P_0) + c(P_0 + \Delta P + P_0)(P - P_0)}{\Delta P} \\ &= \frac{b\Delta P + c(2P_0 + \Delta P)\Delta P}{\Delta P} \\ &= b + c(2P_0 + \Delta P) \\ &= b + 2c\,P_0 + c\Delta P\end{aligned} \tag{4.55d}$$

If we define the marginal cost at the bilateral contract point $(P = P_0)$ as

$$\rho_0 = b + 2c\,P_0 \tag{4.55e}$$

the marginal cost curve would be

$$\rho = \rho_0 + 2c\Delta P \tag{4.55f}$$

and the incremental cost curve would be

$$\rho = \rho_0 + c\Delta P \tag{4.55g}$$

We may generalize the bidding price curve as

$$\rho = \rho_0 + (1+m)c\Delta P \tag{4.56}$$

where ρ is the bidding price, ρ_0 is the price at bilateral point, ΔP is the bidding quantity, c is the quadratic coefficient in the cost function, and $(1+m) \cdot c$ is the bidding slope. To bid at incremental cost, we set $m = 0$ and to bid at marginal cost, we set $m = 1$.

The starting point of energy bid is the bilateral contract (if any, otherwise 0), and the bidding price is $\rho_g = b + 2cP_0 + (1+m_g)c(P-P_0)$. The bid curve is shown in Figure 4.3 where the slope defines the bidding strategy. If we forecast the market price properly, then we could bid at the market price. If we set $\rho_g = \rho_{gm}$, after calculating P, we have

$$m_g = \frac{\rho_{gm} - b - 2cP_0}{c(P - P_0)} - 1 \tag{4.57}$$

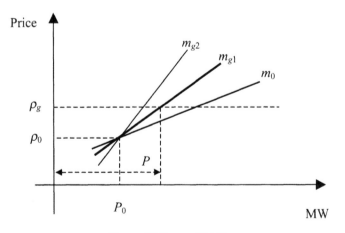

Figure 4.3 Energy Bid Curve

PRICE-BASED UNIT COMMITMENT

The spinning reserve is the unloaded synchronized generation that can ramp up in 10 minutes. The starting point for calculating the spinning reserve bid is the current energy bid as shown in Figure 4.4, and the bidding price is $\rho_r = \rho_g + (1 + m_r)cR$. In the bid curve of Figure 4.4, the slope defines the bidding strategy. If we set $\rho_r = \rho_{rm}$, after calculating R, we have

$$m_r = \frac{\rho_{rm} - \rho_{gm}}{cR} - 1 \tag{4.58}$$

The non-spinning reserve is a generating capacity that is not synchronized to the grid and can ramp up in 10 minutes. The starting point of non-spinning reserve bid is the current spinning reserve bid as shown in Figure 4.5, and the bidding price is $\rho_n = \rho_r + (1 + m_n)cN$. In the bid curve of Figure 4.5, the slope defines the bidding strategy. If we set $\rho_n = \rho_{nm}$, after calculating N, we have

$$m_n = \frac{\rho_{nm} - \rho_{rm}}{cN} - 1 \tag{4.59}$$

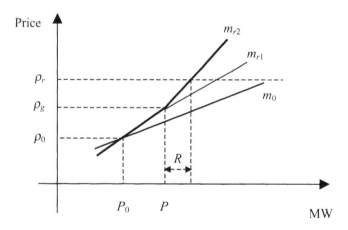

Figure 4.4 Spinning Reserve Bid Curve

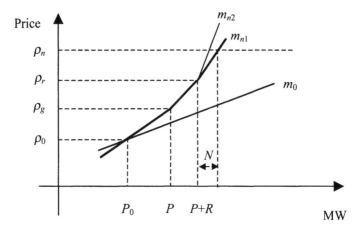

Figure 4.5 Non-spinning Reserve Bid Curve

The market price would represent the maximum bid if we plan to sell to the market and the minimum bid if we plan to buy from the market. If we could forecast the market price, we would bid our product at the market price. The following figures present a few bidding scenarios:

1. Selling energy, spinning reserve, and non-spinning reserve (unit is ON). See Figure 4.6.

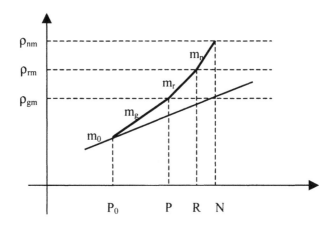

Figure 4.6 Bidding Curve (Scenario 1)

PRICE-BASED UNIT COMMITMENT

2. Purchasing energy, and selling spinning reserve and non-spinning reserve (unit is ON). See Figure 4.7.

3. Purchasing energy and selling non-spinning reserve (unit is OFF). See Figure 4.8.

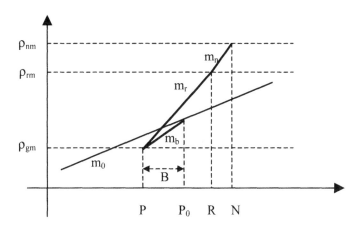

Figure 4.7 Bidding Curve (Scenario 2)

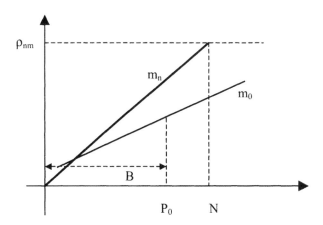

Figure 4.8 Bidding Curve (Scenario 3)

4.6 CASE STUDIES

4.6.1 Case Study of 5-Unit System

The PBUC formulation and solution methodology is implemented as software and applied to a couple of systems, first to a 5-unit system and then to a 36-unit system. See Appendix D.1 for unit characteristics and energy market prices of the 5-unit system. In our implementation, energy and ancillary services (e.g., spinning and non-spinning reserves) are considered simultaneously in the formulation. However, the case studies in this section will only consider the sale of energy.

Using case studies, we will test the basic functions of the software and show the impacts of:

1. Energy market price
2. Ramp rate
3. Fuel price variation (fuel constraints)
4. Identical units
5. Price variation (congestion)

The dispatch results of all cases are listed in Appendix D.1.

4.6.1.1 Case 1: Impact of the Energy Market Price

There are no system constraints such as energy, fuel and emission constraints, considered in our study of the impact of energy market price. Two cases (1-1 and 1-2) are presented here.

Case 1-1. The fuel price is 3 $/MBtu. The market prices for energy are 48 $/MWh at peak hours (hours 6–23) and 20 $/MWh at off-peak hours (hours 1–5 and hour 24).

The dispatch results show that all five units are committed at peak hours and shut down during low market price periods (hours 1–5 and hour 24).

Case 1-2. Fuel and energy market prices are similar to those in Case 1-2, except that energy market prices are ramped up between hours 6 and 15 from 24 $/MWh to 42 $/MWh.

The dispatch results show that all five units are committed between hours 9 and 23, when the energy market price is high. Depending on

PRICE–BASED UNIT COMMITMENT

market prices for energy and the corresponding incremental heat rates, the generating units could be committed at other hours as well. For instance, unit 1 is committed at hour 8, unit 3 is committed at hours 7 and 8, and units 2 and 5 are committed at hours 6, 7, and 8 to satisfy the prevailing constraints.

4.6.1.2 Case 2: Impact of Ramp Rates

Taking Case 1-1 as the base case, we show the impact of ramp rates on generation unit scheduling. In Case 2, we lower the ramp rate of unit 5 to 1 MW/min. Two cases (2-1 and 2-2) are presented here.

Case 2-1. The fuel price is the same as that in the base case (3 $/MBtu).

The dispatch results show that unit 5 is committed at hours 3, 4, 5 and 24 compared to the base case. In this case, the fuel price is relatively low. To reach the full capacity at hours when the energy market price is high, unit 5 is committed at hours 3, 4, and 5, when market price is still low. However, once the ramp rate constraint is relaxed, unit 5 is committed at these hours. It is also noted that because of the ramp rate constraint, the generation of unit 5 cannot be shut down at hour 24.

Case 2-2. The fuel price is increased to 5 $/MBtu.

The dispatch results show that unit 5 is committed at hour 24 compared to the base case. In this case, the fuel price is higher. So it would not be profitable to start the unit up at earlier hours, as was done in Case 2-1. So unit 5 would not be operated at hours 3, 4, and 5, and would reach its partial capacity at hour 6, 7, and 8. Also, because of ramp rate constraint, unit 5 cannot be shut down at hour 24.

4.6.1.3 Case 3: Impact of Fuel Price Variations

This case is used to show the impact of a variable fuel price as a function of the total MBtu consumption on the generation scheduling solution. The base case is Case 1-2. A non-homogeneous relationship between the fuel price and MBtu consumption is illustrated in Figure 4.9.

Table 4.1 lists the optimization results of all price segments for the studied case. The optimization indicates that the best solution is to purchase 250,000 MBtu of fuel, which would result in $175,574 profit. Note that in both the 1 $/MBtu and 7 $/MBtu segments, we get the same best solution, which is not easy to understand at the first glance. The key to understanding this is to look at the fuel purchase cost curve (Figure 4.9b), which is the integral of the fuel purchase price curve (Figure 4.9a). As the cost curve shows the costs to purchase 250,000 MBtu (the best solution) are the same ($950,000) for both the 1 $/MBtu and 7 $/MBtu segments. These are the two points on the price curve: (250,000 MBtu, 1 $/MBtu) and (250,000 MBtu, 7 $/MBtu) and they correspond to the same point on the cost curve: (250,000 MBtu, $950,000). So the two best solutions in 1 $/MBtu and 7 $/MBtu segments are clearly the same.

Note that in Table 4.1 where the fuel consumption corresponds to the 2 $/MBtu segment, the optimal profit is negative. This means that for fuel consumption at 300,000 to 350,000 MBtu, it is impossible to make profit.

Table 4.1 Optimal Results for All Fuel Price Segments

Fuel Price Segment ($/MBtu)	Optimal Consumed Fuel (MBtu)	Profit ($)
1	250000	175574
2	309557	-53936
3	200000	80991
4	50000	85657
5	92362	106559
6	124210	71695
7	250000	175574

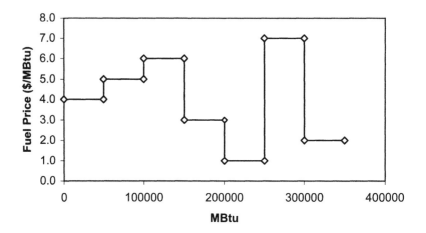

Figure 4.9a Fuel Purchase Price Curve

PRICE–BASED UNIT COMMITMENT

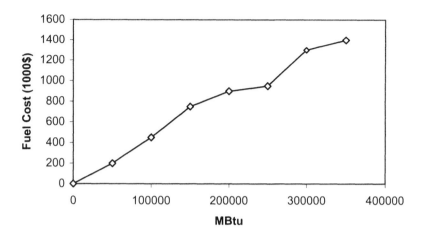

Figure 4.9b Fuel Purchase Cost Curve

4.6.1.4 Case 4: Impact of Identical Generation Units

This case study shows the impact of identical units. Once again, the base case is Case 1-2. The fuel price is 3 $/MBtu, and the fuel constraint is set to 16,000 MBtu. The identical units are 4 and 5 (i.e., unit 4 is set to be identical to unit 5). Four possible cases are presented here.

Case 4-0. There are no identical units. The solution indicates that only unit 5 is committed at hours 14–23 and the fuel constraint is satisfied.

Case 4-1. Unit 4 is set to be identical to unit 5. Both units could be committed at once when only one of them is needed, which is when the fuel constraint would be violated. Likewise, both units could be shut down at once when only one of them is need, which is when no profit would be possible.

Case 4-2. In order to remedy the problem discussed in Case 4-1, the start-up MBtu of unit 4 is increased slightly to make it look different from that of unit 5. If the incremental change in the start-up MBtu is less than 5 MBtu (the original start-up MBtu is 500), both units are shut down (which is the same as Case 4-1). If the incremental change in start-up MBtu is larger than or equal to 5 MBtu, unit 5 is committed at hours 14–23 and the fuel constraint is met. So we see that by a slight modification in the start-up MBtu of one of the identical units (about 1% in this case), the Case 4-1 problem is cured, the fuel constraint is satisfied, and the profit is maximized.

Case 4-3. A different option is considered with a slight modification in the characteristics of unit 4. The incremental heat rate of unit 4 is multiplied by a factor. The results show that if the factor is smaller than 1.0001 (0.01% increase), the same problem as in Case 1-4 is encountered. If the factor is larger than or equal to 1.0001 (0.01% increase), a reasonable schedule can be achieved, the fuel constraint is satisfied, and the profit is maximized.

4.6.1.5 Case 5: Impact of Different LMPs (Transmission Congestion)

The base case is Case 1-2. It is assumed that units 1, 2, and 3 are in zone 1 and units 4 and 5 are in zone 2.

Case 5-1. Units in zone 2 have very cheap costs, while units in zone 1 are expensive. If there is enough transmission capacity between zones 1 and 2, units in zone 2 can generate additional power and make the excessive power available to zone 1. In this case, market prices in zones 1 and 2 would be the same, since no congestion exists, and the profit would be $374,353.

Case 5-2. A lower capacity is assumed for the inter-zonal transmission line between zones 1 and 2, which would result in transmission congestion. The resulting congestion would cause different market prices in two zones. We assume that market prices in zone 2 are the same as those in zone 1, except at hours 22 and 23 when market prices in zone 2 are dropped to 20 $/MWh. The congestion causes less generation to be transferred from zone 2 to zone 1 and the market price in zone 1 (congested zone) to increase. As a result, units 4 and 5 are shut down at those hours and additional generation is supplied locally in zone 1. In this case, the profit is $343,667, which is less than that of Case 5-1.

4.6.2 Case Study of 36-Unit System

The software implementing the proposed formulation and solution methodology is tested using a larger 36-unit system. See Appendix D.2 for unit characteristics and energy market prices. Energy and ancillary services (spinning reserve and non-spinning reserve) are considered in the case study. No system constraints, such as energy, fuel, and emission constraints, are imposed.

PRICE-BASED UNIT COMMITMENT

Four cases are studied. Case 1 is the base case. In Case 2, the base prices for energy and reserve prices are increased by 5. In Case 3, the base prices for energy and reserve prices are increased by 10. In Case 4, the base prices for energy and reserve prices are increased by 15. In all cases, the fuel price is 1.0 $/MBtu. The dispatch results for all cases are listed in Appendix D.2.

4.6.2.1 Base Case (Case 1) Analysis

Generating units are not committed at hours 2–5 and 23–24, when market prices are very low. Units 2–7 have the largest unloaded marginal generation cost (coefficient b in the cost function) and are not committed at all. Unit 1, which also has a high unloaded marginal generation cost, is committed at hours 15–18 because it has no start-up cost. The unloaded marginal generation costs of unit 22–27 are also high and they are committed only at high price periods (hours 15–19).

Units 28–36 are base load units whose unloaded marginal generation costs are small. They are committed almost for the entire scheduling horizon except at hours 1–5 and 23–24 when market prices are low. Units 32–36 should have been committed at hour 1 when the market price is higher than their unloaded marginal generation costs. However, they are decommitted at hour 1 because of minimum OFF time constraints. Units 12 to 18 have similar parameters. However, units 13 and 14 are initially committed and, because of their minimum ON time constraints, remain committed at hour 1, though the market prices are low. Furthermore, units 8–11 are committed at hours 7–21 and units 19–21 are committed at hours 6–22.

Considering energy and reserve sales, we find that the higher the price, the higher is the sales volume. The energy is not sold at hours 2–5 and 23–24 and the reserve is not sold at hours 1–5 and 23–24. Furthermore, at hour 1, units 13 and 14 run at minimum capacities.

4.6.2.2 Case 4 Analysis

In Case 4, as compared to Case 1, units 2–7 are committed at hours 16 and 17 when market prices are high enough that these units can make profits. Another distinction between Cases 4 and 1 is that at hours 2–5 and 23–24, units 8–11, 19–21 and 28–36 are committed. Here market prices are high enough for these units to make profits.

4.6.2.3 Sensitivity to Market Prices

As shown in Appendix D.2, the PBUC results change according to market prices, from Case 1 to Case 2, Case 2 to Case 3, and Case 3 to Case 4. These changes are emphasized in bold letters. Table 4.2 and Figure 4.10 show the progression of the number of ON units based on hourly market prices. As market prices increase, more units are committed at any given hour. Changes are emphasized in bold.

Table 4.2 Sensitivity Analysis: Number of ON Units

Hour	1	2	3	4	5	6	7	8	9	10	11	12	13	14	15	16	17	18	19	20	21	22	23	24
Case 1	2	0	0	0	0	12	16	16	16	23	23	23	23	23	30	30	30	30	29	23	16	12	0	0
Case 2	18	12	9	9	9	16	17	23	26	29	30	30	30	30	30	30	30	30	30	29	23	16	12	0
Case 3	18	16	12	12	16	23	27	29	30	30	30	30	30	30	30	30	30	30	30	29	23	16	12
Case 4	19	19	17	17	23	28	28	30	30	30	30	30	30	30	36	36	30	30	30	30	29	23	16

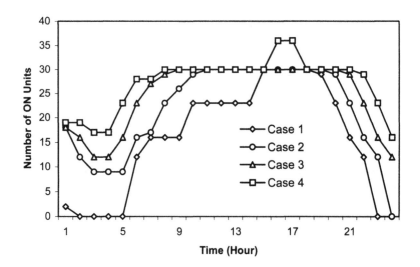

Figure 4.10 Sensitivity Analysis: Number of ON Units

PRICE–BASED UNIT COMMITMENT

Figure 4.11 and Table 4.3 show the sensitivity of energy sales to market prices. As market prices increase, more energy is sold at any given hour. Changes are emphasized in bold.

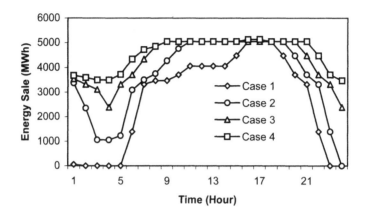

Figure 4.11 Sensitivity Analysis: Energy Sale

Table 4.3 Sensitivity Analysis: Energy Sale

Hour	Case 1	Case 2	Case 3	Case 4
1	50.00	**3372.17**	3540.83	3674.17
2	0.00	**2356.08**	3322.17	3599.17
3	0.00	**1060.00**	3112.77	3499.17
4	0.00	**1060.00**	2389.42	3499.17
5	0.00	**1226.67**	3322.17	3719.78
6	1389.42	**3097.42**	3707.50	4341.30
7	3322.17	**3507.50**	4333.30	4722.17
8	3474.17	**3757.50**	4852.07	4860.07
9	3474.17	**4264.35**	5050.50	5050.50
10	3707.50	**4756.85**	5050.50	5050.50
11	4057.50	**5050.50**	5050.50	5050.50
12	4057.50	**5050.50**	5050.50	5050.50
13	4057.50	**5050.50**	5050.50	5050.50
14	4057.50	**5050.50**	5050.50	5050.50
15	4473.60	**5050.50**	5050.50	5050.50
16	5050.50	5050.50	5050.50	5130.50
17	5050.50	5050.50	5050.50	5130.50
18	5044.90	**5050.50**	5050.50	5050.50
19	4471.20	**5050.50**	5050.50	5050.50
20	3707.50	**4471.20**	5050.50	5050.50
21	3322.17	3707.50	4471.20	**5050.50**
22	1389.42	3322.17	3707.50	**4471.20**
23	0.00	**1389.42**	3322.17	3707.50
24	0.00	0.00	2389.42	3474.17

Figure 4.12 and Table 4.4 show the sensitivity of spinning reserve sales to market prices. As market prices increase, more spinning reserve is sold at any given hour. Changes are emphasized in bold.

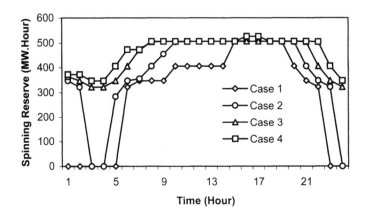

Figure 4.12 Sensitivity Analysis: Spinning Reserve Sale

Table 4.4 Sensitivity Analysis: Spinning Reserve Sale

Hour	Case 1	Case 2	Case 3	Case 4
1	0.00	**347.42**	**364.08**	**372.42**
2	0.00	**322.08**	**347.42**	**372.42**
3	0.00	0.00	**322.08**	**347.42**
4	0.00	0.00	**322.08**	**347.42**
5	0.00	**283.33**	**347.42**	**405.75**
6	322.08	**347.42**	**405.75**	**473.42**
7	347.42	**355.75**	**471.42**	**473.42**
8	347.42	**405.75**	**504.25**	**506.25**
9	347.42	**455.00**	**506.25**	**506.25**
10	405.75	**504.25**	**506.25**	**506.25**
11	405.75	**506.25**	506.25	506.25
12	405.75	**506.25**	506.25	506.25
13	405.75	**506.25**	506.25	506.25
14	405.75	**506.25**	506.25	506.25
15	506.25	506.25	506.25	506.25
16	506.25	506.25	506.25	526.25
17	506.25	506.25	506.25	526.25
18	506.25	506.25	506.25	506.25
19	504.25	**506.25**	506.25	506.25
20	405.75	**504.25**	**506.25**	**506.25**
21	347.42	**405.75**	**504.25**	**506.25**
22	322.08	**347.42**	**405.75**	**504.25**
23	0.00	**322.08**	**347.42**	**405.75**
24	0.00	0.00	**322.08**	**347.42**

PRICE–BASED UNIT COMMITMENT

Figure 4.13 and Table 4.5 show the sensitivity of non-spinning reserve sales to market prices. As market prices increase, more non-spinning reserve is sold. Changes are emphasized in bold.

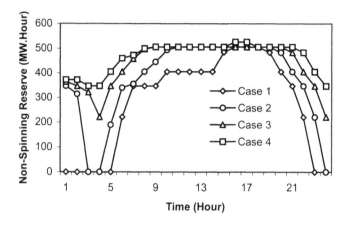

Figure 4.13 Sensitivity Analysis: Non-spinning Reserve Sale

Table 4.5 Sensitivity Analysis: Non-spinning Reserve Sale

Hour	Case 1	Case 2	Case 3	Case 4
1	0.00	**347.42**	**364.08**	**372.42**
2	0.00	**314.33**	**347.42**	**372.42**
3	0.00	0.00	**322.08**	**347.42**
4	0.00	0.00	**221.00**	**347.42**
5	0.00	**190.00**	**347.42**	**405.75**
6	221.00	**339.67**	**405.75**	**460.28**
7	347.42	**355.75**	**458.28**	**473.42**
8	347.42	**405.75**	**497.68**	**499.68**
9	347.42	**445.15**	**506.25**	506.25
10	405.75	**494.40**	**506.25**	506.25
11	405.75	**506.25**	506.25	506.25
12	405.75	**506.25**	506.25	506.25
13	405.75	**506.25**	506.25	506.25
14	405.75	**506.25**	506.25	506.25
15	486.55	**506.25**	506.25	506.25
16	506.25	506.25	506.25	**526.25**
17	506.25	506.25	506.25	**526.25**
18	506.25	506.25	506.25	506.25
19	484.55	**506.25**	506.25	506.25
20	405.75	**484.55**	**506.25**	506.25
21	347.42	**405.75**	**484.55**	**506.25**
22	221.00	**347.42**	**405.75**	**484.55**
23	0.00	**221.00**	**347.42**	**405.75**
24	0.00	0.00	**221.00**	**347.42**

4.7 CONCLUSIONS

The formulation and the solution methodology for the PBUC problem in a restructured market structure can be used by GENCOs in some restructured markets where GENCOs are responsible for unit commitment as in California market. The solution methodology allows GENCOs to commit and schedule their units for selling power, purchasing power, selling spinning and non-spinning reserves in order to maximize their profits. The tests on the 5 and 36 unit system show the effectiveness of the proposed formulation and solution methodology.

Chapter 5

Arbitrage in Electricity Markets

5.1 INTRODUCTION

Arbitrage refers to making profit by a simultaneous purchase and sale of the same or equivalent commodity with net zero investment and without any risk. Arbitrage is a new concept in the power industry which is in the midst of its restructuring. However, arbitrage itself is not new in financial markets. The expanded usage of arbitrage includes any activity that attempts to buy a relatively under-priced commodity and to sell a similar and relatively over-priced commodity for profit.

This chapter will present arbitrage in the general sense, and then discuss arbitrage opportunities in power markets. The two types of arbitrage that are considered in a power market: same-commodity arbitrage and cross-commodity arbitrage. Cross-commodity arbitrage will be the subject of this chapter and examples will be presented to illustrate how arbitrage is executed in the power market.

5.2 CONCEPT OF ARBITRAGE

5.2.1 What Is Arbitrage?

The act of arbitrage depends on three elements: objective, opportunity, and means. The objective of arbitrage is obvious–making profit from price discrepancies by a simultaneous purchase and sale of the same or equivalent commodity with net zero investment and without any risk. The profit opportunity for arbitrage (source) is due to price discrepancies between the same or equivalent commodities. The means of realizing arbitrage is the simultaneous purchase and sale of the same or equivalent commodities. Since we do not want to assume any risks, it would be necessary to purchase and sell commodities at the same time. Otherwise, the chance of risk may be further incurred.

More recently, arbitrage has been expanded to include any activity that attempts to buy a relatively under-priced item and sell a similar and relatively over-priced item for profit. Accordingly, additional changes have taken place corresponding to the three elements of arbitrage to reflect the new definition. Making profit is still the objective in the new paradigm. However, initial investments may be needed. An arbitrageur (i.e., an individual or institution who would engage in arbitrage) may assume some risk (although the risk of arbitrage is relatively low). The profit opportunity still comes from price discrepancies. However, it is not restricted to the same or equivalent commodity. For example, gas and electricity are two different forms of energy. With a gas-fired generating unit, gas may be converted into electricity. If the gas purchase cost and the gas-to-electricity conversion cost are less than the revenue from electricity sales, there is an opportunity for an arbitrage. When implementing an arbitrage opportunity, simultaneity is no longer a factor. In the gas-to-electricity example, the purchase of gas and the sale of electricity cannot happen at the same time.

5.2.2 Usefulness of Arbitrage

The act of arbitrage can benefit arbitrageurs as well as the power market. For arbitrageurs, arbitrage provides a low-risk payoff. It is a reward for the arbitrageur due to its effort to find price discrepancies. For the market, arbitrage provides additional pressure on prices to move to rational or normal levels, maintains liquidity in the market, and provides a pricing mechanism, namely no-arbitrage pricing.

Should an arbitrage opportunity exist temporarily, investors' trading to earn this riskless or low-risk profit would tend to move the price of the asset in a direction that would eliminate the arbitrage. However, the volatility of the market would create new arbitrage opportunities. When such arbitrage opportunities emerge, arbitrageurs would attempt to make it disappear as market is liquidated.

In competitive asset markets, it may be reasonable to assume that equilibrium asset prices are such that no arbitrage opportunity could exist. By assuming the absence of arbitrage, the price of one asset can often be derived from another asset. This would lead to the well-known no-arbitrage pricing theory. For example, we can apply the no-arbitrage theory to price an electricity contract [Bjo00].

Since electricity is a commodity that, in general, cannot be easily stored, the application of the no-arbitrage theory to electricity contract pricing is different from the application to easy-to-store commodities. However, if we follow some basic principles, the no-arbitrage theory can still be applied. The principles include two steps: (1) Find a policy that could lead to arbitrage; (2) set the price of the contract so that there is no arbitrage. Suppose that we would like to define the price for a flexible contract, where the buyer of the contract decides on how the electricity is delivered at a future time and the seller has the obligation to supply the electricity. A possible arbitrage opportunity is that the buyer of the contract may resell the electricity to the spot market. According to the no-arbitrage theory, the price of this type of contract should be its expected value in the spot market. So, to determine the price of the contract, we may find the maximum revenue that the buyer of the contract can get by reselling it to the spot market.

5.3 ARBITRAGE IN A POWER MARKET

In the power industry, there could be many inconsistencies in electricity pricing, which that provide arbitrage opportunities. However, many arbitrage opportunities are not perfect (riskless); that is, even though an arbitrage position looks like a "sure win," significant speculative elements (risks) could be involved. Two types of arbitrage opportunities may be exploited in the power industry: same-commodity arbitrage and cross-commodity arbitrage.

5.3.1 Same-Commodity Arbitrage

When arbitrage is aimed at the same product (e.g., electricity), it is called same-commodity arbitrage. Same-commodity arbitrage in the power market would include both spatial and temporal arbitrage.

Spatial arbitrage refers to exploiting lateral variations in electricity spot prices. Price discrepancy among different regions within a market, or between different markets such as New York and New England, may provide such arbitrage opportunities.

Temporal arbitrage refers to exploiting temporal differences between markets with different time horizons. Consider the arbitrage between futures and spot markets. A futures contract is usually monthly, while a spot market is hourly. If there is a price discrepancy between these two markets, arbitrage opportunities could exist. However, this is not a "sure win" arbitrage since spot prices are very volatile.

5.3.2 Cross-Commodity Arbitrage

Arbitrage that is aimed at different products within a market or different markets is called cross-commodity arbitrage. For example, the arbitrage between gas market and electricity market is a cross-commodity arbitrage. This is the most recognized arbitrage opportunity which is also referred to as spark spread. Other examples of cross-commodity arbitrage include arbitrage between power and steam, and between energy and an ancillary service. The most distinct feature of cross-commodity arbitrage is that generational assets are needed for arbitrage to occur.

5.3.3 Spark Spread and Arbitrage

Generally spark spread is defined as the price of electricity minus the product of generating efficiency (usually represented by heat rate) and the price of fuel. Thus the lower the heat rate and the fuel price, and the higher the electricity price, the larger is the spark spread. Spark spread can be used as an indication for arbitrage, and is sometimes referred to as the spark arbitrage.

Spark spread is useful for monitoring the movements in the energy marketplace and benchmarking relative energy costs. Thus, it is widely used in the power industry. One application of spark spread is for identifying arbitrage opportunities among different energy products and markets such as gas and energy. Consider the following example [Fte02].

On November 29, 2000, Electricity prices at ERCOT (that is, the Electric Reliability Council of Texas) were averaged at 54.58 $/MWh, and gas prices at the Houston Ship Channel were averaged at 5.83 $/MBtu. There were four types of gas turbines at various efficiencies that were used to generate electricity, starting with a highly efficient turbine which only required 7 MBtu to generate 1 MWh of energy, that is, a heat rate of 7 MBtu/MWh, down to turbines that required 8, 10 and 12 MBtu to generate the same amount of energy. Using the most efficient turbine, the cost of generating 1 MWh of energy was 7 MBtu x 5.83 $/MBtu = $40.81. When we compared that cost in ERCOT to the cost of buying energy at $54.58/MWh from elsewhere, it turned out to be cheaper (by 54.58 - 40.81 = 13.77) to generate energy in ERCOT, using the high-efficiency turbine, than to buy it off the grid. Thus, at a heat rate of 7 MBtu/MWh, the spark spread from the gas at the Houston Ship Channel to the electricity at ERCOT was 13.77 $/MWh. Here, the spark spread was positive which meant that it was more profitable to buy gas for generating electricity and selling electricity off to the market. At a heat rate of 8 MBtu/MWh, it would still be profitable, though the spark spread drop to 7.94 $/MWh. However, the spark spreads would be -3.72 $/MWh and -15.38 $/MWh at heat rates of 10 MBtu/MWh and 12 MBtu/MWh, respectively. Negative spark spread means that it would be more economical to purchase electricity from the market than generating it locally.

It should be emphasized that spark spread in our example is computed for a simplified version of a market and should be used with caution. In our example, electricity prices were averaged. However, in an practical case, daily electricity prices could be very volatile, and peak-hours and off-peak hours could be used to improve the usefulness of spark spread in the market.

We also know that the heat rate of a generating unit is a function of generation level. A more accurate model for representing heat rate would be to use a piecewise linear curve as in the economic dispatch model. However, using a more cumbersome representation of system variables may result in a complicated spark spread curve which could be more difficult to analyze. A compromise would be to use several typical values of heat rates as referred to in our example.

In general, spark spread can only be used as an indication of arbitrage in a market and would not be enough for a detailed analysis of arbitrage opportunities. In this chapter, we apply the PBUC that was discussed in Chapter 4 for a more detailed analysis of arbitrage opportunities.

5.3.4 Applications of Arbitrage Based on PBUC

In the following, we discuss arbitrage opportunities based on the PBUC model. First, in our PBUC model, energy and ancillary services are optimized simultaneously, and the PBUC results provide a portfolio of energy and ancillary services bids. These results are used for exploring arbitrage opportunities between energy and ancillary services.

Second, if there are any bilateral contracts, a GENCO may satisfy them either by local generation or by purchases from the market. Since bilateral contracts are included in PBUC, the optimization procedure can discover the prevailing arbitrage opportunities.

Third, PBUC can discover gas-to-electricity arbitrage opportunities. Given a certain amount of gas, a GENCO may opt to sell gas directly to the market, use gas as fuel in generating different electricity products (energy, reserve), or sell a fraction of the available gas to the market and burn the rest to generate electricity for the power market. The arbitrage depends on market prices for gas and energy.

Fourth, PBUC can discover arbitrage opportunities among gas, energy and emission. Depending on market prices, a GENCO may decide to sell the emission compliance and generate less electricity, or sell more gas to the market. This can be done since PBUC has the capability of handling emission constraints.

Last, arbitrage between steam and power may be exploited. Once steam is produced, a GENCO may sell a fraction of it directly to the market if the price of steam is relatively higher than the price of energy.

5.4 ARBITRAGE EXAMPLES IN POWER MARKET

A 6-unit system is used to show the application of PBUC to cross-commodity arbitrage. See Appendix D.3 for unit characteristics and market prices. A few types of arbitrage opportunities are discussed as follows:

1. Arbitrage between energy and ancillary service
2. Arbitrage of bilateral contract
3. Arbitrage between gas and power
4. Arbitrage of emission allowances
5. Arbitrage between steam and power

5.4.1 Arbitrage between Energy and Ancillary Service

The arbitrage across energy and ancillary services commodities may provide a higher profit for a GENCO than selling energy alone as a commodity. In [Deb01b], it is pointed out that "a generator will seek to maximize income by seeking profits and advantage across all available markets. In order to reveal the full earnings potential of an asset, valuation must also include revenues that operators will earn from proactive participation in the lucrative ancillary service and spot markets." In [Hir01], a model is presented for maximizing profits across energy and ancillary services markets. If the price of an ancillary service is higher than that of energy, we may reduce the sale of energy and opt to sell more

ARBITRAGE IN ELECTRICITY MARKETS

ancillary services. The numerical example in this section demonstrates that by applying PBUC, arbitrage can be realized between energy and ancillary services.

First, we consider the energy sale exclusively for the 6-unit system (see Appendix D.3). Table 5.1 shows the scheduling summary with a profit of $240,995.53. As expected, more energy is offered for sale at hours when the spot market price is high (i.e., hours 1, 6–21), a smaller amount of energy is sold when the spot market price is low (i.e., hours 2, 22–23), and energy is not sold at hour 24 when the spot market price is very low. At hours 3-5, the spot market prices are also very low when units are supposed to be OFF. However, because of minimum up/down time constraints, units 4–6 would run at those hours at their minimum capacities (considering ramp rate constraints).

If we consider both energy and ancillary services as trading commodities, a cross-commodity arbitrage would be scheduled as shown in Table 5.2, with expected profit of $245,244.63. In this case, ancillary services are sold at all hours except hours 3–4 and 24, when spot market prices for ancillary services are very low. Although units 4–6 are ON at hours 3–4, they do not offer any ancillary services for sale.

Comparing Tables 5.1 and 5.2, we learn that the total capacity offered as energy and ancillary services is the same in the two cases above. However, in the second case, less energy and more spinning and non-spinning reserves are scheduled. This is because the prices of ancillary services are higher than those of energy. The total profit in the second case is increased by 245,244.63 – 240,995.53 = $4,249.10 (i.e., 1.76% increase). This example shows that the arbitrage between energy and ancillary services can improve the profitability.

Table 5.1 Summary of Energy Schedule

Hour	Energy (MW)	Spinning Reserve	Non-spinning Reserve	Total (MW)
1	950.00	0.00	0.00	950.00
2	723.64	0.00	0.00	723.64
3-4	260.00	0.00	0.00	260.00
5	475.00	0.00	0.00	475.00
6-7	950.00	0.00	0.00	950.00
8	1025.00	0.00	0.00	1025.00
9-10	1100.00	0.00	0.00	1100.00
11-12	1150.00	0.00	0.00	1150.00
13-14	1100.00	0.00	0.00	1100.00
15	1150.00	0.00	0.00	1150.00
16-17	1200.00	0.00	0.00	1200.00
18-19	1150.00	0.00	0.00	1150.00
20	1100.00	0.00	0.00	1100.00
21	1025.00	0.00	0.00	1025.00
22	675.00	0.00	0.00	675.00
23	200.00	0.00	0.00	200.00
24	0.00	0.00	0.00	0.00

Table 5.2 Summary of Energy and Ancillary Services Schedule

Hour	Energy (MW)	Spinning Reserve	Non-spinning Reserve	Total (MW)
1	791.67	79.17	79.17	950.01
2	565.31	79.17	79.17	723.64
3-4	260.00	0.00	0.00	260.00
5	340.00	55.83	79.17	475.00
6-7	791.67	79.17	79.17	950.00
8	841.67	91.67	91.67	1025.00
9-10	916.67	91.67	91.67	1100.00
11-12	950.00	100.00	100.00	1150.00
13	916.67	91.67	91.67	1100.00
14-15	950.00	100.00	100.00	1150.00
16-17	983.33	108.33	108.33	1200.00
18-19	950.00	100.00	100.00	1150.00
20	916.67	91.67	91.67	1100.00
21	841.67	91.67	91.67	1025.00
22	540.00	55.83	79.17	675.00
23	133.33	33.33	33.33	200.00
24	0.00	0.00	0.00	0.00

5.4.1.1 Impact of Price

If the spot market price of ancillary services is lower than that of energy, the schedule will be the same as the first case above. If the spot market price of ancillary services is fluctuating with respect to that of energy, PBUC will determine those periods when it is more economical to offer ancillary services for sale instead of energy.

Table 5.3 shows the schedule for unit 5 at certain hours. Recall from Section 4.4.3 that at least two of P, R, and N would be at their limits when we reach the optimal solution for PBUC. Considering Table 5.3, at hour 1, P, R, and N are all at their upper limits. At hour 2, both R and N are at their upper limits. At hour 3, P, R, and N are at their lower limits. At hour 5, P is at its lower limit, N is at its upper limit.

Table 5.3 Detailed Schedule of Unit 5 at Some Hours

Hour	Energy (MW)	Spinning Reserve	Non-spinning Reserve	Total (MW)
1	291.67	29.17	29.17	350
2	256.67	29.17	29.17	315
3	140	0	0	140
5	140	5.83	29.17	175

5.4.1.2 Impact of Ramp Rates

In the example, the ramp rate of unit 5 is 175 MW/h (i.e., 175/60 MW/min). So the maximum quantities for spinning and non-spinning reserves would be 175/60 x 10 = 29.17 with a profit of $78,360.15 for unit 5. If we increase the ramp rate to 240 MW/h (i.e., the maximum quantity of spinning and non-spinning reserves is 40 MW, the unit 5 profit would be $78,869.47 and the incremental profit of unit 5 would be 78,869.47 - 78,360.15 = $509.32 (i.e., 0.65% increase). If we reduce the ramp rate to 120 MW/h (i.e., maximum spinning and non-spinning reserves is 20), the unit 5 profit would be $77,517.50, and the incremental profit of unit 5 would be 77,517.50 - 78,360.15 = -$842.65 (i.e., 1.075% decrease). Table 5.4 shows the impact of different ramp rates.

The ramp rate sensitivity analysis shows that higher ramp rates can improve the profitability of arbitrage between energy and ancillary services. This type of analysis would help make better decisions on capital investment for increasing the efficiency of generating units.

Table 5.4 Impact of Ramp Rate

Ramp Rate (MW/min)	Maximum Spinning Reserve	Maximum Non-spinning Reserve	Unit 5 Profit ($)	Profit Change
2.92	29.17	29.17	78360.15	Base
4.00	40.00	40.00	78869.47	0.65%
2.00	20.00	20.00	77517.50	-1.075%

5.4.1.3 Impact of Constraints on Energy and Reserve Arbitrage

In the preceding analyses, energy and ancillary services were not constrained. Table 5.5 shows the upper limits imposed on energy and ancillary services with the resulting schedule in Table 5.6. In Table 5.6, profit is $236,079.47 and constrained hours are illustrated in bold letters. Here the profit is reduced by 245,244.63 - 236,079.47 = $9165.16 (i.e., 3.74% decrease) as compared with the unconstrained solution.

Table 5.5 Limits on Energy and Ancillary Services

Hour	Energy (MW)	Spinning Reserve	Non-spinning Reserve
1	800	160	80
2	600	120	60
3	400	80	40
4	300	60	30
5	400	80	40
6	600	120	60
7	700	140	70
8	800	160	80
9	900	180	90
10-11	1000	200	100
12	900	180	90
13	800	160	80
14	700	140	70
15	800	160	80
16-17	1000	200	100
18-19	900	180	90
20	800	160	80
21	700	140	70
22	600	120	60
23	500	100	50
24	400	80	40

ARBITRAGE IN ELECTRICITY MARKETS

Table 5.6 Constrained Energy and Ancillary Services Schedule

Hour	Energy (MW)	Spinning Reserve	Non-spinning Reserve	Total (MW)
1	500.00	50.00	50.00	600.00
2	308.64	50.00	50.00	408.64
3–4	120.00	0.00	0.00	120.00
5	159.97	42.08	**40.00**	242.04
6	**600.00**	57.05	**60.00**	717.04
7	**700.00**	79.17	**70.00**	849.17
8	791.67	79.17	79.17	950.01
9	843.34	91.67	**90.00**	1025.00
10	916.67	91.67	91.67	1100.01
11	950.00	100.00	**100.00**	1150.00
12	866.67	91.67	**90.00**	1048.34
13	**800.00**	91.67	**80.00**	971.67
14	**700.00**	91.67	**70.00**	861.68
15	**800.00**	91.67	**80.00**	971.66
16–17	991.67	108.33	**100.00**	1200.00
18–19	**900.00**	91.67	**90.00**	1081.65
20	**800.00**	91.67	**80.00**	971.66
21	**700.00**	79.17	**70.00**	849.17
22	557.96	57.05	**60.00**	675.00
23	133.33	33.33	33.33	200.00
24	0.00	0.00	0.00	0.00

5.4.2 Arbitrage of Bilateral Contract

Bilateral contract can be fulfilled either by local generation at GENCO or by energy purchases from the power market. The example in this section would show that the arbitrage between local generation and purchases from the market would be more profitable than the local generation alone.

First, the bilateral contract is fulfilled by local generation exclusively. Table 5.7 shows the scheduling summary with a $129,217.91 profit. Since the bilateral contract is fulfilled by local generation, the generated power at each hour would be larger than or equal to the respective bilateral contract.

Table 5.8 shows the arbitrage results with a $133871.30 profit. In Table 5.8, market prices are low at hours 2–5 and 21–24. So GENCO would purchase power from the market to fulfill its contractual obligation. When the market price is high, as in the previous case, bilateral contracts would be satisfied by local generation. The incremental profit due to arbitrage is 133,871.30 - 129,217.91 = $4,653.39 (i.e., 3.6% increase).

Table 5.7 Summary of Schedule
(Bilateral Contracts Satisfied Only by Local Generation)

Hour	Energy (MW)	Spinning Reserve	Non-spinning Reserve	Bilateral (MW)	Purchase (MW)
1	791.67	79.17	79.17	400.00	0.00
2	680.00	79.17	79.17	400.00	0.00
3–4	480.00	0.00	0.00	400.00	0.00
5	480.00	3.33	41.67	400.00	0.00
6–8	791.67	79.17	79.17	400.00	0.00
9–10	916.67	91.67	91.67	400.00	0.00
11–12	950.00	100.00	100.00	560.00	0.00
13	916.67	91.67	91.67	560.00	0.00
14–15	950.00	100.00	100.00	560.00	0.00
16–17	983.33	108.33	108.33	560.00	0.00
18–19	950.00	100.00	100.00	560.00	0.00
20	916.67	91.67	91.67	560.00	0.00
21	911.37	91.67	91.67	560.00	0.00
22	785.00	65.83	79.17	560.00	0.00
23	483.33	33.33	62.50	400.00	0.00
24	400.00	0.00	0.00	400.00	0.00

Table 5.8 Summary of Schedule
(Bilateral Contracts Satisfied by Local Generation and Purchase)

Hour	Energy (MW)	Spinning Reserve	Non-Spinning Reserve	Bilateral (MW)	Purchase (MW)
1	791.67	79.17	79.17	400.00	0.00
2	565.31	79.17	79.17	400.00	16.67
3–4	260.00	0.00	0.00	400.00	160.00
5	340.00	55.83	79.17	400.00	126.67
6–7	791.67	79.17	79.17	400.00	0.00
8	841.67	91.67	91.67	400.00	0.00
9-10	916.67	91.67	91.67	400.00	0.00
11–12	950.00	100.00	100.00	560.00	0.00
13	916.67	91.67	91.67	560.00	0.00
14–15	950.00	100.00	100.00	560.00	0.00
16–17	983.33	108.33	108.33	560.00	0.00
18–19	950.00	100.00	100.00	560.00	0.00
20	916.67	91.67	91.67	560.00	0.00
21	841.67	91.67	91.67	560.00	30.00
22	540.00	55.83	79.17	560.00	103.33
23	133.33	33.33	33.33	400.00	266.67
24	0.00	0.00	0.00	400.00	400.00

Some comments on Tables 5.7 and 5.8 follow. In the case of no arbitrage (i.e., bilateral contract is satisfied exclusively by local generation), the total generation at hour 2 is 680 MW. It is expected in the case of arbitrage (i.e., bilateral contract is satisfied by local generation as well as purchases from the market), the local generation plus purchases from the market would add up to 680 MW. However, the total in this case is limited to 565.31 MW, which is due to ramping constraints.

At hour 3, when the spot market price is low, all generating units should be OFF. However, because of minimum ON/OFF constraints, the units are ON. In the case of no arbitrage, units 4, 5, and 6 would satisfy bilateral contracts, which are 80, 150, and 250 MW respectively. In the case of arbitrage, the same units would generate at their minimum capacities, which are 20, 140, and 100 MW, respectively.

At hour 2, when the market price is high, all units are ON. In the case of no arbitrage, units would generate power based on economic dispatch results. In the case of arbitrage, the maximum generation of individual units is constrained by the generation at hour 3 because of ramping constraints, which are 100, 175, and 200 MW, respectively. Economic dispatch shows that the generation of unit 4 is less than its maximum capacity while units 5 and 6 would reach their maximum capacity.

The details are shown in the Tables 5.9 and 5.10. According to the two tables, the arbitrage between local generation and purchases from the market for satisfying bilateral contracts can significantly impact economic dispatch and alter the commitment status of generating units.

Table 5.9 Analysis of Schedule at Hour 2
(Bilateral Contracts Satisfied by Local Generation)

Unit	4	5	6	Total
Generation at hour 3 (MW)	80	150	250	480
Maximum capacity at hour 2 (MW)	180	325	400	905
Spinning reserve (MW/10min)	16.67	29.17	33.33	79.17
Non-spinning reserve (MW/10min)	16.67	29.17	33.33	79.17
Maximum generation (MW)	146.66	266.66	333.34	746.66
Minimum generation (MW)	80	150	250	260
Economic dispatch generation (MW)	80	266.66	333.34	680

Table 5.10 Analysis of Schedule at Hour 2
(Bilateral Contracts Satisfied by Local Generation and Purchase)

Unit	4	5	6	Total
Generation at hour 3 (MW)	20	140	100	260
Maximum capacity at hour 2 (MW)	120	315	300	735
Spinning reserve (MW/10min)	16.67	29.17	33.33	79.17
Non-spinning reserve (MW/10min)	16.67	29.17	33.33	79.17
Maximum generation (MW)	86.66	256.66	233.34	576.66
Minimum generation (MW)	20	140	100	260
Economic dispatch generation (MW)	75.31	256.66	233.34	565.31
Bilateral contract (MW)	0	150	250	400
Energy purchase (MW)	0	0	16.67	16.67

5.4.3 Arbitrage between Gas and Power

Given gas purchase and sale opportunities, a GENCO can apply PBUC to arbitrage between gas and electric power sales. We first consider a simple example in which the fuel sale price is 0.5 $/MBtu higher than the fuel purchase price. Suppose that the GENCO has 100,000 MBtu of gas at its possession.

If the purchase price is 1 $/MBtu, the relationship between fuel consumption and generation profit is as shown in Figure 5.1. Figures 5.2 and 5.3 show the cases with purchase prices of 2 $/MBtu and 3 $/MBtu, respectively. Accordingly, when the purchase price is relatively low (1 $/MBtu or 2 $/MBtu), the GENCO can make more profit by consuming additional fuel for generating electricity. When the purchase price is relatively high (3 $/MBtu), the profit first increases and then decreases with additional fuel consumption.

Since the fuel sale price is 0.5 $/MBtu higher than the fuel purchase price, the relationship is linear between fuel sale profit and MBtu as shown in Figure 5.4. Table 5.11 shows the results of a profit analysis of the arbitrage between gas and power. Table 5.12 shows the differences between arbitrage and no arbitrage cases. As these tables indicate, depending on the fuel purchase price, the arbitrage between gas and power is more profitable than the exclusive sales of power or fuel.

ARBITRAGE IN ELECTRICITY MARKETS

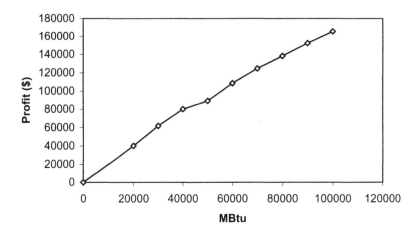

Figure 5.1 Generation Profit versus Fuel Consumption (Purchase Price: 1 $/MBtu)

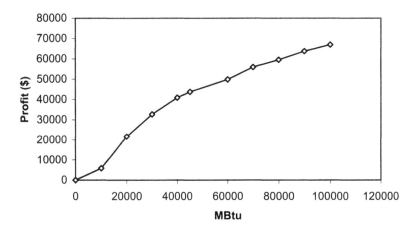

Figure 5.2 Generation Profit Vs. Fuel Consumption (Purchase Price: 2 $/MBtu)

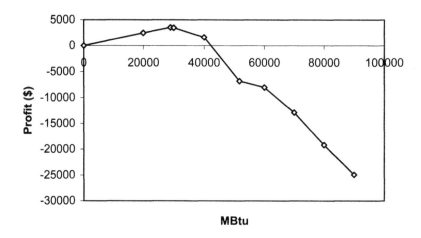

Figure 5.3 Generation Profit versus Fuel Consumption (Purchase Price: 3 $/MBtu)

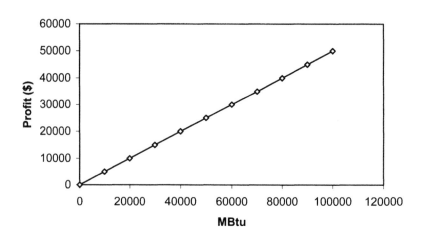

Figure 5.4 Sale Profit versus Sale MBtu
(Sale Price is 0.5 $/MBtu Higher Than Purchase Price)

Table 5.11 Arbitrage between Gas and Power

Fuel Purchase Price ($/MBtu)	Purchase MBtu	Purchase Cost ($)	Gen MBtu	Gen Profit ($)	Sale MBtu	Sale Profit ($)	Total Profit ($)
1	100000	100000	100000	165541	0	0	165541
2	100000	200000	45081	43709	54919	27460	71169
3	100000	300000	0	0	100000	50000	50000

ARBITRAGE IN ELECTRICITY MARKETS

Table 5.12 Comparison of Arbitrage and No Arbitrage (Gas Arbitrage Case)

Fuel Purchase Price ($/MBtu)	Generation Only Profit ($)	Fuel Sale Only Profit ($)	Arbitrage Profit ($)	Arbitrage Strategy
1	165541	50000	165541	Only generation
2	66899	50000	71169	Part generation, Part sale
3	-28000	50000	50000	Only sale

We turn now to a more complicated case where the fuel purchase price is a function of fuel consumption. Figure 5.5 shows the purchase price curve for gas as a function of fuel consumption. If we consider segment z for example, the minimum available fuel would be $(x + y)$ and the maximum available fuel would be $(x + y + z)$. Note that the GENCO can use any portion of the total available MBTU (i.e., $x + y + z$) to generate electricity and sell the rest of the available gas.

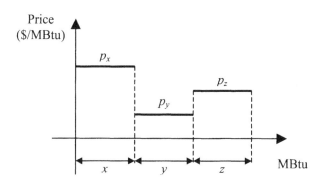

Figure 5.5 Fuel Purchase Price Curve

Suppose that w is the amount that is used for generating electricity and v is the amount of fuel that is sold to the market. Two cases are considered:

- If w is between $(x + y)$ and $(x + y + z)$, then v could be between 0 and $(x + y + z - w)$.

- If w is less than $(x + y)$, then v could be between $(x + y - w)$ and $(x + y + z - w)$ so that $(w + v)$ would be between $(x + y)$ and $(x + y + z)$.

To simplify the arbitrage calculation, we discretize the $(0, x + y + z)$ segment into n subsegments. For subsegment i, the lower limit for fuel consumption would be $(i - 1) \times (x + y + z) / n$, and the upper limit for fuel consumption would be $i \times (x + y + z) / n$. Imposing lower and upper limits on PBUC, we calculate the optimal w between $(i\text{-}1) \times (x + y + z) / n$ and $i \times (x + y + z) / n$. Given w, we determine the limits of v, and based on the fuel purchase price curve, we determine the optimal v. Once we have the optimal solutions for the subsegments, we determine the best solution for the desired segment. Figure 5.6 shows the procedure.

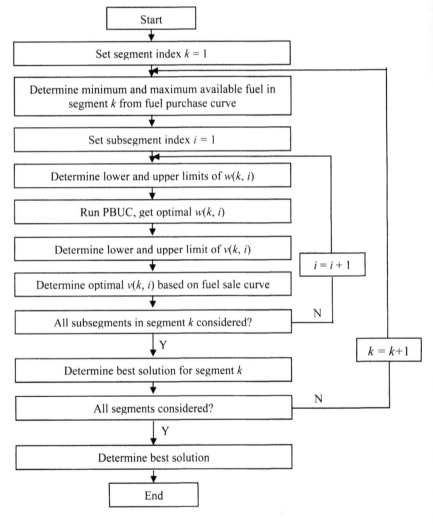

Figure 5.6 Procedure of Arbitraging between Gas and Power

ARBITRAGE IN ELECTRICITY MARKETS

Let us consider an example. The fuel purchase price and cost curves are shown in Figures 5.7 and 5.8, respectively. The cost of supplying a certain amount of fuel is the integral of the fuel purchase price curve for that amount of fuel. We assume that the fuel sale price and cost curves are the same as those in Figures 5.7 and 5.8. We are interested in the second segment of Figure 5.7 (i.e., the minimum available MBtu is 50,000 and the maximum available MBtu is 100,000). We set the steps at 10,000 MBtu increments for a total of 10 segments. The arbitrage results are shown in Table 5.13.

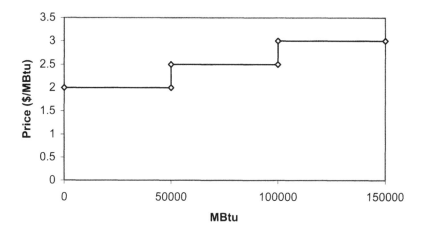

Figure 5.7 Fuel Purchase Price Curve

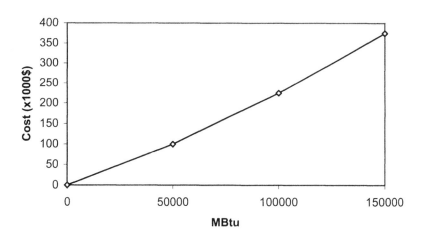

Figure 5.8 Fuel Purchase Cost Curve

Table 5.13 Arbitrage between Gas and Power (Second Segment)

Sub. Seg. No	Min Gen MBtu	Max Gen MBtu	Opt. Gen MBtu	Opt. Gen Profit	Min Sale MBtu	Max Sale MBtu	Opt. Sale MBtu	Opt. Sale Profit	Add. Cost $	Total Profit $
1	0	10000	0	0	50000	100000	50000	-25000	-25000	0
2	10000	20000	20000	11952	30000	80000	30000	-15000	-25000	21951
3	20000	30000	30000	18757	20000	70000	20000	-10000	-25000	33757
4	30000	40000	40000	22119	10000	60000	10000	-5000	-25000	42118
5	40000	50000	41403	22502	8597	58597	8597	-4298	-25000	43203
6	50000	60000	60000	21559	0	40000	0	0	-25000	46559
7	60000	70000	67732	23021	0	32268	0	0	-25000	**48021**
8	70000	80000	70000	22943	0	30000	0	0	-25000	47943
9	80000	90000	80083	21784	0	19917	0	0	-25000	46784
10	90000	100000	90000	21254	0	10000	0	0	-25000	46254

Consider subsegment 5 for example. The minimum and the maximum available MBtu for generating electricity are 40,000 and 50,000. According to PBUC, a GENCO consumes a total of 41,403 MBtu with $22,502 profit. Hence, the GENCO has an opportunity to sell the difference in fuel for at least (50,000 - 41,403 =) 8,597 MBtu and at most (100,000 - 41,403 =) 58,597 MBtu. According to the fuel sale price curve, the optimal MBtu sale occurs at 8,597 with -$4,298 profit (i.e., loss). The loss is inevitable in order to meet the minimum and maximum available MBtu. The additional cost is -$25,000; see Section 4.5.2 where $\Delta C = (x \cdot p_x + y \cdot p_y) - (x + y) \cdot p_z$ is defined as the additional cost). So the total profit is (22,502 – 4,298 + 25,000 = $43203).

The optimal arbitrage solution is subsegment 7. Here, the minimum and the maximum available MBtus are 60,000 and 70,000, respectively. According to PBUC, a GENCO consumes 67,732 MBtu with a $23,021 profit. Since the minimum and the maximum available MBtu are 50,000 and 100,000, we sell at most 100,000 - 67,732 = 32,268 MBtu. Checking the fuel sale price curve, we learn that the optimal strategy would be to sell nothing. Here the additional cost is - $25000 with a total profit of (23,021 + 25,000 = $48,021).

The optimization results for the first and third segments are shown in Tables 5.14 and 5.15 in which the optimal solution for each segment is shown in bold. Note that for the third segment, there are 15 subsegments with 10,000 MBtu steps. Generally, smaller steps will result in smaller errors caused by discretization.

ARBITRAGE IN ELECTRICITY MARKETS

Comparing the three preceding tables, the optimal strategy would be to purchase 67,732 MBtu and apply all of it to generate electricity with a $48021 profit.

Table 5.14 Arbitrage between Gas and Power (First Segment)

Seg. No	Min MBtu	Max MBtu	Opt. Gen MBtu	Gen Profit	Min Sale MBtu	Max Sale MBtu	Opt. Sale MBtu	Sale Profit	Add. Cost	Total Profit
1	0	5000	0	0	0	50000	0	0	0	0
2	5000	10000	Infeasible							
3	10000	15000	15000	14889	0	35000	0	0	0	14889
4	15000	20000	20000	21551	0	30000	0	0	0	21551
5	20000	25000	25000	28005	0	25000	0	0	0	28005
6	25000	30000	30000	33357	0	20000	0	0	0	33357
7	30000	35000	35000	37893	0	15000	0	0	0	37893
8	35000	40000	40000	41718	0	10000	0	0	0	41718
9	40000	45000	45000	44648	0	5000	0	0	0	**44648**
10	45000	50000	50000	41362	0	0	0	0	0	41362

Table 5.15 Arbitrage between Gas and Power (Third Segment)

Seg. No	Min MBtu	Max MBtu	Opt. Gen MBtu	Gen Profit	Min Sale MBtu	Max Sale MBtu	Opt. Sale MBtu	Sale Profit	Add. Cost	Total Profit
1	0	10000	0	0	100000	150000	100000	-50000	-75000	25000
2	10000	20000	20000	3051	80000	130000	80000	-40000	-75000	38050
3	20000	30000	28948	4213	71052	121052	71052	-35526	-75000	43688
4	30000	40000	30000	4158	70000	120000	70000	-35000	-75000	44158
5	40000	50000	40000	2519	60000	110000	60000	-30000	-75000	**47519**
6	50000	60000	58852	-12214	41148	91148	41148	-41148	-75000	21637
7	60000	70000	60000	-7791	40000	90000	40000	-40000	-75000	27209
8	70000	80000	70000	-11408	30000	80000	30000	-30000	-75000	33593
9	80000	90000	80000	-17464	20000	70000	20000	-20000	-75000	37536
10	90000	100000	90000	-23072	10000	60000	10000	-10000	-75000	41929
11	100000	110000	100000	-29716	0	50000	0	0	-75000	45284
12	110000	120000	110000	-36997	0	40000	0	0	-75000	38003
13	120000	130000	120000	-45904	0	30000	0	0	-75000	29096
14	130000	140000	130000	-55603	0	20000	0	0	-75000	19397
15	140000	150000	140000	-67232	0	10000	0	0	-75000	7768

5.4.4 Arbitrage of Emission Allowance

We run an emission-constrained PBUC for emission arbitrage. We first consider a simple case with a fixed emission allowance trading price.

The information included in Table 5.16 is the basis for emission allowance arbitrage. First, we exclude emission constraints, and the PBUC solution shows a total emission of 585 tons and $245,244.63 profit. That is, whether the emission allowance is larger or less than 585 tons, the PBUC profit will be less than $245,244.63. Next we include emission allowance constraints in PBUC ranging from 100 tons to 500 tons. Table 5.16 shows the profits corresponding to each value of the constraint.

Table 5.16 PBUC Solution with Emission Constraints

Emission Constraint (tons)	Profit ($)
none	245244.63
500	243819.28
400	234635.75
350	227320.03
300	217425.89
275	209655.50
250	200003.00
225	189289.92
200	175884.64
175	164902.23
150	149950.63
125	132474.77
100	93963.67

If the emission allowance is more than the unconstrained emission solution (585 tons), for example, 600 tons, we could sell the difference of 600 - 585 = 15 tons to the market for $15k$ profits, in which k represents the market price for the emission allowance and the unit of k is $/ton. If the emission allowance is less than 585 tons, we might purchase the additional allowance from the market or sell an additional emission allowance to the market, based on the market's purchase price for an emission allowance.

Suppose that our emission allowance is denoted by e, where $100 \leq e \leq 585$ tons. As we consider emission allowance trades, the total profit could be represented by $F(w) = y(w) + (e-w) \times k$, where w is the optimal emission requirement calculated by PBUC and $y(w)$ is the PBUC profit

ARBITRAGE IN ELECTRICITY MARKETS

without any emission allowance trading. While $w > e$ indicates that we would need to purchase additional emission allowance, $w < e$ indicates that an excess allowance is available that can be sold in the market. If $k = 0$, then $w = 585$ tons which is the original case, since emission allowance would be free of charge. If $k > 0$, Tables 5.17 to 5.19 show different arbitrage scenarios for certain values of (e). In these tables, the optimal strategies are presented in bold. Note that the total emission and profit (without emission trade) in these tables refer to the PBUC results.

Table 5.20 shows the differences between arbitrage and no arbitrage when the original emission allowance is 250 tons. This is the case shown in Table 5.18.

From Tables 5.17 to 5.19, for constant emission allowance trading price (k), the optimal strategy is irrelevant to (e). That is, the optimal solution for (w) in $F(w) = y(w) + (e - w) \times k = y(w) - w \times k + e \times k$ would be irrelevant to (e), since $(e \times k)$ would be constant as long as (e) is specified and (k) is constant. If (k) is not constant, this conclusion may change. Consider an example.

Table 5.17 Arbitrage of Emission Allowance ($e = 200$ tons)

Total Emission (ton)	Profit without Trade ($)	Emission Trades (ton)	Net Profit ($)					
			$k = 10$	$k = 100$	$k = 200$	$k = 300$	$k = 400$	$k = 500$
585	245245	385	**241395**	206747	168249	129751	91253	52755
500	243819	300	240850	214126	184433	154740	125047	95354
400	234636	200	232636	**214636**	194637	174637	154638	134638
350	227320	150	225820	212320	197320	182320	167320	152320
300	217426	100	216426	207426	**197427**	**187427**	177427	167428
275	209656	75	208905	202155	194655	187155	179655	172155
250	200003	50	199503	195003	190003	185003	**180003**	175003
225	189290	25	189040	186790	184290	181790	179290	176790
200	175885	0	175885	175885	175886	175886	175887	175887
175	164902	-25	165152	167403	169903	172403	174904	**177404**
150	149951	-50	150451	154951	159951	164951	169951	174951
125	132475	-75	133225	139975	147476	154976	162477	169977
100	93964	-100	94964	103964	113964	123964	133964	143964

Table 5.18 Arbitrage of Emission Allowance (e = 250 tons)

Total Emission (ton)	Profit without Trades ($)	Emission Trades (ton)	Net Profit ($)					
			$k = 10$	$k = 100$	$k = 200$	$k = 300$	$k = 400$	$k = 500$
585	245245	335	**241895**	211747	178249	144751	111253	77755
500	243819	250	241350	219126	194433	169740	145047	120354
400	234636	150	233136	**219636**	204637	189637	174638	159638
350	227320	100	226320	217320	207320	197320	187320	177320
300	217426	50	216926	212426	**207427**	**202427**	197427	192428
275	209656	25	209405	207155	204655	202155	199655	197155
250	200003	0	200003	200003	200003	200003	**200003**	200003
225	189290	-25	189540	191790	194290	196790	199290	201790
200	175885	-50	176385	180885	185886	190886	195887	200887
175	164902	-75	165652	172403	179903	187403	194904	**202404**
150	149951	-100	150951	159951	169951	179951	189951	199951
125	132475	-125	133725	144975	157476	169976	182477	194977
100	93964	-150	95464	108964	123964	138964	153964	168964

Table 5.19 Arbitrage of Emission Allowance (e = 300 tons)

Total Emission (ton)	Profit without Trade ($)	Emission Trades (ton)	Net Profit ($)					
			$k = 10$	$k = 100$	$k = 200$	$k = 300$	$k = 400$	$k = 500$
585	245245	285	**242395**	216747	188249	159751	131253	102755
500	243819	200	241850	224126	204433	184740	165047	145354
400	234636	100	233636	**224636**	214637	204637	194638	184638
350	227320	50	226820	222320	217320	212320	207320	202320
300	217426	0	217426	217426	**217427**	**217427**	217427	217428
275	209656	-25	209905	212155	214655	217155	219655	222155
250	200003	-50	200503	205003	210003	215003	**220003**	225003
225	189290	-75	190040	196790	204290	211790	219290	226790
200	175885	-100	176885	185885	195886	205886	215887	225887
175	164902	-125	166152	177403	189903	202403	214904	**227404**
150	149951	-150	151451	164951	179951	194951	209951	224951
125	132475	-175	134225	149975	167476	184976	202477	219977
100	93964	-200	95964	113964	133964	153964	173964	193964

Table 5.20 Comparison of Arbitrage and No-Arbitrage (Emission Allowance Arbitrage)

Emission Price ($/ton)	No-Arbitrage Profit ($)	Arbitrage Profit ($)	Arbitrage Strategy	Arbitrage Amount (ton)
10	200003	241895	Purchase	335
100	200003	219636	Purchase	150
200	200003	207427	Purchase	50
300	200003	202427	Purchase	50
400	200003	200003	No Trades	0
500	200003	202404	Sale	75

ARBITRAGE IN ELECTRICITY MARKETS

The price curve for the emission trade is as shown in Figure 5.9. Also, the arbitrage is shown in Table 5.21 in which the optimal strategies are in bold. If the price of emission allowance were not constant in Table 5.21, the arbitrage would depend on the original amount of emission allowance. Table 5.22 shows the differences between arbitrage and no arbitrage cases.

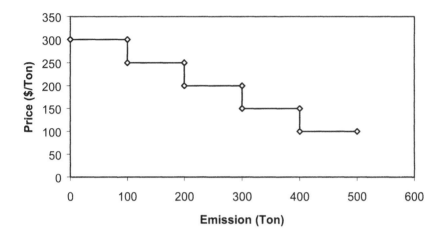

Figure 5.9 Emission Trading Price Curve

Table 5.21 Arbitrage of Emission Allowance (Considering Emission Trades)

Total Emission (PBUC) (Tons)	Profit without Trades ($)	Emission Allowance (Tons)								
		200			300			400		
		Emission Trades (Tons)	Cost ($)	Net Profit ($)	Emission Trades (Tons)	Cost ($)	Net Profit ($)	Emission Trades (Tons)	Cost ($)	Net Profit ($)
585	245245	385	87747	157498	285	71996	173249	185	51245	194000
500	243819	300	75000	168819	200	55000	188819	100	30000	213819
400	234636	200	55000	179636	100	30000	204636	0	0	234636
350	227320	150	42500	184820	50	15000	212320	-50	-15000	242320
300	217426	100	30000	**187426**	0	0	**217426**	-100	-30000	**247426**
275	209656	75	22500	187156	-25	-7500	217156	-125	-36250	245906
250	200003	50	15000	185003	-50	-15000	215003	-150	-42500	242503
225	189290	25	7500	181790	-75	-22500	211790	-175	-48750	238040
200	175885	0	0	175885	-100	-30000	205885	-200	-55000	230885
175	164902	-25	-7500	172402	-125	-36250	201152	-225	-60000	224902
150	149951	-50	-15000	164951	-150	-42500	192451	-250	-65000	214851
125	132475	-75	-22500	154975	-175	-48750	181225	-275	-70000	202475
100	93964	-100	-30000	123964	-200	-55000	148964	-300	-75000	168564

Table 5.22 Comparison of Arbitrage and No Arbitrage for Emission Allowance

Emission Allowance (ton)	No-Arbitrage Profit ($)	Arbitrage Profit ($)	Arbitrage Strategy	Arbitrage Amount (ton)
200	175885	187426	Purchase	100
300	217426	217426	No Trades	0
400	234636	247426	Sale	100

5.4.5 Arbitrage between Steam and Power

A cogeneration unit can sell both steam and electric power. So, depending on market prices for steam and power, it would be possible to arbitrage between steam and power. We use a simplified model for the cogeneration unit to demonstrate the arbitrage between steam and power.

Assume that the fuel consumption is given by the function

$$C_f(P,H) = a + b_1 P + c_1 P^2 + b_2 H + c_2 H^2 + dPH \tag{5.1}$$

where

P	=	power production
H	=	steam production
a	=	constant fuel consumption coefficient
b_1, c_1	=	fuel consumption coefficients for power production
b_2, c_2	=	fuel consumption coefficients for steam production
d	=	fuel consumption coefficient coupling power and steam

Also assume that the ratio of steam to power is fixed at all time, that is,

$$H/P = \gamma \tag{5.2}$$

The arbitrage between steam and power is determined by the optimal value of γ. In the following, we show how the original PBUC formulation and solution methodology is modified to implement the arbitrage between steam and power.

First, we modify the fuel consumption function. Substituting (5.2) into the fuel consumption function (5.1), we have

ARBITRAGE IN ELECTRICITY MARKETS

$$C_f(P,H) = a + b_1 P + c_1 P^2 + b_2(\gamma P) + c_2(\gamma P)^2 + dP(\gamma P)$$
$$= a + (b_1 + \gamma b_2)P + (c_1 + \gamma^2 c_2 + \gamma d)P^2 \quad (5.3)$$

If we define,

$$b' = b_1 + \gamma b_2 \quad (5.4a)$$

$$c' = c_1 + \gamma^2 c_2 + \gamma d \quad (5.4b)$$

the fuel consumption function can be modified by replacing b with b', c with c'. The corresponding modification of the profit formulation is given as follows:

The profit for selling power is

$$F = \rho \times P \quad (5.5)$$

where, ρ is the price of power.

The profit for selling power and steam is

$$F = \rho \times P + \rho_t \times H \quad (5.6)$$

where, ρ_t is the price of steam.

Manipulating (5.6), we have

$$F = \rho \times P + \rho_t \times H = \rho \times P + \rho_t \times (\gamma P) = (\rho + \gamma \rho_t)P \quad (5.7)$$

If we define

$$\rho' = \rho + \gamma \rho_t \quad (5.8)$$

then (5.5) can be used to represent the profit for selling power and steam (5.6) by replacing ρ with ρ'. Thus, by replacing b with b', c with c' and ρ with ρ', the original PBUC formulation and solution methodology can be used to calculate the arbitrage between steam and power.

For the 6-unit system, assume that unit 4 is a cogeneration unit and that its fuel consumption function is given by

$$C_f(P,H) = 142.7348 + 10.694P + 0.00463P^2$$
$$+ 4.2H + 0.004H^2 + 0.0042PH$$

Table 5.23 shows profit (only for cogeneration unit 4) variations with different γ and the steam price. The ratio γ varies between 0 and 1 with a step of 0.1 and optimal strategies are given in bold. Table 5.24 compares difference between arbitrage and no arbitrage.

Table 5.23 Arbitrage between Steam and Power

Ratio γ	Steam Price ($/MW)		
	5	7	10
0	**42927**	42927	42927
0.1	42566	43184	44112
0.2	42148	43384	45240
0.3	41672	43526	46324
0.4	41146	43642	47442
0.5	40576	**43718**	48528
0.6	39941	43713	49509
0.7	39251	43650	50429
0.8	38509	43387	51289
0.9	37713	43202	52096
1	36864	42955	**52912**

Table 5.24 Comparison of Arbitrage and No-Arbitrage (Steam Arbitrage)

Steam Price ($/MW)	No-Arbitrage Profit ($)	Arbitrage Profit ($)	Arbitrage Strategy
5	42927	42927	Sell power
7	42927	43718	Sell steam equal to 0.5 of power
10	42927	52912	Sell equal amounts of steam and power

5.5 CONCLUSIONS

As the preceding examples show, many arbitrage opportunities in the power market can be found using PBUC. In some cases, no change is made to the original formulation and solution methodology. While in other cases, appropriate changes need to be made to accommodate a new situation. We end this chapter with a discussion of the arbitrage between leasing a plant and building a plant.

The rapid development of the power market has attracted a myriad of investments into the electric power industry, mostly in the generation sector. Those who expect to make profits in this industry may either build a new plant or lease an existing plant. Depending on the leasing conditions

ARBITRAGE IN ELECTRICITY MARKETS

and contracts for an existing plant and the cost of building a new plant, arbitrage opportunities could arise.

To lease a plant, we would pay a leasing fee, for example X dollars for one year. With the leased plant, we purchase fuels, generate electricity, and sell the electricity to the market. We may engage a PBUC to realize the maximum value of the plant. Suppose that the revenue for operating the plant (i.e., revenue from selling electricity minus the fuel cost) is Z dollars for one year's operation. Then, the profit for leasing the plant is $Z - X$ dollars in one year.

To build a plant, we would invest a large sum. Suppose that money is borrowed from a bank and should be returned in 10 years for a total of Y dollars (considering interest and depreciation). Here, the annual debt would be $Y/10$ dollars. Similar to leasing a plant, we apply a PBUC to run the plant. We assume the revenue for one year is Z'. Then, the net profit from building the plant and operating it in one year is $Z' - Y/10$ dollars for one year. Comparing $Z - X$ and $Z' - Y/10$, we can arbitrage between leasing a plant and building a plant.

More realistically, leasing a plant would involve the valuation of the plant. However, building a new plant would involve decisions on the location and the type of the plant. It could also involve a great deal of risk since it is a relatively long-term commitment. Further discussion of these topics would be beyond the scope of this book. However, we should point out that no matter whether the decision is leasing a plant or building a new plant, efficient PBUC is an indispensable part of the decision making.

Chapter 6
Market Power Analysis Based on Game Theory

6.1 INTRODUCTION

It is claimed that the increasing competition in the energy market can help maximize customers' payoffs. This claim is conceivably supported by the results obtained from the application of game theory. In general, the bigger the game, the greater the variety of coalitions is. However, the grand coalition of the game will be dominating as long as the market is not giving any participants[1] or group of participants a relative advantage over the remaining participants. Hence, in the case of a game played by numerous participants there will be a narrower margin for possible coalitions against grand coalition.

In this chapter, we present game theory-based methodologies that can be used to:

- Identify non-competitive situations in energy marketplaces (transaction analysis from the market coordinator's point of view)

- Provide support for minimizing risks involved in price decisions in energy marketplaces (transaction analysis from a participant's point of view).

Power market coordinators must identify and correct situations in which some companies possess "market power." In such situations, companies acting alone or colluding with their counter-parts, may have enough share of the generation (or the load) to be able to act as price-setters; hence, some companies' payoffs exceed those obtained in competitive situations while systemwide payoffs are reduced. In order to

---

[1] Also referred to as agents.

prevent collusions in a power market, market coordinators would follow a methodology which will:

- Identify possible collusions among participants
- Identify participants of possible collusions and their strategies
- Compute transactions and payoffs associated with those collusions
- Discourage collusions that could minimize market payoffs
- Identify coalitions that are likely to be formed among participants
- Encourage coalitions that could maximize market payoffs.

6.2 GAME THEORY

Although game theory has been associated with parlor games, most research in game theory has focused on how groups of people interact. Game theory has been extensively used in microeconomic analysis, where its prediction record has been remarkable in areas such as industrial organization theory.

Game theory is the mathematical theory of bargaining, the essentials of which were developed by [Neu47]. The authors restricted their attention to zero-sum games, that is, to games in which no participant can gain except at another's expense. However, this restriction was overcome by the work of John Nash during the 1950s.

In game theory, beliefs are formulated against risky alternatives for the maximization of expected revenue (a numerical utility function). Probability theory is heavily used in order to represent the uncertainty of outcomes and Bayes' law is used to revise beliefs. The following example discusses some of the features of the game theory.

6.2.1 An Instructive Example

In game theory, we search for so-called Nash equilibria, that is, sets of strategies used by multiple participants such that each participant would have no incentive to change its strategy given the strategies of other participants. Nash equilibria are stable, but not necessarily desirable. For example, in what is undoubtedly the most-discussed instance of a game, the

MARKET POWER ANALYSIS BASED ON GAME THEORY

Prisoner's Dilemma, the unique Nash equilibrium is a state in which both participants are as badly off, given their utility functions, as possible.

One way to describe a game is by listing the participants in the game, and for each participant, listing the alternative choices (called strategies) available to that participant. This is similar to the Prisoner's Dilemma case. The Prisoner's Dilemma concerns two participants who are partners in a crime and are captured by the police. Each suspect is placed in a separate cell and offered the opportunity to confess to the crime. The point of game theory is not prescriptive but descriptive: analysis of a game permits us to locate the equilibria, and thus to predict those states of play that will be stable, barring exogenous interference.

Consider a system of two electric utilities. This system will be discussed thoroughly in the following sections of this chapter. Here, we review one of the conditions prevailing on the two-utility system in order to familiarize the reader with game theory. The two utilities supply their own load by generating power and trading with a neighboring utility. Utility 1 may sell power to utility 2 if the spot price is larger than its cost of generating additional power. The lambdas for these two utilities are $\lambda_1 = 21$ \$/MWh and $\lambda_2 = 26$ \$/MWh, respectively. It is expected that if the spot price is between these two lambdas, the two utilities will be willing to exchange power, where the flow will be from utility 1 to utility 2. The transaction flow in the opposite direction is not likely, since utility 2 cannot sell its power at a price less than its marginal cost. Figure 6.1 depicts the situation where transmission losses are neglected.

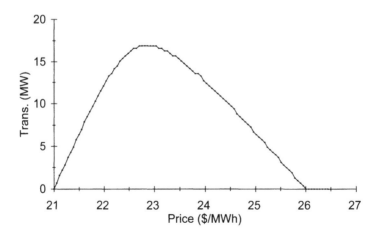

Figure 6.1 Spot Price and Power Transaction without Losses

The price range is from 21 to 26 $/MWh, and the transaction flow is from 0 to 16.667 MW. This curve corresponds to positive payoffs in both utilities. Prices beyond the range of $21 and $26 may be refused by either utility. In Figure 6.1, the maximum transaction is achieved at 23.083 $/MWh. This flow corresponds to the maximum payoff which may be obtained by the cooperation of two utilities. The corresponding power flow from utility 1 to utility 2 is 16.667 MW (Nash equilibrium) and payoff (total saving) is 2 × $23.083 which is divided equally between the utilities. Apart from the optimal price, there are two prices corresponding to each transaction. On the left side, where the price would be lower, utility 2 is requesting a transaction but utility 1 is not willing to sell more (i.e., utility 1 is expecting a higher spot market price). On the right side, where the price is higher, utility 1 is willing to sell but utility 2 is not interested in buying at the given price (i.e., utility 2 is expecting a lower spot market price). Thus, the transaction would be zero if the spot price is higher than 26 $/MWh or lower than 21 $/MWh (i.e., marginal costs of the two utilities).

Similar to the Prisoner's Dilemma, if the two utilities cooperate, they receive the highest payoff and split the payoff of their cooperation. However, if one utility sets the price at less than 23.083 $/MWh, though utility 2 is willing to buy, utility 1 is not willing to sell because its payoff will be less than that of utility 2.

This type of two-participant game has fascinated power engineers for several reasons. First, it is a simple representation of numerous important situations in a power market. Utility 1 would like to set the price at higher than the Nash equilibrium to gain most of the payoff, whereas utility 2 would like to set the spot price at lower than the Nash equilibrium to gain most of the payoff. If the two utilities collude, it is best for them if both set high prices. However, individually, it is best for each utility to set a low price while the opposition sets a high price.

The second feature of this game is that it illustrates how a utility should behave in a competitive power market. No matter what a utility believes his market opponent is going to do, it is always best to cooperate. This conflict between the pursuit of individual goals and the common good is at the heart of many game-theoretic problems.

Repetition in a power market opens up the possibility of being rewarded or punished in the future for current behavior. Game theorists have provided a number of applications to explain that if the game is repeated often enough, the market participants ought to cooperate.

6.2.2 Game Methods in Power Systems

In the past, Nash game, cooperative game and hyper-game methods were used for non-commercial power plants operation analysis [Mae92]. Other applications of game theory to power systems were presented in [Eht89]. The cogeneration problem [Hau92] was presented using Stackelberg equilibrium method and a successive approximation algorithm was applied to solve the problem. The cogeneration minimizes its net cost given the selling and buying prices, and the utility minimizes its net production cost based on the exchanged power. The selling price is the utility's average cost, while the buying price is the same as utility's marginal cost. The allocation of cost savings in an energy brokerage system is discussed in [Cha95], and linear programming is used to determine optimal transactions. The Shapley value criterion is used for the allocation of savings. In [Ruu91], the Nash bargaining scheme is used for electricity exchange in a power market. The objective is to calculate optimal power exchanges in a power market for different time periods to adjust the load shape and reduce the operation cost of utilities in the market. Transmission pricing policies are analyzed in [Hob92], where static cooperative models are used to calculate the possible outcomes of short-run transmission game. Long-run games, in which the amount of transmission capacity is a decision variable, are modeled as dynamic non-cooperative Stackelberg game. A detailed discussion of the applications of game theory to power systems is presented in [Sin99] and [Sto02].

In this chapter, the transaction analysis by the Nash game theory will be discussed. Note that in a restructured power market, the participants' cost functions are considered confidential. Therefore, we consider a three-utility case in which the market power is analyzed with and without the complete information.

6.3 POWER TRANSACTIONS GAME

Figure 6.2 illustrates a power market where generators and loads submit bids and the ISO (or PX) sets the market spot price based on the last generator dispatched in order to balance the power market's generation and demand. If the market demand is higher than the offered power by participants, the spot price will increase. Conversely, if the offered power is higher than the market demand, the spot price will decrease. [Car82] refers to this process as "real-time closed-loop control law."

Game theory is used here by the ISO, the market coordinator, to understand participants' behaviors. In a power transaction game, participants' transactions are modeled as a game of strategies in which participants compete to maximize their payoffs. Here, economic payoffs constitute *payoffs* and participants' options are treated as *strategies*. Any subset of participants is called a *coalition,* with a *grand coalition* representing all participants. A coalition may be constituted by *one* participant maximizing its payoffs. See Appendix E for a brief introduction to game theory.

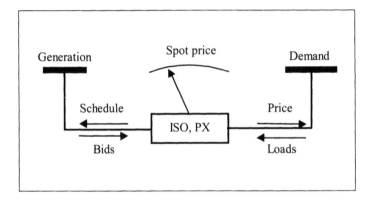

Figure 6.2 The Power Market

Two types of games are considered in this chapter: non-cooperative and cooperative. A *non-cooperative* game takes place when any coalition is interested in maximizing its own payoffs regardless of payoffs for the counter-coalition. Within a coalition, strategies are coordinated among participants in a *cooperative* game. In other words, in non-cooperative games, participants make no commitments to coordinate their strategies. Conversely, in cooperative games, participants can make *credible* commitments to coordinate their strategies. In a non-cooperative game, some participants choose a strategy while other participants try to identify their best response to that strategy. In a cooperative game, participants define strategies that would lead to the best outcome for the coalition.

There are two sets of agreements discussed by participants in a coalition. The first one is to agree on the coordination of strategies, for example, *we all bid high.* The second one is to agree on the payoff distribution. In this chapter, we do not analyze the second agreement. We

MARKET POWER ANALYSIS BASED ON GAME THEORY

presume that when a strategy is coordinated for a coalition, participants manage to find a way to share the payoffs in order to preserve the coalition. The situation is simpler for a one-participant coalition: it has to anticipate a bid that maximizes its payoffs when transactions are defined by the market coordinator.

6.3.1 Coalitions among Participants

In Figure 6.3, the pricing strategy is given in a power market where participation is not certain. Individuals may decide to either *participate* in the market or to make *bilateral agreements* with other individuals. They may, in essence, join the power market if payoffs are higher than those of bilateral agreements.

Figure 6.3 Participant's Pricing Decision

In making the bilateral agreements, an individual could choose to *act alone* (hence, not coordinate pricing policies with other participants) or to *coordinate* its pricing strategy with market participants. In deciding to participate in the pool, a participant may choose either to *cooperate* with the market (taking the spot price) or *collude* with other participants. Market regulations should prevent any coalitions except the grand coalition and encourage *cooperation* in the power market.

6.3.2 Generation Cost for Participants

A power transaction and its spot price are negotiated by buyers and sellers, and the total savings in supplying the load is the payoff for cooperation in a power market. Suppose that the generation cost function in a utility is represented by Figure 6.4.

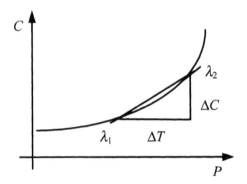

Figure 6.4 Cost Function and Transaction

In the Figure, P and C represent system generation and operation cost, λ_1 and λ_2 represent the system's marginal costs before and after a transaction, ΔT and ΔC represent a transaction increment and the corresponding cost increment. The incremental cost is defined as $\Delta C/\Delta T$, and it may represent the transaction price, which depends on the utility's initial generation level and the amount of power transaction. In this chapter, we will evaluate the relationship between transaction and price based on a game method.

Figure 6.5 illustrates the generation cost as a function of generation. The generation cost C for a power P generated by generator i is represented as a quadratic polynomial $a + bP + cP^2$ (though, the analysis is not limited to the quadratic function), where a, b, and c are cost coefficients.

MARKET POWER ANALYSIS BASED ON GAME THEORY

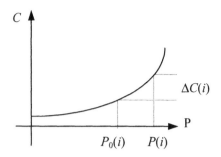

Figure 6.5 Generation Cost for Generator *i*

The cost increment incurred by the generation increment $\Delta P(i) = P(i) - P_0(i)$ is calculated as:

$$\begin{aligned}\Delta C(i) &= C(P(i)) - C(P_0(i)) \\ &= b(i)\Delta P(i) + c(i)P(i)\Delta P(i) + c(i)P_0(i)\Delta P(i)\end{aligned} \tag{6.1}$$

The incremental cost is determined as

$$\pi(i) = \frac{\Delta C(i)}{\Delta P(i)} = b(i) + c(i)P(i) + c(i)P_0(i) \tag{6.2}$$

where $\pi(i)$ is also referred to as the transaction price. The transaction price is the minimum price at which one can accept an increment $\Delta P(i)$, because this price multiplied by the incremental power represents the incremental cost to the utility. An exporting utility will offer $\Delta P(i) > 0$ to the market at a price equal to or larger than $\pi(i)$ ($/MWh) and collect a minimum of $\Delta P(i) \times \pi(i)$ ($/h) from the market for the export. Likewise, the utility may buy $\Delta P(i) < 0$ from the market if the transaction cost is lower than or at most equal to $\Delta C(i)$. When $\Delta P(i)$ approaches 0, the incremental cost would represent the generator's marginal cost:

$$\lambda(i) = \frac{dC(i)}{dP(i)} = b(i) + 2c(i)P(i) \tag{6.3}$$

In Figure 6.6, λ and π are depicted as a function of the generation level in a utility. In this figure, the utility may increase its generation up to its maximum capacity P_{max} and construct its trading price curves either from the marginal cost or from the incremental cost curve. The incremental cost curve will provide a utility with the maximum price to pay for a purchase (or minimum price to receive from the sale) of power equal to $T(i)$.

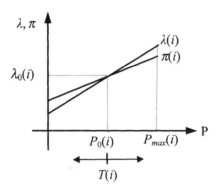

Figure 6.6 Marginal Cost and Incremental Price Curves

The curves in Figure 6.6 are interpreted as follows: the utility will increase its generation level beyond $P_0(i)$ if the selling price for each unit of extra power is greater than $\lambda_0(i)$. The lowest price to accept for the export is given by the incremental cost curve. If the spot price is lower than $\lambda_0(i)$, the utility will import power to supply its local load. The highest price to pay for the import is given by the incremental cost curve.

From the participant's point of view, the problem is to define the price levels for maximizing the payoffs associated with individual utilities. From a market's point of view, the aim in the optimal scheduling of power transactions is to define the amount and price of power transactions among utilities in order to maximize the payoffs corresponding to the market operation. These payoffs are maximized for a particular electricity market and a given amount of generation offered to the power system, by minimizing the total operation cost of the system.

Depending on competitive conditions in an electricity market, utilities drop the offered price to its lowest limit, meaning the incremental cost. The utilities in this environment try to buy electric power from the network at a price that is at most equal to their incremental costs. If the price of power offered by other utilities is higher than the buyer's incremental cost, it will be more reasonable for the buyer utility to generate its own power.

Since incremental cost and marginal cost are linear functions of generation, we assume that participants' bids are also linear functions of generation:

$$\lambda(i) = \lambda_0(i) + m(i)P(i)\lambda(i) \tag{6.4}$$

where

$\lambda(i)$ = price (bid) per unit of power at generation level $P(i)$

$\lambda_0(i)$ = marginal cost of electricity at $P_0(i)$

$m(i)$ = slope of bid curve

The market coordinator receives participants' bids as (6.4) and matches the lowest bid with the load. In Figure 6.7, the participant intends to increase the generation beyond $P_0(i)$ if the spot market price is greater than $\lambda_0(i)$. Here, $T(i)$ is the net interchanged power; if $T(i)$ is positive (negative), the participant is selling power to (buying power from) the market, $-P_0(i) \leq T(i) \leq P_{max}(i) - P_0(i)$. Each participant will produce $P(i) = P_0(i) + T(i)$ to supply the market's load.

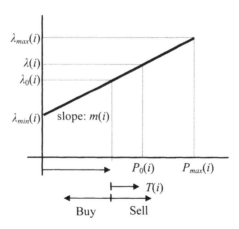

Figure 6.7 Participant *i*'s Bid

6.3.3 Participant's Objective

For a given spot market price ρ, the participant's payoff is

$$B(i) = -\Delta C(i) + \rho T(i) \tag{6.5}$$

Other pricing schemes [Rud97] would result in different allocations of payoffs among participants; however, our approach is general enough to be used in other scenarios as well. We focus on generating units and transmission charges in this chapter. The participant's objective is to maximize its payoff; hence, using (6.5), we obtain the condition for payoff maximization as

$$\frac{dB(i)}{dP(i)} = \frac{d}{dP(i)}(-b(i)(P(i)-P_0(i))-c(i)(P(i))^2+c(i)(P_0(i))^2+\rho T(i))$$

$$= -b(i) - 2c(i)P(i) + \frac{d(\rho T(i))}{dP(i)} = 0 \quad (6.6)$$

Hence, in the absence of binding constraints,

$$\frac{dC(i)}{dP(i)} = \frac{d(\rho T(i))}{dP(i)} = \rho \frac{d(T(i))}{dP(i)} + T(i)\frac{d(\rho)}{dP(i)} = \rho + T(i)\frac{d(\rho)}{dP(i)} \quad (6.7)$$

In perfect competition, sellers and buyers are very small compared to the market size; no participant can significantly affect the existing market price; hence, the spot price at which each participant can sell or buy electric power is essentially given. In (6.7), for a fixed spot market price ρ, a participant adjusts its generation level in order to match its marginal cost with the given price. Electricity market's and participant's payoffs are maximized simultaneously, and the market price is the optimal spot market price defined in [Sch88].

6.4 NASH BARGAINING PROBLEM

In this chapter, the transaction problem is considered as a bargaining model and analyzed by the Nash bargaining method. The main features of the method are:

- The objective function is the sum of product of individual payoff functions.
- In formulating the payoff functions, local power demands are considered proprietary information.
- The price that a utility is willing to pay for importing power is considered an important variable for local decision making. An

arbitrator may assist both participants to coordinate their plans for optimizing revenues.

- The final result should be pareto optimal; that is no participant will be better off financially by withdrawing from the game.

Here both buyer and seller are considered participants that buy and sell power according to a spot market price. If the two participants want to reach an agreement on transactions, they need to negotiate a flow pattern that will be acceptable to both parties. This scenario corresponds to coordination among participants in a decentralized power system, which is different from that in a centralized system; in a decentralized system, each participant presents an objective or price policy while mastering its own control devices and payoffs from association with other power systems in a bilateral manner.

A detailed discussion of the Nash bargaining problem is provided in [Aub82]. The main point are summarized here. Let T denote a possible trade; associated with each trade T is a pair of utility payoffs (μ, v) representing participants 1 and 2. The Nash bargaining problem for two participants is denoted by $[\Re, (\mu, v)]$ characterized by a region $\Re$, and a payoff point (μ, v) in $\Re$. If no trade occurs, the payoff is $(0, 0)$; a trade would take place if and only if both participants agree upon a solution in $\Re$ which represents a fair outcome for the participants. The problem is stated as follows: given $(\mu_0, v_0) = (0, 0)$ in $\Re$ as the initial solution, find a different solution $(\mu_1, v_1,)$ that satisfies the conditions

- (μ_1, v_1) is a point of $\Re$.
- $\mu_1 v_1 \geq \mu_0 v_0$, for all (μ_0, v_0) in $\Re$, where $\mu_1 > 0$ and $v_1 > 0$.

The point (μ_1, v_1) is the Nash equilibrium solution to the bargaining game $[\Re, (\mu, v)]$.

6.4.1 Nash Bargaining Model for Transaction Analysis

The objective function, based on the preceding discussion, is stated as

$$\max \sum_{k \in K} \prod_{p_{ij}, T_{ij}} R_{ij}^k (p_{ij}, T_{ij}) \tag{6.8}$$

where $R_{ij}^k(p_{ij}, T_{ij}) = p_{ij}T'_{ij} - C_i(P_{gi} + T_{ij}) + C_i(P_{gi})$

$T_{ij} = T'_{ij} + \Delta T'_{ij}$

and
- R_{ij}^k = payoff function for contract k
- K = set of contracts
- T_{ij} = transaction power generated by participant i
- T'_{ij} = transaction Power received by participant j
- $\Delta T'_{ij}$ = transmission loss
- p_{ij} = transaction price per unit of transaction T_{ij}
- P_{gi} = power output of participant i
- C_i = production cost of participant i

T_{ij} and p_{ij} correspond to a transaction contract between the two participants. There are additional constraints considered for each participant that affect the level of transaction, including load balance constraints as well as the transmission system constraints. In Figure 6.8, the payment is based on the amount of power received by the buyer; that is, transmission losses are accounted for in the operation cost of the seller.

The Nash bargaining solution is obtained next by stating the problem as a two-participant problem and searching an optimal solution for (6.8).

Figure 6.8 A Two-Participant System

6.4.2 Two-Participant Problem Analysis

In this study, a two-participant system is analyzed based on the Nash bargaining method for seeking the optimal price for each transaction. According to (6.8), the two-participant bargaining problem is described as

MARKET POWER ANALYSIS BASED ON GAME THEORY

$$L = \max \{R_1 \times R_2\} \quad (6.9)$$

where

$$R_1 = p_T T'_{12} + (-C_1(P_{g1} + T_{12}) + C_1(P_{g1}))$$
$$R_2 = -p_T T'_{12} + (-C_2(P_{g2} - T'_{12}) + C_2(P_{g2}))$$

where p_T is the transaction price. The first term in R_1 or R_2 is the payment for a given transaction, and the second term is the change in the participant's operation cost owing to the transaction. Each utility has a constrained generation capacity, and payoffs are constrained by $R_1 \geq 0$ and $R_2 \geq 0$, indicating that the negative payoff is excluded. When there are no transactions, the payoff for each participant is zero. In order to maximize (6.9), we use

$$\frac{\partial L}{\partial p_T} = T'_{12}\left(-p_T T'_{12} - C_2(P_{g2} - T'_{12}) + C_2(P_{g2})\right)$$
$$- T'_{12}\left(p_T T'_{12} + C_1(P_{g1}) - C_1(P_{g1} + T_{12})\right) = 0 \quad (6.10)$$

and

$$\frac{\partial L}{\partial T'_{12}} = \left(p_T - \frac{\partial C_1(P_{g1} + T_{12})}{\partial T_{12}} \cdot \frac{\partial T_{12}}{\partial T'_{12}}\right) \cdot \left(-p_T T'_{12} + C_2(P_{g2}) - C_2(P_{g2} - T'_{12})\right)$$
$$+ \left(-p_T + \frac{\partial C_2(P_{g2} - T'_{12})}{\partial T'_{12}}\right) \cdot \left(p_T T'_{12} + C_1(P_{g1}) - C_1(P_{g1} + T_{12})\right) = 0 \quad (6.11)$$

Equation (6.9) is a non-linear problem with a non-linear objective function and constraints, and multiple extreme points exist in the optimization problem. One of these extreme points is the zero transaction defined by (6.10), which of course does not represent the maximum objective function. Three important results may be deducted from (6.10) and (6.11), for $p_T > 0$ and $|T| > 0$:

1. Transaction payoffs are divided equally between two participants.

2. If we assume a lossless case, i.e., $T_{ij} = T'_{ij}$, then the marginal costs of the two participants will be the same, indicating that the operation cost is minimized.

3. The optimal value of p_T is derived as (6.12), which represents the incremental system operation cost divided by the amount of transaction,

$$p_T = \frac{\left(-C_2(P_{g2} - T'_{12}) + C_2(P_{g2})\right) + \left(-C_1(P_{g1} + T_{12}) + C_1(P_{g1})\right)}{2T'_{12}} \quad (6.12)$$

Hence, the optimal price is not decided by one participant; rather it depends on both participants' transmitted power and incremental cost of operation.

6.4.3 Discussion on Optimal Transaction and Its Price

The transaction price (6.12) obtained by the game model is different from the spot price. Here we discuss the difference. When there is a surplus generation in a participant and no network constrains are imposed, the spot price will be the same as the system's marginal cost. The marginal cost relates to the system's optimal status. However, it may not provide equal savings for the two participants when used for spot pricing as there is no consideration of payoff assignments in the optimization procedure. In contrast, (6.12) may be written as

$$p_T = \left(\frac{-C_2(P_{g2} - T'_{12}) + C_2(P_{g2})}{T'_{12}} + \frac{-C_1(P_{g1}) + C_1(P_{g1} + T_{12})}{T'_{12}} \right)/2 \quad (6.13)$$
$$= (\bar{p}_1 + \bar{p}_2)/2$$

where $\bar{p}_1$ and $\bar{p}_2$ represent the incremental costs of the two participants, and the transaction price is denoted as the average of these two costs. Furthermore, let the cost function be represented as a quadratic function,

$$C_i(P_i) = a_i + b_i P_i + c_i P_i^2 \quad (6.14)$$

also let the marginal costs in the two participants before trading power be $\lambda_{10} < \lambda_{20}$. Then the optimal operation of two interconnected participants is given by

$$\lambda_{OP} = \frac{\partial C_1(P_{g1} + T_{12})}{\partial P_{g1}} = \frac{\partial C_2(P_{g2} - T'_{12})}{\partial P_{g2}}$$

If we disregard transaction losses (i.e., $T_{12} = T'_{12}$) and use (6.13), the optimal transaction price, and λ will be related by

$$p_T = \lambda_{OP} + (c_2 - c_1) \cdot T'_{12}/2 \quad (6.15)$$

MARKET POWER ANALYSIS BASED ON GAME THEORY

in which

$$\begin{aligned} p_T > \lambda_{OP} \quad &\text{if} \quad c_2 > c_1 \\ p_T = \lambda_{OP} \quad &\text{if} \quad c_2 = c_1 \\ p_T < \lambda_{OP} \quad &\text{if} \quad c_2 < c_1 \end{aligned} \qquad (6.16)$$

where λ_{OP} is the spot price of the system. Though these results are presented for a lossless system, a lossy system can be analyzed similarly. The first condition in (6.16) shows that in order to divide the payoffs fairly between the two participants (equal division in this chapter), the transaction price should be higher than the optimal system's lambda. In other words, using the optimal price as transaction price may result in an unfair (less than optimal) division of payoffs between the two participants. Other conditions in (6.16) can be analyzed similarly. The quadratic cost function is used here, although the method works equally well for other types of production cost curves (such as piecewise linear).

Based on (6.11),

$$\frac{\partial C_1(P_{g1}+T_{12})}{\partial T_{12}} \cdot \frac{\partial T_{12}}{\partial T'_{12}} = -\frac{\partial C_2(P_{g2}-T'_{12})}{\partial T'_{12}} = 0 \qquad (6.17)$$

which in a lossless case corresponds to $T_{12} = T'_{12}$ and $\dfrac{\partial T_{12}}{\partial T'_{12}} = 1$. In a lossy transmission, (6.17) represents equal marginal costs after trading power, which results in minimum operation costs for the two participants.

In general, many factors including security and environmental or political considerations may affect transaction flow limits in a decentralized system. As discussed in the following section, the production cost may be affected by transaction flow constraints. However, the proposed method for price and transaction optimization is valid with or without the transaction flow limits.

6.4.4 Test Results

6.4.4.1 Two-Utility System

The two-utility system illustrated in Figure 6.8 is used here for discussion, with data given in Table 6.1. In this example, we don't consider transmission loss. Later, we will study the effect of transmission loss.

By solving the optimization problem (6.9) for different prices, we obtain the optimal transaction between the two utilities, as shown in Figure 6.1. The summary before and after the transaction is given in Table 6.2.

From Figure 6.1 and Table 6.2, the optimal transaction power is 16.667 MW. The transaction price (23.083 $/MWh) is between the two λ's for utility 1 and utility 2 before the transaction (21 and 26 $/MWh). The two λ's after the transaction are the same (22.667 $/MWh), which is also the system λ since no loss is considered. Note that the transaction price satisfies (6.15), that is, 23.083 = 22.667 + (0.1- 0.05) × 16.667/2, and that price is higher than that of the system λ after the transaction, which corresponds to the first condition of (6.16) since $c_2 > c_1$. The total operation cost of the system decreases from $1085 (= 205 + 880) to $1043.334 (= 568.889 + 474.445), and the difference represents the system's saving (1085 - 1043.334 = 2 × 20.833 = $41.666), which is equally divided by the two utilities.

From utility 1, the seller's point of view, its payoff is represented by utility 2, the buyer's payment minus the increase in its operation cost, that is, 20.833 = 16.667 × 23.083 - [(0.0 + 20 × 26.667 + 0.05 × 26.667^2) - (0.0 + 20 × 10 + 0.05 × 10^2)]. From utility 2, the buyer's point of view, its payoff is represented by the reduction in its operation cost minus its payment to utility 2, the buyer, that is 20.833 = [(0.0 + 18 × 40 + 0.1 × 40^2) -(0.0 + 18 × 23.333 + 0.1 × 23.333^2)] - 16.667 × 23.083. In summary, both utilities have positive payoffs through cooperation.

Table 6.1 Test Data for the Two-Utility System

Utility	a	b	c	Load (MW)	Generation Limit (MW)
1	0.0	20	0.05	10	50
2	0.0	18	0.10	40	50

Table 6.2 Test Results without Losses

		Generation (MW)	λ_{OP} ($/MWh)	Operation Cost ($/h)	Savings ($/h)
Before transaction	Utility 1	10.0	21	205	-
	Utility 2	40.0	26	880	-
After transaction	Utility 1	26.667	22.667	568.889	20.833
	Utility 2	23.333	22.667	474.445	20.833
Transaction power T'_{12}		16.667 MW			
Transaction price		23.083 $/MWh			
Transaction payment		$384.722			

6.4.4.1.1 Effect of Transmission Losses. Transmission losses are mostly neglected in transaction analyses. Here, transmission losses are incorporated in utility 1, the seller's generation using $T = \alpha T' + (\beta T')^2$. Other loss formulations, e.g., B coefficients, may also be used. Three simple cases defined in Table 6.3 are used to analyze our results, where Case I corresponds to the lossless case discussed above.

Figure 6.9 shows the relationship between price and transaction power for the three cases based on (6.12), where the effect of transmission losses is considered in calculating the seller utility's operation cost. It is anticipated that the inclusion of transmission losses will reduce the net transaction for a given power generation at the seller's utility level, or increase the transaction price for a given power transaction as compared with the lossless condition. In Figure 6.9, the optimal power transmission and its price shifts from curve I to II to III are depicted as transmission losses increase. That is, the transaction is decreased, and the price is increased corresponding to an increase in system losses.

Table 6.3 Loss Coefficients for the Test System

Cases	α	β
I	1.00	0.000
II	1.04	0.001
III	1.08	0.001

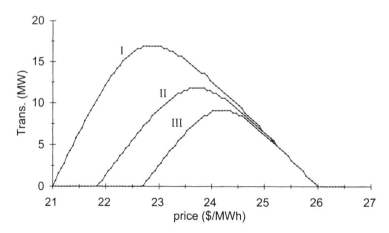

Figure 6.9 Transaction Price and Power Losses

Taking a closer look, we can see that the results of Case I are the same as those in Table 6.2, and that Table 6.4 corresponds to the results of Case III. Comparing Table 6.4 (with loss) with Table 6.2 (without loss), we can see that the optimal transaction is reduced from 16.667 to 9.180 MW, and that the transaction price is increased from 23.083 to 24.249 $/MWh. The total system operation cost increases from $1043.334 to $1069.718, thus the system payoff decreases from $41.666 (=2 × 20.833) to $15.282 (=2 × 7.641). The net payoff to each utility is reduced to $7.641.

Table 6.4 Test Results with Losses

		Generation (MW)	λ_{OP} ($/MWh)	Operation Cost ($/h)	Savings ($/h)
Before transaction	Utility 1	10.0	21	205	-
	Utility 2	40.0	26	880	-
After transaction	Utility 1	19.999	22.000	419.986	7.641
	Utility 2	30.819	24.164	649.732	7.641
Transaction power T'_{12}		9.180 MW			
Transaction price		24.249 $/MWh			
Transaction payment		$222.627			

6.4.4.1.2 Production Cost Curve and Optimal Price. As was discussed in Section 6.4.3, the optimal transaction price might be higher or lower than the marginal cost. Three different production cost functions in each utility are tested, with data given in Table 6.5 and results in Table 6.6. Note that Cases 1, 2, and 3 correspond to the first, the second, and the third conditions of (6.16), respectively. For instance, when coefficient c is the same for the two utilities (Case 2), the optimal transaction price is equal to the marginal cost.

6.4.4.1.3 Discussion on Transaction Flow Limits. Several factors in addition to the thermal limit of a line may affect the transmission capability; among these factors are dynamic stability limit, voltage stability limit, and static security constraint. Transmission capability may even be constrained by long-term economics or strategic considerations. The marginal costs may be different if transactions are limited. A limited transaction case is tested and results are shown in Table 6.7.

In Table 6.7, the system's data are the same as those of Case III in Table 6.6 except that the transaction at the receiving end (utility 2) is limited to 5 MW. The transaction price is 24.29 $/MWh, which is between the two lambdas after transaction. The transaction power on the seller's side is 5.425 MW, which includes losses.

Table 6.5 Data for Different Production Cost Curves

Case	Utility	a	b	c
I	1	0.0	20	0.05
	2	0.0	18	0.10
II	1	0.0	20	0.05
	2	0.0	18	0.05
III	1	0.0	20	0.05
	2	0.0	20	0.025

Table 6.6 Optimal Price and Transaction

Case	Initial $\lambda_{10}, \lambda_{20}$	After Transaction $\lambda_{OP1}, \lambda_{OP2}$	Transaction Price ($/MWh)	Transaction Power (MW)	Saving ($/h)
I	21, 26	22.667	23.083	16.667	20.833
II	21, 22	21.500	21.500	5.000	1.250
III	21, 22	21.667	21.583	6.667	1.667

Table 6.7 Optimal Price and Transaction with Limit

Initial $\lambda_{10}, \lambda_{20}$	After Transaction $\lambda_{OP1}, \lambda_{OP2}$	Transaction Price ($/MWh)	Transaction Power (MW)	Saving ($/h)
21, 26	21.54, 25.0	24.290	5.0	6.052

6.4.4.2 Three-Utility System

The power market coordinator may use the procedures described in this section to identify market power imposed by some of the participants. The coordinator may accordingly take adequate corrective measures to eliminate unanticipated problems.

The power system used in this section is a three-bus power market in which each utility can either supply its local load or sell its power depending on the market price. These utilities are tied with lines that have equal reactance with negligible resistance. For simplicity, each utility is represented by a single generator. Here, no capacity limits are imposed on tie lines, and participants are not facing any additional constraints (e.g., interruptible power from contracts, contingencies in local resources) in defining offered prices.

Before transactions are defined, each utility supplies its local load by applying an economic dispatch. Then, each utility offers prices for its power to the market using curves similar to those in Figure 6.7. The characteristics of utilities are listed in Table 6.8. Based on price coefficients in Table 6.8, utility A and B will sell power while utility C will buy power in the market. The information in Table 6.8 is only available to the power market coordinator.

Table 6.8 Example System

Utility	Load (MW)	P_{max} (MW)	Price Coefficients			λ_0 ($/MWh)
			a	b	c	
A	50	100	0	15	0.025	17.5
B	40	100	0	12	0.05	16
C	100	100	0	17	0.01	19

Utilities' Strategies. Utilities are able to change their prices by adjusting the slope in Figure 6.7. Using economic dispatch, market's payoffs will be maximized when utilities trade power at marginal cost, $m(i) = 2c(i)$ in equation (6.4). However, as utilities try to maximize their own payoffs, they may opt to either decrease their bids in order to sell more power or increase the price in order to earn more. Among the infinite set of feasible alternatives for each participant, we analyze the following three strategies:

H: Trade power at 1.15 times the marginal cost, $m(i) = 2.3c(i)$. The utility's strategy is to bid high.

M: Trade power at marginal cost, $m(i) = 2c(i)$. The utility's strategy is to cooperate with the market.

L: Trade power at 0.85 times the marginal cost, $m(i) = 1.7c(i)$. The utility's strategy is to bid low.

In each case, payoffs are assumed positive, and they represent the sum of payments and incremental costs. Here, we identify the possible coalitions: a coalition between any two participants, the grand coalition, and any representation of the three one-utility coalitions. Correspondingly, there are four possible non-cooperative game strategies between coalitions and counter-coalitions.

MARKET POWER ANALYSIS BASED ON GAME THEORY

The situation when all utilities cooperate with the power market (i.e., grand coalition) conduces to the maximum payoff. The grand coalition's solution is the same as that of a traditional centralized economic dispatch with minimizing cost functions. The optimum dispatch in this case corresponds to the minimum operation cost of 3138.75 $/h, generations of 57.5, 51.25, and 81.25 MW for the three utilities, and payoffs of 7.03, 23.20, and 3.52 $/h for the three utilities, respectively.

The competition under the condition that no limit is imposed on the tie lines represents *perfect competition* as participants' size or location does not pose a biased control over the market price and participants cannot take advantage of external constraints. However, various conditions could deter perfect competition, including the economic pre-eminence of certain participants, the mix of generating resources available to each participant, and the geographical location of participants. When the market is under *imperfect competition*, some participants may find that it is possible to obtain higher payoffs by colluding with other participants.

Consider an example where the line connecting B to C is limited to 10 MW, which could constrain the economic dispatch among utilities. In this game, A and C could collude against B for higher payoffs. The set of available strategies in each coalition is represented by multiplying the number of strategies in the coalition. In Table 6.9, the payoff matrix of non-cooperative game is shown for coalition $S = \{A, C\}$ and its counter-coalition $S^c = \{B\}$.

Table 6.9 Payoff Matrix ($/h)

$\{A,C\}\{B\}$ ↓ ↦	H	L	M
HH	13.40, 17.23	12.63, 21.05	11.43, 21.39
HL	10.70, 22.93	4.77, 28.20	6.34, 27.05
HM	11.67, 22.00	5.69, 27.15	7.28, 26.02
LH	19.05, 13.51	18.85, 11.90	14.50, 19.12
LL	14.85, 15.38	14.62, 16.14	11.41, 22.05
LM	15.58, 15.18	9.67, 24.04	11.94, 21.69
MH	18.04, 14.52	17.66, 14.40	13.57, 19.87
ML	14.00, 17.74	13.72, 18.34	9.90, 23.83
MM	14.78, 17.28	8.36, 25.17	10.55, 23.20

Coalition $\{A, C\}$ have nine possible strategies as listed in Table 6.9. The value of each entry in the payoff matrix is computed using the spot price to define transactions among utilities for that particular strategy (set of prices). Each entry in the table signifies a pair of payoffs for coalition members and the counter-coalition, respectively. The first value in each pair represents the sum of payoffs for A and C. The second value is B's payoff for the same set of strategies.

The *characteristic function* (v) represents an estimate of the best possible outcome in the worst situation of a coalition. In this example, the characteristic function of coalition $\{A,C\}$ is found by first locating the minimum payoff in each row in Table 6.9 (i.e., 11.43 \$/h in row 1, 4.77 \$/h in row 2, ..., 8.36 \$/h in row 9) and then choosing the maximum payoff for B in the selected minimums which results in $v(\{A,C\}) = 14.50$ \$/h. The chosen strategy is the max-min strategy. In this regard, A would bid below marginal cost and C would bid above marginal cost, because the coalition would offer a higher payoff than that obtained in the grand coalition (10.55 \$/h). Note that C is buying; hence, when bidding high, it is buying at a lower price. It is seen that no matter what B's bid is, coalition $\{A,C\}$ would make more money than that of grand coalition.

The characteristic function is a pessimistic estimate because it assumes that the counter-coalition is playing to minimize the coalition's payoff when in fact the counter-coalition is trying to maximize its own payoff. For instance, in Table 6.9, if B decides to sell above its marginal cost (i.e., H instead of M), the coalition payoff would be higher than that of the characteristic function.

If all bids are based on max–min strategies, the market's payoff will not be maximum (i.e., 33.75 \$/h), total operation cost will increase to 3138.88 \$/h, and B's payoffs will drop to 19.12\$/h. In this case, A will bid L while C will bid H, while the best bid for B will be M. The game is in equilibrium because these strategies are the best according to the opponent's strategy. Other equilibrium states may also be identified in the game.

Other criteria than the max–min criterion may also be used to simulate participants' decision process. For instance, both pessimism–optimism criteria and the criterion based on the principle of insufficient reasons [Lut57] would concentrate the decision on a weighted combination of the best and the worst states of participants. The proper decision criteria will depend on participants' characteristics.

6.5 MARKET COMPETITION WITH INCOMPLETE INFORMATION

In this section, we analyze the competition (with incomplete information) among participants in an electricity market. Each participant's payoff in this market is a function of production costs, power transactions, and spot market price. Hence, individual participant's payoff is a function of bids offered by other participants, and each participant intends to estimate the other participants' bids in order to maximize its own payoff.

6.5.1 Participants and Bidding Information

In the electricity market, it is perceived that each participant has the complete information on its own payoffs but lacks much information on other participant's payoffs. Hence, the competition between participants for the market's load is modeled as an (incomplete) i-game. Here, a participant's unknown characteristics are modeled as a participant's type. The type of participant embodies any information that is not common to all participants. This information may include, in addition to the participant's payoff function, its beliefs about other participants' payoff functions, its beliefs about other participants' perception of its beliefs, fuel prices, availability of transmission installations, and so on. In this section, a participant's type corresponds to its cost structure, that is, coefficients a, b, and c.

Each participant would have a full knowledge of its own costs, but only an estimate of the remaining participants' costs. Each participant adjusts m in (6.4) in order to maximize its payoff B in (6.5). We consider the choice of m as a strategy (bid low, bid high, etc.) in the game. The problem is defined as: what value of m should a participant offer to the market in order to maximize its payoff when the participant would not have much knowledge of other participants' parameters.

This game is based on the assumption that participants would use a Bayesian approach for dealing with incomplete information. That is, they assign a basic joint probability distribution Π to unknown variables. Once this is done, they maximize the mathematical expectation of their payoffs in terms of Π. In estimating the probability distribution, each participant uses the information common to all participants. Here, not only a participant's perspective is taken into account but also other participants' perspectives of the participant are considered.

The basic probability of game is assumed as being based on random fuel prices. This topic is discussed next.

6.5.2 Basic Probability Distribution of the Game

In order to discuss the proposed methodology, we present a scenario from participant B's perspective when B is competing with participant A for selling power to participant C. We assume that participants' types are drawn at random from hypothetical populations Φ_A and Φ_B containing types t_A^q and t_B^r respectively. The superscripts $q = 1, ..., Q$ and $r = 1, ..., R$ stand for the types of participants A and B, respectively. For instance, to model the uncertainty in A's cost, B would assume that there are Q possible types of A, each with its corresponding cost. Generally, B knows its type r but does not know its opponent's type q (i.e., it does not know A's costs).

Participants estimate the opponents' probability distributions Π based on the published information such as known fuel contracts, availability of transmission lines, and participants' parameters. Element π_{qr} in the basic probability distribution Π corresponds to the probability that A is type q and B is type r.

The amount of power that individual participants could trade is a function of spot market price. The higher the price, the lower will be the amount of power that A and B will sell to C. For simplicity, we suppose that C's cost function is known to A and B. From the data in Table 6.8, we see that C would purchase 0 MW when the spot price is as high as 19 \$/MWh and 100 MW when the spot price is as low as 17 \$/MWh. Given the participants' bids in (6.4), when network constraints and losses are ignored, power transactions are derived as

$$\lambda_A = \lambda_B = \rho \qquad \text{s.t.} \quad P_A + P_B = L_C(\rho) \qquad (6.18)$$

where

P_A = excess power over participant A's load sold to participant C

P_B = excess power over participant B's load sold to participant C

L_C = purchased power by participant C

ρ = spot market price

6.5.3 Conditional Probabilities and Expected Payoff

The conditional probability $\theta_A^q(r)$ signifies the probability that A's type q would face B's type r:

$$\theta_A^q(r) = Pr(t_B^r | t_A^q) = \frac{\pi_{qr}}{\sum_{r=1}^{R} \pi_{qr}} \qquad (6.19)$$

Likewise,

$$\theta_B^r(q) = Pr(t_A^q | t_B^r) = \frac{\pi_{qr}}{\sum_{q=1}^{Q} \pi_{qr}} \qquad (6.20)$$

A participant's strategy (bid) will depend on its type. A possible vector of strategies could include: bid high, bid at marginal cost, and bid low. Let s_A^q be the vector of strategies for A's type q. A's conditional payoff $G_A^q = G_A^q(s_A^q, s_B^r, q, r)$ would depend not only on its strategy but also on its opponent's strategies s_B^r. In order to maximize G_A^q, A should know its opponent's type, and since this information is unavailable to A in the i-game, A will try to maximize its *expected payoff* as

$$E_A^q = \sum_{r=1}^{R} \theta_A^q(r) G_A^q(s_A^q, s_B^r, q, r) \qquad (6.21)$$

where E_A^q depends on its opponents' strategies t_B^r. Likewise,

$$E_B^r = \sum_{q=1}^{Q} \theta_B^r(q) G_B^r(s_B^r, s_A^q, q, r) \qquad (6.22)$$

In (6.21) and (6.22), participants would consider all types of opponents. The original i-game is now interpreted as a $(Q + R)$-participant (complete) c-game with Q types of participant A and R types of participant B. The transformed game is with complete but imperfect information. Participants know the mathematical structure of the game (payoff functions E_A^q and E_B^r, basic probability distribution of the game, etc.) but do not know the opponent's type. In this c-game, the Nash equilibrium is the solution.

6.5.4 Gaming Methodology

We now illustrate the proposed methodology by applying the following steps to the market:

Step 1: *Define participants' types.*

We would define $Q = 2$ and $R = 2$ with participant A's cost coefficients are

$a_A = [0, 0]$ $/h

$b_A = [15, 15.5]$ $/MWh

$c_A = [0.025, 0.027]$ $/MW^{2}h

The two elements in each vector corresponds to $q = 1, 2$, respectively.

Participant B's cost coefficients are

$a_B = [0, 0]$ $/h

$b_B = [12, 11.7]$ $/MWh

$c_B = [0.05, 0.06]$ $/MW^{2}h

The two elements in each vector corresponds to $r = 1, 2$ respectively.

The participant's type is defined based on scenarios of fuel prices (f). A probability distribution function $\varphi(f)$ is used where $\sum_{f=1}^{F} \varphi(f) = 1$. Two scenarios are considered here for fuel prices. The first scenario corresponds to a low fuel price with $\varphi(1) = 0.6$, while the second scenario corresponds to a higher fuel price with $\varphi(2) = 0.4$.

Probability distributions Ω_A^f and Ω_B^f are used for modeling uncertainties in each participant's operation cost. For the first fuel scenario, we use

$$\Omega_A^1 = [0.84, 0.16]$$
$$\Omega_B^1 = [0.58, 0.42]$$
(6.23)

where the first element in Ω_A^1 corresponds to the probability that A is type 1 for $f = 1$, and the second element corresponds to the probability that A is type 2 for $f = 1$. The same notation applies to Ω_B^1. In (6.23), it is more probable that A and B will have low operation costs when fuel prices are

MARKET POWER ANALYSIS BASED ON GAME THEORY 219

low. These probability distributions may also represent the available information on each participant's equipment characteristics. For the second fuel scenario we use

$$\Omega_A^2 = [0.36, 0.64]$$
$$\Omega_B^2 = [0.37, 0.63]$$

Step 2: *Define the basic probability distribution of the game.*

The probability that (q, r) would represent participants A and B's types, respectively, would depend on fuel price. We define π_{qr} as the expected probability that A is type q and B is type r:

$$\pi_{qr} = \sum_{f=1}^{F} \left(\varphi(f) \Omega_A^f(q) \Omega_B^f(r) \right)$$

In the example:

$$\Pi = \begin{bmatrix} 0.3456 & 0.3024 \\ 0.1504 & 0.2016 \end{bmatrix}$$

Through matrix Π, the original i-game is transformed into c-game with imperfect information. In the new c-game, each participant would know its fuel prices and compute its own operation costs without knowing the opponent's operation cost.

According to (6.19) and (6.20), the conditional probability vectors for participants A and B are

$$\theta_A^1 = [0.53333 \quad 0.46667] \qquad \theta_A^2 = [0.42727 \quad 0.57273]$$
$$\theta_B^1 = [0.69677 \quad 0.30323] \qquad \theta_B^2 = [0.6 \quad 0.4]$$

where participant B assumes a positive correlation between its type and participant A's type. For instance, when B has low costs (i.e., type 1), there is a higher probability that A will also have low costs, $\theta_B^1(1) > \theta_B^1(2)$.

Step 3: *Define participants' strategies.*

A participant's type corresponds to a set of strategies defined by bid slopes. We assume that each participant observes the same three strategies as in (6.4). The slope of a bid curve is computed as $m_i = k_S \times c_i$ where k_S is set to

1.7, 2, and 2.3 for the three strategies. The vector of strategies for each participant's type is:

$$s_A^1 = [0.0425 \quad 0.050 \quad 0.0575] \; \$/MW^2h$$

$$s_A^2 = [0.0459 \quad 0.054 \quad 0.0621] \; \$/MW^2h$$

$$s_B^1 = [0.0850 \quad 0.100 \quad 0.1150] \; \$/MW^2h$$

$$s_B^2 = [0.1020 \quad 0.120 \quad 0.1380] \; \$/MW^2h$$

A strategy vector element represents a bid slope for the combination of a participant's type and strategy. For instance, the second element in s_A^2 represents the bid slope for A's type 2, as A bids marginal (i.e., 2×0.027 $\$/MW^2h$.)

Step 4*: Define participants' conditional payoff.*
We solve (6.18) for a combination of participant types and strategies and calculate each participant's payoff using (6.5). For example, conditional payoffs ($\$/h$) for participant A's type 1 against participant B's type 1 are

$$G_A^1(1) = \begin{bmatrix} 4.381 & 4.999 & 5.501 \\ 4.920 & 5.625 & 6.198 \\ 5.160 & 5.908 & 6.518 \end{bmatrix}$$

In $G_A^1(1)$, each row corresponds to a strategy of A for type 1 and each column corresponds to a strategy of B for type 1. For instance, the element at row 3 and column 2 in $G_A^1(1)$ corresponds to A's type 1 payoff if A decides to bid above marginal cost against a situation where B is type 1 and bids marginal. Other conditional payoff matrices such as $G_A^1(2)$, $G_A^2(1)$, $G_A^2(2)$, $G_B^1(1)$, $G_B^1(2)$, $G_B^2(1)$, and $G_B^2(2)$ could be computed similarly.

Step 5*: Define expected payoff matrices.*
We compute expected payoff matrices E_A^1, E_A^2, E_B^1, and E_B^2 using (6.21) and (6.22) and the conditional payoff matrices computed in Step 4. Each row in E_A^1 or E_A^2 is participant A's strategy corresponding to type 1 or

MARKET POWER ANALYSIS BASED ON GAME THEORY

type 2, respectively. Each column in E_A^1 or E_A^2 corresponds to the presumed strategy of participant A's opponent. For instance, column z_{23} in E_A^1 is A's type 1 payoff when B uses strategy 2 against A's type 1 and strategy 3 against A's type 2. The same notation applies to E_B^1 and E_B^2.

$$\begin{array}{ccccccccc} z_{11} & z_{12} & z_{13} & z_{21} & z_{22} & z_{23} & z_{31} & z_{32} & z_{33} \end{array}$$

$$E_A^1 = \begin{Bmatrix} 4.944 & 5.124 & 5.266 & 5.338 & 5.518 & 5.660 & 5.657 & 5.837 & 5.979 \\ 5.558 & 5.765 & 5.929 & 6.007 & 6.213 & 6.377 & 6.371 & 6.578 & 6.742 \\ 5.834 & 6.054 & 6.229 & 6.309 & 6.530 & 6.705 & 6.697 & 6.918 & 7.093 \end{Bmatrix}$$

$$E_A^2 = \begin{Bmatrix} 0.415 & 0.482 & 0.538 & 0.534 & 0.601 & 0.657 & 0.647 & 0.713 & 0.769 \\ 0.465 & 0.540 & 0.603 & 0.598 & 0.673 & 0.736 & 0.725 & 0.800 & 0.863 \\ 0.486 & 0.565 & 0.632 & 0.626 & 0.705 & 0.772 & 0.759 & 0.838 & 0.904 \end{Bmatrix}$$

$$\begin{array}{ccccccccc} y_{11} & y_{12} & y_{13} & y_{21} & y_{22} & y_{23} & y_{31} & y_{32} & y_{33} \end{array}$$

$$E_B^1 = \begin{Bmatrix} 19.212 & 19.536 & 19.584 & 20.321 & 20.645 & 20.693 & 20.547 & 20.871 & 20.919 \\ 19.786 & 20.132 & 20.188 & 20.954 & 21.300 & 21.356 & 21.209 & 21.555 & 21.611 \\ 20.250 & 20.613 & 20.676 & 21.467 & 21.829 & 21.893 & 21.746 & 22.109 & 22.172 \end{Bmatrix}$$

$$E_B^2 = \begin{Bmatrix} 23.432 & 23.628 & 23.781 & 24.222 & 24.419 & 24.571 & 24.840 & 25.037 & 25.189 \\ 24.539 & 24.574 & 24.740 & 25.215 & 25.429 & 25.595 & 25.885 & 26.099 & 26.266 \\ 24.082 & 24.301 & 24.472 & 24.954 & 25.173 & 25.344 & 25.638 & 25.857 & 26.028 \end{Bmatrix}$$

Step 6: *Obtain Nash equilibrium of strategies.*

Using E_A^1, E_A^2, E_B^1, and E_B^2, we try to find Nash equilibrium pairs. We look for the collection of strategies in which each participant's strategy would be represented by the best response to other participant's strategies. By inspecting E_A^1 and E_A^2, we learn that rows $s_A^1(3)$ and $s_A^2(3)$ dominate other rows; in other words, no matter what strategy participant B uses, participant A obtains a higher payoff by bidding above its marginal costs. Hence, a rational participant A will always choose to bid above marginal costs regardless of its costs. Participant A's optimal strategy is represented by column y_{33} in E_B^1 and E_B^2.

Accordingly, participant B has learned that A would bid above its marginal costs in all cases. Hence, B would only analyze the last column in E_B^1 and E_B^2. By inspecting column y_{33}, we learn that a rational participant B would bid $s_B^1(3)$ (above its marginal cost) when it is type 1 and $s_B^2(2)$ (at marginal cost) when it is type 2. This strategy is represented by column z_{32} in E_A^1 and E_A^2.

The pair of strategies y_{33} and z_{32} is the Nash equilibrium of the game. The Nash equilibrium is a "consistent" prediction of how the game will be played. A participant's optimal bid is derived for this equilibrium point. In essence, all participants predict that a particular Nash equilibrium will occur and there is no incentive to play differently. The strategy pairs in Nash equilibrium are the participants' maximum strategies [Aub82, Fud91, Mor94], which maximize participants' conditional payoffs. That is, a participant could obtain at least the payoff at the equilibrium point (or it may obtain more depending on his opponent's strategy).

6.6 MARKET COMPETITION FOR MULTIPLE ELECTRICITY PRODUCTS

In this section, we extend the discussion of Section 6.5 in optimizing the participant's payoff by including multiple products (i.e., energy and ancillary services). The participant applies generation scheduling and provides the ISO with bids for supplying the load and ancillary services as discussed in the PBUC chapter (Chapter 4). The ISO will analyze the offered bids and define power transactions among participants, by minimizing the spot prices for energy and ancillary services in the power market while preserving network constraints.

6.6.1 Solution Methodology

This method in general considers any number of participants within the market. It resembles the method discussed in Section 6.5 except that we would run PBUC here in order to distinguish different products and their bids offered to the market. The procedure is explained as follows:

1. Depending on fuel prices, various types are assigned to individual participants.

MARKET POWER ANALYSIS BASED ON GAME THEORY

2. PBUC is applied to each participant's type and marginal bids of generators are calculated. Each participant simulates the market behavior in order to calculate its own bids. The procedure is presented next for participant A.

3. Participant A simulates the market operation using the available information based on the following steps:

    3.1 For Energy

    I. Participant A assumes that other participants can play different strategies against it.

    II. Participant A applies the game and obtains its own dominant strategies and bids by applying Nash equilibrium.

    III. Participant A identifies the operating point x shown in Figure 6.10, for every unit in A (see Chapter 4 for additional details on PBUC).

    3.2 For Ancillary Services

    Participant A repeats the procedure above based on point x in order to define the dominant strategy for its ancillary services bids.

4. The above procedure is followed by all participants in order to calculate their bids.

5. All bids are then submitted by participants to the ISO.

6. The ISO analyzes the submitted bids, determines the winners and accordingly sends the transaction information back to each winner.

6.6.2 Study System

Figure 6.11 shows a modified IEEE 30-bus system with 9 generating units, 20 loads, 30 buses, and 41 transmission lines. The system is divided into two participants: A and B. Participant A has three base load units (1, 5, and 8), one intermediate unit (2), and three peaking units (11, 24, and 30). Participant B has two intermediate units (13,15). A study period of 24 hours is considered. The spinning reserve is 5% of the load and the peak load occurs at hour 18. The corresponding data are given in Appendix D.4.

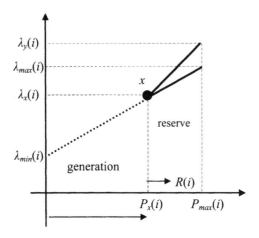

Figure 6.10 Spinning Reserve Bid

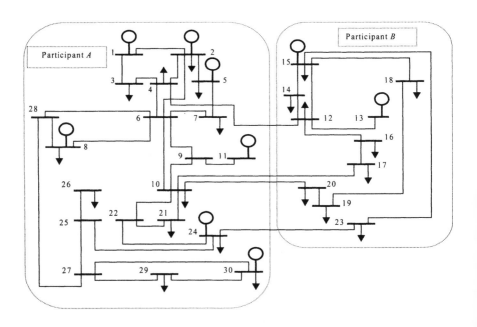

Figure 6.11 Modified IEEE 30-Bus System

MARKET POWER ANALYSIS BASED ON GAME THEORY

6.6.3 Gaming Methodology

Step 1: *Define participant types.*

Participants may define two scenarios for fuel prices. Assume that each participant's type depends on fuel prices and that the fuel price of type 2 is higher than that of type 1. This condition is represented by increasing the value of coefficient c in generation cost functions by 20%. The participant's type 1 cost coefficients are shown in Table D.20 of Appendix D.4. The two types are estimated by the participants using the information available to them. Local loads are independent of the participant's type.

A probability distribution $\varphi(f)$ is used to model the fuel price uncertainty. Two scenarios are introduced as $\varphi(1) = 0.6$ and $\varphi(2) = 0.4$ to represent low and high fuel prices, respectively. Probability distributions Ω_A^f and Ω_B^f are for modeling uncertainties in each participant's actual cost. For example, experience provides the following probabilities.

$$\Omega_A^1 = [0.85, 0.15] \qquad \Omega_A^2 = [0.36, 0.64]$$
$$\Omega_B^1 = [0.70, 0.30] \qquad \Omega_B^2 = [0.37, 0.63]$$

The first (second) element in Ω_A^1 corresponds to the probability that type 1 (2) represents participant A when fuel price is low $f = 1$. Others could be explained similarly. These probability distributions may be derived from the information available on the equipment characteristics of each participant.

Step 2: *Define basic probability distributions of the game.*

Define π_{qr} as the expected value, over all fuel scenarios, that participant A is type q and participant B is type r:

$$\pi_{qr} = \sum_{f=1}^{2} \left(\varphi(f) \Omega_A^f(q) \Omega_B^f(r) \right)$$

For this example, we have

$$\Pi = \begin{bmatrix} 0.41028 & 0.24372 \\ 0.15772 & 0.18828 \end{bmatrix}$$

Matrix Π is used to transform the original incomplete information game into a complete game with imperfect information. Each participant has partial information on its opponent but does not know the opponent's operation cost. The conditional probability vectors in participants A and B are

$$\theta_A^1 = [0.6273395 \quad 0.3726605]$$

$$\theta_A^2 = [0.4558382 \quad 0.5441618]$$

$$\theta_B^1 = [0.7223239 \quad 0.2776761]$$

$$\theta_B^2 = [0.5641667 \quad 0.4358333]$$

The first (or second) element in θ_A^1 corresponds to the conditional probability that participant A is playing against participant B type 1 (or 2) subject to the condition that participant A is type 1. Others could be explained similarly.

Step 3: *Unit commitment before gaming.*

Consider the base case unit commitment executed by each participant for determining optimal transactions before gaming. The base case unit commitment for local loads is shown in Table 6.10. Note that at off-peak hours (1– 9, 24), the intermediate unit at bus 2 is decommitted.

Table 6.10 Base Case Unit Commitment before the Game

Participant	Generator At Bus No.	Hour (1-24)
A	30	0 0 0 0 0 0 0 0 0 0 0 0 0 0 0 0 0 0 0 0 0 0 0 0
	24	0 0 0 0 0 0 0 0 0 0 0 0 0 0 0 0 0 0 0 0 0 0 0 0
	11	0 0 0 0 0 0 0 0 0 0 0 0 0 0 0 0 0 0 0 0 0 0 0 0
	2	0 0 0 0 0 0 0 0 1 1 1 1 1 1 1 1 1 1 1 1 1 1 1 0
	8	1 1 1 1 1 1 1 1 1 1 1 1 1 1 1 1 1 1 1 1 1 1 1 1
	5	1 1 1 1 1 1 1 1 1 1 1 1 1 1 1 1 1 1 1 1 1 1 1 1
	1	1 1 1 1 1 1 1 1 1 1 1 1 1 1 1 1 1 1 1 1 1 1 1 1
B	13	0 0 0 0 0 0 0 0 0 0 0 0 0 0 0 0 0 0 0 0 0 0 0 0
	15	1 1 1 1 1 1 1 1 1 1 1 1 1 1 1 1 1 1 1 1 1 1 1 1

Step 4: *Define participants' generation strategies.*

Each participant may play different strategies represented by different bid slopes. Some basic bid parameters are shown in Table 6.11. $P_0(i)$ represents

MARKET POWER ANALYSIS BASED ON GAME THEORY

bilateral contract which is calculated for each participant, and marginal cost $\lambda_0(i)$ is evaluated at $P_0(i)$. Other parameters are also computed, including $\lambda_{min}(i)$ and $\lambda_{max}(i)$, which are defined as $\lambda_{min}(i) = \lambda_0(i) - P_0(i) \times c(i)$, and $\lambda_{max}(i) = \lambda_0(i) + (P_{max}(i) - P_0(i)) \times c(i)$.

Table 6.11 Marginal Generation Bids at Hour 18

Participant	Bus No.	Type 1 $\lambda_0(i)$ $\lambda_{min}(i)$ $\lambda_{max}(i)$ ($/MWh)			$P_0(i)$ (MW)	Type 2 $\lambda_0(i)$ $\lambda_{min}(i)$ $\lambda_{max}(i)$ ($/MWh)			$P_0(i)$ (MW)
A	30	49.32	49.32	49.81	0.0	49.32	49.32	49.91	0.0
	24	40.05	39.97	40.29	5.0	40.08	39.98	40.37	5.0
	11	38.03	37.96	38.24	5.0	38.06	37.97	38.31	5.0
	2	18.86	18.48	18.97	62.3	19.01	18.55	19.14	62.3
	8	14.24	13.80	14.24	50.0	14.42	13.89	14.42	50.0
	5	14.20	13.76	14.20	50.0	14.37	13.85	14.37	50.0
	1	11.20	10.94	11.41	54.9	11.30	10.99	11.55	54.9
B	13	19.32	19.32	20.05	0.0	19.32	19.32	20.19	0.0
	15	19.10	18.70	19.12	56.2	19.26	18.78	19.29	56.2

Step 5: *Compute participants' parameters for trading energy.*

For each combination of type-strategy, participant A plays a game against participant B using its generation bids. Each participant's objective is to maximize its own payoff by modifying the slope of its generation bid curve for each generator in Table 6.11. While modifying bids, a participant would check its trade options by monitoring its hourly revenues and calculating operating point x in Figure 6.10. This point identifies each unit's power generation and the corresponding spinning reserve available for trading. Table 6.12 shows each participant's generation before and after gaming at hour 18. For type 1, participant A is buying 11 MW from participant B since participant B's offer is cheaper.

Table 6.12 Generation before and after Gaming at Hour 18 without Generation Reserve

Generation before Game (MW)		Generation after Game (MW)	
Participant A	Participant B	Participant A	Participant B
227.2	56.2	216.2	67.2

Step 6: *Compute participants' parameters for spinning reserve.*

In Figure 6.10, once point x is calculated for a participant, the participant can play a game of offering reserve bids. The participant's objective is to maximize its own payoff by modifying the slope of the reserve bid curve for each generator. While modifying bids, the participant checks its trades by monitoring hourly revenues.

Step 7: *Determine expected optimal parameters.*

By substituting the results of Steps 5 and 6 into the conditional probability formula, the participant obtains the results for types 1 and 2.

Step 8: *Find the Nash equilibrium.*

Using the results of Step 7, the participant determines the Nash equilibrium point. Tables 6.13 and 6.14 show the dominant strategies without and with spinning reserve for hours 6 and 18. In these tables, the modified load refers to the amount of load that a participant has to supply due to its gaming strategy. Note that when participant A's fuel price is low (i.e., type 1), its modified load is higher since it can trade more power. Also, the optimal reserve price (38.3 $/MWh) is higher than the corresponding spot price for energy (25.41 $/MWh).

Table 6.13 Payoffs without Spinning Reserve at Hours 6 and 18

Type	Hour	Participant A Total Payoff ($/h)	Participant A Modified Load (MW)	Participant B Total Payoff ($/h)	Participant B Modified Load (MW)	Spot Price for Energy ($/MWh)
1	6	44.0	178.5	23.1	5.7	17.86
1	18	134.4	216.2	67.0	67.2	25.41
2	6	37.3	173.3	19.9	10.9	17.9
2	18	134.3	216.1	67.7	67.3	25.46

Table 6.14 Payoffs with Spinning Reserve at Hours 6 and 18

Type	Hour	Participant A Total Payoff ($/h)	Participant A Modified Load (MW)	Participant B Total Payoff ($/h)	Participant B Modified Load (MW)	Optimal Reserve Price ($/MWh)
1	6	46.0	180.0	27.5	13.4	19.2
1	18	173.0	219.0	222.3	78.57	38.3
2	6	38.6	174.8	22.4	18.6	19.2
2	18	182.2	218.9	217.6	78.67	38.4

MARKET POWER ANALYSIS BASED ON GAME THEORY

The total load at hour 18 is 283.4 MW and the required spinning reserve by the ISO is 5% (14.17 MW). The reserve allocation after gaming for type 1 at hour 18 is shown in Table 6.15. Participant B because of its lower generation cost, will generate most of the spinning reserve.

As was discussed in Chapter 4, the spinning reserve and power generation can have different slopes (bids). Table 6.16 is analogous to Table 6.14, except that Table 6.16 represents the case where the spinning reserve and generation are forced to bid the same. We decided to use Table 6.14 (which is based on unit commitment) rather than Table 6.16 because the payoffs are higher. Note that the optimal spinning reserve price is higher than the energy spot price. However, the difference is smaller at off-peak hours.

Table 6.15 Reserve Allocation after Gaming at Hour 18

Participant	Unit	Reserve (MW)
A	2	2.8
B	13	11.37

Table 6.16 Payoffs with Spinning Reserve

| Type | Hour | Participant A | | Participant B | | Optimal Reserve Price ($/MWh) |
		Total Payoff ($/h)	Modified Load (MW)	Total Payoff ($/h)	Modified Load (MW)	
1	6	45.5	180.0	23.8	13.40	18.88
1	18	160.4	218.8	122.5	78.77	29.54
2	6	38.3	179.9	20.9	13.41	19.00
2	18	164.7	218.7	128.8	78.87	31.62

Step 10: *Generate schedule with spinning reserves.*

After calculating the modified hourly load, each participant applies unit commitment and the results are shown in Table 6.17. Table 6.18 shows comparative results without spinning reserves. In comparison, participant B commits the unit at bus 13 for a longer period to sell spinning reserve. Participant A does not have to commit any more units, since the available capacity of unit 2 is enough to sell the required reserve.

Table 6.17 Schedule with Spinning Reserve

Participant	Generator At Bus No.	Hour (1-24)
A	30	0 0 0 0 0 0 0 0 0 0 0 0 0 0 0 0 0 0 0 0 0 0 0 0
	24	0 0 0 0 0 0 0 0 0 0 0 0 0 0 0 0 0 0 0 0 0 0 0 0
	11	0 0 0 0 0 0 0 0 0 0 0 0 0 0 0 0 0 0 0 0 0 0 0 0
	2	1 1 1 1 1 1 1 1 1 1 1 1 1 1 1 1 1 1 1 1 1 1 1 1
	8	1 1 1 1 1 1 1 1 1 1 1 1 1 1 1 1 1 1 1 1 1 1 1 1
	5	1 1 1 1 1 1 1 1 1 1 1 1 1 1 1 1 1 1 1 1 1 1 1 1
	1	1 1 1 1 1 1 1 1 1 1 1 1 1 1 1 1 1 1 1 1 1 1 1 1
B	13	0 0 0 0 0 0 0 0 0 0 1 1 1 0 0 0 1 1 1 1 1 1 0 0
	15	1 1 1 1 1 1 1 1 1 1 1 1 1 1 1 1 1 1 1 1 1 1 1 1

Table 6.18 Schedule without Spinning Reserve

Participant	Generator At Bus No.	Hour (1-24)
A	30	0 0 0 0 0 0 0 0 0 0 0 0 0 0 0 0 0 0 0 0 0 0 0 0
	24	0 0 0 0 0 0 0 0 0 0 0 0 0 0 0 0 0 0 0 0 0 0 0 0
	11	0 0 0 0 0 0 0 0 0 0 0 0 0 0 0 0 0 0 0 0 0 0 0 0
	2	1 1 1 1 1 1 1 1 1 1 1 1 1 1 1 1 1 1 1 1 1 1 1 1
	8	1 1 1 1 1 1 1 1 1 1 1 1 1 1 1 1 1 1 1 1 1 1 1 1
	5	1 1 1 1 1 1 1 1 1 1 1 1 1 1 1 1 1 1 1 1 1 1 1 1
	1	1 1 1 1 1 1 1 1 1 1 1 1 1 1 1 1 1 1 1 1 1 1 1 1
B	13	0 0 0 0 0 0 0 0 0 0 0 0 0 0 0 0 1 1 1 1 0 0 0 0
	15	1 1 1 1 1 1 1 1 1 1 1 1 1 1 1 1 1 1 1 1 1 1 1 1

6.7 CONCLUSIONS

Deregulation in the electric power industry is expected to increase payoffs associated with the operation of interconnected power systems. In this chapter, we use game theory concepts to simulate the behavior of participants in the restructured energy markets. We assume a completely restructured power market in which each participant defines prices to interchange power with the remaining participants of the electricity marketplace.

From the results obtained in the chapter, we learn that all participants try to maximize their payoffs by cooperating with the power market to obtain the maximum systemwide payoffs when the system constraints do not limit the interchanges. The notion that increasing competition will help decrease operational costs appears to be supported by the results obtained in the chapter. Mathematically, as competition increases (through entry of more participants, additional tie lines to avoid flow congestion, etc.) the grand coalition strategy becomes more appealing to all participants.

We also reviewed the case where a network imposes additional constraints on the bids, and participants increase their payoffs by coordinating bidding strategies and sharing payoffs. The analysis may be used by market coordinators to identify non-competitive situations and to encourage pricing policies that lead to maximum systemwide payoffs.

Competition among participants for the market's load is modeled as a non-cooperative game. Each participant has incomplete information of the game. In our example, participants know their own operation costs, but they do not know the operation costs of their opponents. The game is solved using the Nash equilibrium idea for a transformed game with complete but imperfect information. The optimal price decision is derived for the Nash equilibrium point.

The proposed methodology is geared toward providing support for price decisions in the electricity market. We do not look for a specific value in the price that maximizes the participant's payoff; this "best value" can be later defined using the results obtained from the methodologies in this chapter and other mathematical tools.

The obtained strategies for the Nash equilibrium of the game maximize the expected conditional payoff in participants; moreover, the obtained strategies are the participants' maximum strategies. This fact makes the obtained Nash equilibrium strategies even more appealing to participants.

The proposed approach is a more general case than that when participants model their opponents' costs with uncorrelated probability distributions. The approach presented in this chapter allows a participant in the electricity market to make optimal pricing decisions that take into account not only his own perspective of the energy market but also the beliefs that other participants have of him.

In a restructured power market, participants can bid for generation and spinning reserve. In this uncertain environment, generation scheduling is done by participants in order to maximize payoffs while satisfying transmission constraints provided by the TRANSCOs. For each participant, unit commitment depends on the generation spot price, spinning reserve price, and transmission capabilities, and an hourly scheduling may be optimized by security-constrained optimal power flow. We model the competition in the market as a non-cooperative game with incomplete information; then we transform the incomplete information game into a complete game with imperfect information and seek Nash equilibrium. The unit commitment solution determines the modified load, committed units, as well as dominant strategies for each committed unit. This method can be used by the ISO to calculate the expected spot price and the spinning reserve price, and by participants in risk management.

Chapter 7

Generation Asset Valuation and Risk Analysis

7.1 INTRODUCTION

7.1.1 Asset Valuation

Asset valuation is the process of calculating the profit that occurs due to the utilization of a certain asset. The value of the asset is the difference between the revenue from and the cost incurred by that asset. In generation scheduling, a GENCO commits certain generating units and transforms various types of fuel into electricity for sale to the power market. The profit in this process is the value of committed generating units.

There are two types of valuation for generating units. One is the valuation based on the daily scheduled generation. The other is the valuation based on the available capacity of generating units.

Generation Asset Valuation. In the first application, we operate generating units on a daily basis and the issue would be to value generating units based on the current market information. As we commit units and submit bids to the market based on the forecasted market price, the value of the generating units is realized by the accepted bids. So valuation is based on the spot market price and not on a forecasted market price. In addition to the market price, the bidding strategy (i.e., commitment and bidding of units) has a major impact on the value of generating units.

Generation Capacity Valuation. In the second application, valuation is based on the available capacity for trading in the market. Hence, the physical characteristics of the units such as maximum/minimum capacity, force outage rate (i.e., availability), fuel consumption function (i.e., efficiency), and ramp rate (i.e., response capability) are among factors used in determining the value of generating units. Besides the physical characteristics, market price and commitment procedure can affect the value of the generating capacity. Obviously, higher market prices for electricity and lower fuel prices would increase the value of generation capacity. In addition, a weak commitment strategy could decrease the value of generating capacity.

There are two main differences between the two applications. First, the value of the generation asset in a daily operation is mostly realized in the form of accepted bids through market settlement, while generation capacity is mainly settled bilaterally. The second difference concerns long-term and short-term operation. The daily generation asset valuation is basically short term, while generation capacity valuation is long term, since it signifies the future availability of units.

7.1.2 Value at Risk (VaR)

Risk plays a major role in valuing generating units in a competitive power market. Risk is both objective and subjective. It is objective because all market participants would face the same market uncertainties. It is subjective because different attitudes to the market operation could incur different degrees of risk. Risk measurement then deserves rigorous indices to encompass the underlining issues. VaR is a viable measure for risk analysis that considers many risk factors synthetically and provides a single number for evaluating the effect of risk.

VaR is an estimate that shows how much a portfolio could lose due to market movements at a particular time horizon and for a given probability of occurrence. The given probability is called confidence level, which represents the degree of certainty of VaR. The common confidence level is 0.95 (or 95%) which means that 95% of the time participants' losses will be less than VaR and 5% of the time losses will be more than VaR. In mathematical terms, VaR corresponds to a percentile of a portfolio Profit & Loss (P&L) distribution, and can be expressed either as a potential loss from the current value of the portfolio, or as the loss from the expected value at the horizon [Min00].

Figure 7.1 shows a portfolio's P&L distribution. The expected P&L is 20 and the 5% percentile is minus 60. Hence, we can either express 95% VaR as a loss from the current value (VaR = 60), or as a loss from the expected value (VaR = 80). The decision to anchor the VaR calculation at the current value or the expected value is arbitrary, and as most people do, we define VaR as the difference between the corresponding percentile of the P&L distribution and the current value of a portfolio.

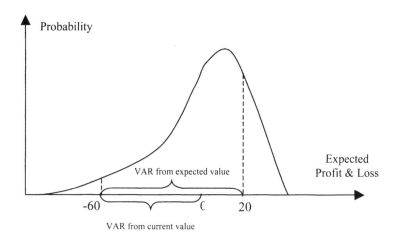

Figure 7.1 Illustration of VaR

7.1.3 Applications of VaR to Asset Valuation in Power Markets

As described earlier, the value of generating units depends on the generating units' efficiency and on market prices, which can be very uncertain in a restructured power market. In this chapter, we apply the concept of VaR to value the generation assets and assess the risk of generation capacity profitability, based on the daily operation and trading point of view.

According to Section 7.1.2, VaR is the estimated *loss* of a portfolio due to market movements at a particular time horizon and for a given probability of occurrence. So the larger the VaR, the *higher* the risk is.

In generation asset valuation, the value of generation asset is positive, so the estimated loss (VaR) is a negative number. In order to represent VaR with a positive value in this chapter, we consider a slightly different definition for VaR. We define VaR as an estimated *earning* of a portfolio due to market movements at a particular time horizon and for a given probability of occurrence. So the larger the VaR, the *smaller* the risk is.

The traditional value of VaR is the negative of the VaR considered in this chapter.

7.2 VAR FOR GENERATION ASSET VALUATION

The value of generation assets in a daily operation is realized through bidding in a market. The value is calculated based on accepted bids. In this section, we first propose a framework for calculating VaR. Then, a single time period example is presented for illustrating the procedure. Next, a more practical example with 36 units, 24 time periods, and bilateral contracts is presented. In this section, we also study the impact of market prices and bidding strategies.

7.2.1 Framework of the VaR Calculation

Figure 7.2 shows the framework for calculating the VaR for short-term generation asset valuation. Six modules are represented in Figure 7.2:

- Price forecasting module
- Market price simulation module
- Bids generating module
- Market settling module
- Profits analysis module
- VaR calculation module

These modules are discussed as follows:

- The details of market price forecast are the same as provided in Chapter 3.
- According to the forecasted spot market price, GENCO runs PBUC for calculating energy bids. Here, physical constraints of generating units such as emission and fuel constraints are considered.
- The actual spot market price is simulated, which could be different from the forecasted spot market price. In simulating the price, a GENCO may assume that the spot price distribution follows that of the historical price information and calculate the probability distribution of price difference accordingly (statistical simulation model-1). The results of the price difference analysis are fed to the market price simulation module. The GENCO may otherwise assume a priori that the actual spot market price follows a certain probability distribution with the forecasted spot market price as its expected value (statistical simulation model-2).

GENERATION ASSET VALUATION AND RISK ANALYSIS

- Based on the simulated spot market price, the GENCO runs the PBUC and calculates the bids that are going to be offered to the market. The bids are settled in the market-settling module. The output of the market-settling module includes the generation quantity and its corresponding price.

- The output of the market-settling module is fed to the profit analysis module. The profit analysis module calculates the revenue and the production cost for GENCO, and obtains the profit distribution.

- Based on the profit distribution and a specified confidence level, VaR is calculated in the VaR calculation module.

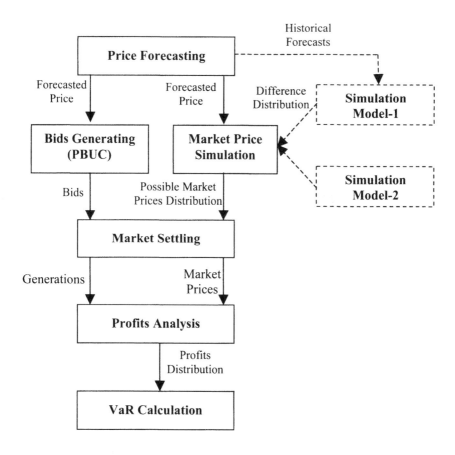

Figure 7.2 VaR Calculation for Short-Term Generation Asset Valuation

Note that the proposed framework for calculating VaR is very general. We can substitute any of the proposed modules with a different module in a specific application. For example, we can specify a different method for simulating the spot market price and study the sensitivity of VaR to the specified method. Within this framework, we can also study the risk of different bidding strategies in terms of VaR. The modules in Figure 7.2 are discussed further in the section that follows.

7.2.2 Spot Market Price Simulation

Recall that we simulate the actual spot market price based on the forecasted price. In doing so, we choose one of two patterns of the market price behavior as follows:

7.2.2.1 Statistical Simulation Model-1

Figure 7.3a compares the typical values for a spot market price forecast with the actual spot market price. We assume that the difference between actual and forecasted market prices follows the historical pattern of the spot price. Based on Figure 7.3a, we derive the price difference (actual price - forecasted price) distribution as shown in Figure 7.3b. Using Figure 7.3b, we add the price difference to any forecasted spot price in order to simulate the actual spot market price.

7.2.2.2 Statistical Simulation Model-2

We assume that the actual spot market price deviation from the forecasted market price follows a probability distribution function with forecasted price as its expected value. The most widely used probability distribution is the normal distribution. However, other distributions including triangular, lognormal, and exponential distributions can be used in our discussion.

Suppose that the spot market price follows a normal distribution. Figure 7.4 shows the distribution of the forecasted spot market price where the mean is 20 $/MWh (forecasted spot price) and the standard deviation is presumed to be 10% of the mean (2 $/MWh).

GENERATION ASSET VALUATION AND RISK ANALYSIS 239

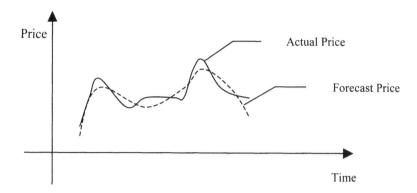

Figure 7.3a Historical Prices

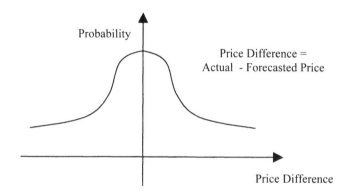

Figure 7.3b Price Difference Distribution

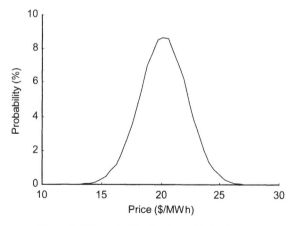

Figure 7.4 Normal Distribution of Price Forecast

7.2.3 A Numerical Example

In this section, we consider an example in which a GENCO follows the framework described in Section 7.2.1 and include a simple calculation of the VaR.

7.2.3.1 Historical Price Difference Analysis

Figure 7.5 shows a typical spot price forecast with relation to the actual spot market price. From Figure 7.5, we obtain the price difference curve shown in Figure 7.6 and then the price difference distribution shown in Figure 7.7. The corresponding numerical values for price difference distribution are given in Table 7.1.

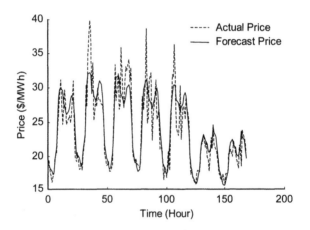

Figure 7.5 Historical Price Forecast

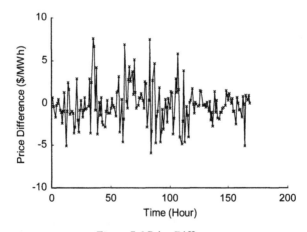

Figure 7.6 Price Difference

GENERATION ASSET VALUATION AND RISK ANALYSIS

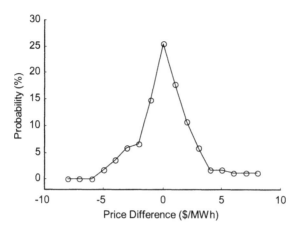

Figure 7.7 Price Difference Distribution

Table 7.1 Price Difference Distribution

Price difference ($/MWh)	-5	-4	-3	-2	-1	0	1
Probability (%)	1.79	3.57	5.95	6.55	14.88	25.60	17.86
Price difference ($/MWh)	2	3	4	5	6	7	8
Probability (%)	10.71	5.95	1.79	1.79	1.19	1.19	1.19

7.2.3.2 Market Price Simulation based on the Price Forecast

Suppose that the forecasted price for a given hour is 30 $/MWh. Based on the price difference distribution (Figure 7.7), we obtain the simulated spot market price distribution (spot market price = forecasted price + price difference). The probability of a possible spot market price would be the same as the probability of the corresponding price difference, as shown in Table 7.2 and Figure 7.8.

Table 7.2 Possible Market Price Distribution

Spot market price ($/MWh)	25	26	27	28	29	30	31
Probability (%)	1.79	3.57	5.95	6.55	14.88	25.60	17.86
Spot market price ($/MWh)	32	33	34	35	36	37	38
Probability (%)	10.71	5.95	1.79	1.79	1.19	1.19	1.19

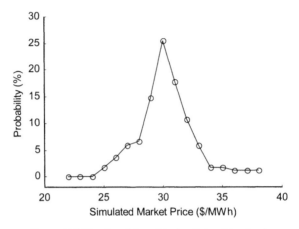

Figure 7.8 Simulated Spot Market Price Distribution

7.2.3.3 Bidding Curve and Cost Curve

We execute PBUC using a specific bidding strategy. Accordingly, the generating capacity cost curve and the bidding curve are shown in Table 7.3 and Figure 7.9.

Table 7.3 Bidding Curve and Capacity Cost Curve

MW	50	100	150	200	250	300	350	400	500
Bid Price ($/MWh)	26.2	27.1	27.8	28.5	29.3	29.5	29.9	31.2	33.5
Unit Cost ($/MWh)	27.5	27.75	28	28.25	28.5	28.75	29	30	32

7.2.3.4 Market Settlement Simulation

Using the simulated spot market prices (Figure 7.8) and the bidding curve (Figure 7.9), a GENCO could simulate the market settlement. The market settlement refers to the amount of energy that the GENCO will be able to sell to the spot market based on the market prices. The market will accept any bidding price offered by the GENCO that is less than the spot market price. Table 7.4 shows results of the market settlement simulation. The accepted bids depend on market prices.

GENERATION ASSET VALUATION AND RISK ANALYSIS

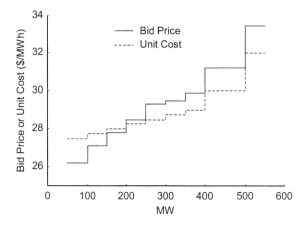

Figure 7.9 Bidding Curve and Capacity Cost Curve

Table 7.4 Market Settlements

Simulated market price ($/MWh)	25	26	27	28	29	30	31
Probability (%)	1.79	3.57	5.95	6.55	14.88	25.60	17.86
Accepted generation (MW)	0	0	50	150	200	350	350
Simulated market price ($/MWh)	32	33	34	35	36	37	38
Probability (%)	10.71	5.95	1.79	1.79	1.19	1.19	1.19
Accepted generation (MW)	400	400	500	500	500	500	500

7.2.3.5 Profit Analysis

From the market settlement and the marginal cost curve, we calculate the profit as shown in Table 7.5. Here, revenue = simulated market price x MW, total cost = capacity cost x MW, and profit = revenue - total cost. Table 7.6, which is derived from Table 7.5, shows the probability distribution of the profit. Note here that if certain market prices would result in the same profit, the probability for that profit would be the sum of corresponding probabilities.

Figure 7.10 shows the revenue distribution. Figure 7.11 shows the probability density function (PDF) of the profit. Figure 7.12 shows the cumulative distribution function (CDF) of the profit.

Table 7.5 Profit Analysis

Spot market price ($/MWh)	25	26	27	28	29	30	31
Probability (%)	1.79	3.57	5.95	6.55	14.88	25.60	17.86
Accepted Generation (MW)	0	0	50	150	200	350	350
Revenue ($)	0	0	1350	4200	5800	10500	10850
Total Cost ($)	0	0	1375	4200	5650	10150	10150
Net Profit ($)	0	0	-25	0	150	350	700
Spot market price ($/MWh)	32	33	34	35	36	37	38
Probability (%)	10.71	5.95	1.79	1.79	1.19	1.19	1.19
Accepted Generation (MW)	400	400	500	500	500	500	500
Revenue ($)	12800	13200	17000	17500	18000	18500	19000
Total Cost ($)	12000	12000	16000	16000	16000	16000	16000
Net Profit ($)	800	1200	1000	1500	2000	2500	3000

Table 7.6 Probability Distribution of Profit

Net Profit ($)	-25	0	150	350	700	800
Probability (%)	5.95	11.91	14.88	25.60	17.86	10.71
Net profit ($)	1000	1200	1500	2000	2500	3000
Probability (%)	5.95	1.79	1.79	1.19	1.19	1.19

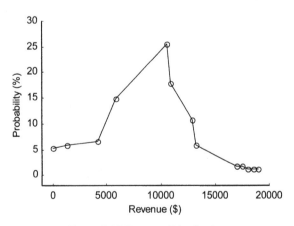

Figure 7.10 Revenue Distribution

GENERATION ASSET VALUATION AND RISK ANALYSIS

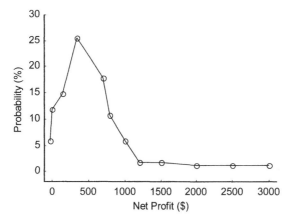

Figure 7.11 Profit Distribution

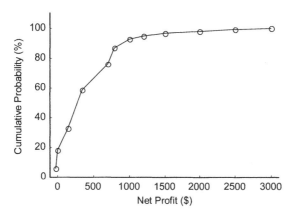

Figure 7.12 Cumulative Profit Distribution

7.2.3.6 Calculating VaR

Suppose that we are interested in calculating VaR with a 94.05% (i.e., 100 - 5.95) confidence level. According to Figure 7.12 and Table 7.6, VaR = -$25, which means that 94.05% of time the profit is larger than -$25, or 5.95% time the profit is worse than -$25. Another example is when we want to calculate VaR with a 95% confidence level. From Table 7.5b, the probability of the first net profit value is over 5% (i.e., 100 - 95) and extrapolation is employed to get the net profit corresponding to 5% probability. The first two points in Figure 7.10 are used:

$$\frac{\text{VaR} - (-25)}{5\% - 5.95\%} = \frac{0 - (-25)}{11.91\% - 5.95\%}$$

We obtain the VaR with 95% confidence level as -28.98$. This means that 95% of the time the participant's profit will be more than -$28.98, and that 5% of the time the profit will be less than -$28.98. Since the profit is negative, it becomes a loss. In other words, 5% of the time the participant's loss will be more than $28.98, and 95% (confidence level) of the time the participant's loss will be less than $28.98.

7.2.4 A Practical Example

In this section, we turn to a practical example of VaR calculation. The GENCO has 36 units with 24 time periods. Some units must serve bilateral contracts.

Since VaR depends on market structure and market settlement rules, we first describe the market model and set up some settlement rules. Then, we discuss the modules in the framework.

7.2.4.1 Market Model

In order to value generation assets in a market, we must know the market rules. The market model considered in this example is shown in Figure 7.13. In this model, the ISO runs an auction market. GENCOs submit sale bids to the market, DISCOs submit purchase bids to the market, and the ISO settles the market and returns settlements to the GENCOs and DISCOs. GENCOs can sign bilateral contracts with customers in this market. Note that in Figure 7.13, a bid includes its price and quantity, and a settlement includes its price and quantity.

7.2.4.2 Market Settlement Rules

Settlement rules in the auction market are given as follows:
- Based on the submitted sale and purchase bids, the ISO determines the market-clearing price (MCP).

GENERATION ASSET VALUATION AND RISK ANALYSIS

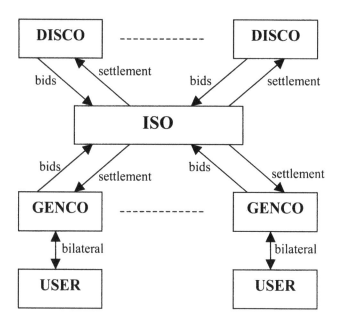

Figure 7.13 Market Model in the Study

- Sale bids that are less than or equal to MCP will be accepted. The settlement price will be the same as MCP for the bidding quantity.
- Purchase bids that are higher than or equal to MCP will be accepted. The settlement price will be the same as MCP for the bidding quantity.

Settlement rules for bilateral contracts are given as follows:

- If the bilateral contract is honored in its entirety (GENCO can provide the entire contracted quantity to the user), the settlement quantity will be the contracted quantity, and settlement price will be the GENCO's marginal price to produce the contracted quantity times one plus a profit factor. The profit factor will be negotiated beforehand to characterize the GENCO's profit margin from the bilateral contract.
- If the bilateral contract is nullified (GENCO provides nothing to the user), the settlement quantity will be the contracted quantity, and the settlement price will be the current market price times one plus a penalty factor. The penalty factor will be negotiated beforehand to characterize the GENCO's defaulted contract with the user.

- If the bilateral contract is partially honored (GENCO can only provide part of the contracted quantity to the user), the settlement will be comprised of two parts. In the first part, the settlement quantity is the quantity that the GENCO can provide, and the settlement price is the GENCO's marginal price to produce the contracted quantity times one plus a profit factor. In the second part, the settlement quantity will be the contracted quantity minus the settlement quantity in the first part of the settlement, and the settlement price will be the current market price times one plus a penalty factor.

7.2.4.3 PBUC and Generation Bids

Based on market prices, a GENCO executes the PBUC that answers the following question: Given market prices, what is the best generation schedule? However, PBUC doesn't answer the following question: What is the best bidding strategy (i.e., how to bid)? Here, we characterize the bids based on the PBUC output. The bidding quantity is the same as generation schedule and the bidding price is the marginal cost corresponding to generation schedule.

The input to PBUC includes:

- Unit characteristics and operating limits
- Market prices
- Bilateral contracts
- System constraints (energy, reserve, emission and fuel constraints)

The output of PBUC includes:

- Generating units status
- Generation schedule

See Appendix D.2 for unit characteristics. In this example, we only consider energy market with prices shown in Figure 7.14. Bilateral contracts are plotted in Figure 7.15. Energy constraints are shown in Table 7.7. No emission and fuel constraints are imposed. Tables 7.8 to 7.11 include all the bidding information.

GENERATION ASSET VALUATION AND RISK ANALYSIS

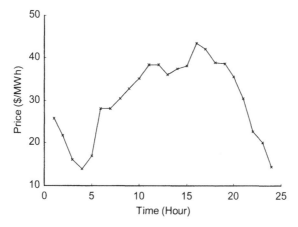

Figure 7.14 Forecasted Energy Market Prices

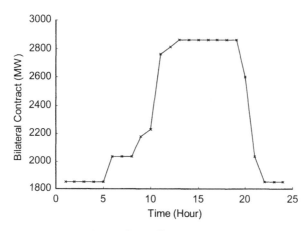

Figure 7.15 Bilateral Contracts

Table 7.7 System Energy Limits

Hour	Lower Limit (MW)	Upper Limit (MW)	Hour	Lower Limit (MW)	Upper Limit (MW)
1	3308.76	2008.76	13	3817.80	2517.80
2	3054.24	1754.24	14	3732.96	2432.96
3	2884.56	1584.56	15	3690.54	2390.54
4	2799.72	1499.72	16	3690.54	2390.54
5	2794.88	1494.88	17	3860.22	2560.22
6	2797.30	1497.30	18	4242.00	2920.00
7	2799.72	1499.72	19	4199.58	2899.58
8	2969.40	1669.40	20	4114.74	2814.74
9	3393.60	2093.60	21	3987.48	2687.48
10	3732.96	2232.96	22	3902.64	2602.64
11	3817.80	2517.80	23	3690.54	2390.54
12	3860.22	2560.22	24	3436.02	2136.02

Table 7.8 PBUC Schedule

Unit \ Hour	1	2	3	4	5	6	7	8	9	10	11	12	13	14	15	16	17	18	19	20	21	22	23	24
1	2	0	0	0	0	2	2	2	2	2	2	2	2	2	2	2	2	2	2	2	0	0	0	0
2-7	0	0	0	0	0	0	0	0	0	0	0	0	0	0	0	0	0	0	0	0	0	0	0	0
8-11	15	15	15	15	15	15	15	15	15	15	15	15	15	15	15	15	15	15	15	15	0	0	0	0
12	0	0	0	0	0	25	25	25	25	25	25	25	53	53	53	53	53	53	52	52	25	0	0	0
13	25	25	0	0	0	25	25	25	25	25	25	53	53	53	53	53	53	53	52	52	25	0	0	0
14	25	25	0	0	0	25	25	25	25	25	53	53	53	53	53	53	53	53	52	50	0	0	0	0
15	0	0	0	0	0	25	25	25	25	53	53	53	53	53	53	53	53	53	52	52	50	0	0	0
16	0	0	0	0	0	25	25	25	53	53	53	53	53	53	53	53	53	53	52	52	50	0	0	0
17	0	0	0	0	0	25	25	25	53	53	53	53	53	53	53	53	53	53	51	52	50	0	0	0
18	0	0	0	0	0	25	25	25	53	54	54	54	54	54	54	54	54	53	53	50	0	0	0	0
19	54	54	54	54	54	62	62	62	64	63	63	63	64	64	64	64	64	64	61	62	62	78	0	0
20	54	54	54	54	54	62	62	62	64	63	64	64	64	64	64	64	64	64	62	62	62	78	0	0
21	54	54	54	54	54	62	62	62	64	64	64	64	64	64	64	64	64	64	62	62	62	78	0	0
22	0	0	0	0	0	69	69	69	69	69	82	82	83	83	83	83	83	82	77	78	77	0	0	0
23	0	0	0	0	0	69	69	69	69	69	82	82	83	83	83	83	83	82	77	78	78	0	0	0
24	0	0	0	0	0	69	69	69	69	69	82	82	83	83	83	83	83	82	77	78	78	0	0	0
25	0	0	0	0	0	0	69	69	69	69	82	82	83	83	83	83	82	77	78	69	0	0	0	0
26	0	0	0	0	0	0	69	69	69	69	82	82	83	83	83	83	82	77	78	69	0	0	0	0
27	0	0	0	0	69	69	69	69	69	82	82	83	83	83	83	82	77	78	78	0	0	0	0	0
28	150	155	156	157	157	155	155	156	158	155	155	155	156	157	157	157	155	147	148	148	350	350	350	175
29	150	155	157	157	157	155	155	156	158	155	156	155	156	157	157	157	156	148	149	149	350	350	350	350
30	148	154	156	157	157	155	155	156	157	154	154	154	155	156	157	157	155	145	147	146	350	350	350	175
31	152	156	157	158	158	155	155	156	158	156	157	156	157	158	158	158	157	149	151	150	350	350	350	350
32	275	274	273	271	271	260	260	261	270	274	274	274	274	273	272	271	271	274	274	274	400	400	400	400
33	275	274	273	272	271	260	260	261	270	274	274	274	273	272	272	272	274	274	275	275	400	400	400	400
34	275	274	273	272	272	260	260	261	270	274	274	274	273	272	272	272	274	275	275	275	400	400	400	400
35	275	274	273	272	272	260	260	261	270	274	274	274	273	272	272	272	274	275	275	275	400	400	400	400
36	275	274	273	272	272	260	260	261	270	274	274	274	273	272	272	272	274	275	275	275	400	400	400	400

GENERATION ASSET VALUATION AND RISK ANALYSIS

Table 7.9 Purchase Schedule

Unit\Hour	1	2	3	4	5	6	7	8	9	10	11	12	13	14	15	16	17	18	19	20	21	22	23	24
1-11	0	0	0	0	0	0	0	0	0	0	0	0	0	0	0	0	0	0	0	0	0	0	0	0
12	0	0	0	0	0	0	0	0	0	0	0	0	0	0	45	40	37	37	47	50	50	0	0	0
13	0	0	0	0	0	0	0	0	0	0	0	0	50	46	41	38	38	48	50	50	0	0	0	0
14	0	0	0	0	0	0	0	0	0	0	49	50	45	40	38	38	48	50	50	50	0	0	0	0
15	0	0	0	0	0	0	0	0	0	50	50	50	46	41	39	39	49	50	50	50	0	0	0	0
16	0	0	0	0	0	0	0	0	32	50	49	50	46	41	38	38	48	50	50	50	0	0	0	0
17	0	0	0	0	0	0	0	0	33	50	50	50	47	42	39	39	49	50	50	50	0	0	0	0
18	0	0	0	0	0	0	0	0	30	47	45	46	42	37	35	35	44	50	50	50	0	0	0	0
19	0	0	0	0	0	15	15	17	40	60	60	60	57	50	47	47	60	60	60	60	0	0	0	0
20	0	0	0	0	0	15	15	17	40	60	60	60	56	50	47	47	59	60	60	60	0	0	0	0
21	0	0	0	0	0	15	15	17	39	60	60	60	56	50	47	47	59	60	60	60	0	0	0	0
22-23	0	0	0	0	0	0	0	0	0	0	80	80	80	80	77	77	80	80	80	80	0	0	0	0
24-26	0	0	0	0	0	0	0	0	0	0	80	80	80	80	76	76	80	80	80	80	0	0	0	0
27	0	0	0	0	0	0	0	0	0	0	80	80	80	80	77	77	80	80	80	80	0	0	0	0
28	150	150	150	137	136	44	44	50	116	150	150	150	150	146	137	137	150	150	150	150	0	0	0	0
29	150	150	150	136	134	43	44	49	114	150	150	150	150	144	135	135	150	150	150	150	0	0	0	0
30	150	150	150	142	141	45	46	52	120	150	150	150	150	150	141	141	150	150	150	150	0	0	0	0
31	150	150	147	130	129	42	42	47	110	150	150	150	150	138	130	130	150	150	150	150	0	0	0	0
32	250	198	168	148	147	47	47	53	125	198	192	195	178	158	148	148	188	250	250	250	0	0	0	0
33	250	197	168	148	147	46	47	53	124	198	191	195	178	157	148	148	188	250	250	250	0	0	0	0
34	250	197	167	147	146	46	47	53	124	197	191	194	177	157	147	147	188	250	250	250	0	0	0	0
35	250	196	167	147	146	46	47	53	124	197	190	194	177	157	147	147	187	250	250	250	0	0	0	0
36	250	196	166	146	145	46	46	53	123	196	189	193	176	156	146	146	186	250	250	250	0	0	0	0

Table 7.10 Sale Bidding Price

Unit\Hour	1	2	3	4	5	6	7	8	9	10	11	12
1	25.669	0	0	0	0	25.669	25.669	25.669	25.669	25.669	25.669	25.669
2-7	0	0	0	0	0	0	0	0	0	0	0	0
8	13.8	13.8	13.8	13.8	13.8	13.8	13.8	13.8	13.8	13.8	13.8	13.8
9	13.594	13.594	13.594	13.594	13.594	13.594	13.594	13.594	13.594	13.594	13.594	13.594
10	13.626	13.626	13.626	13.626	13.626	13.626	13.626	13.626	13.626	13.626	13.626	13.626
11	13.691	13.691	13.691	13.691	13.691	13.691	13.691	13.691	13.691	13.691	13.691	13.691
12	0	0	0	0	0	18.312	18.312	18.312	18.312	18.312	18.312	18.312
13	18.899	18.899	0	0	0	18.899	18.899	18.899	18.899	18.899	18.899	19.409
14	18.406	18.406	0	0	0	18.406	18.406	18.406	18.406	18.406	18.923	18.925
15	0	0	0	0	0	18.574	18.574	18.574	18.574	18.574	19.073	19.074
16	0	0	0	0	0	18.499	18.499	18.499	18.951	19.007	19.007	19.008
17	0	0	0	0	0	17.569	17.569	17.569	18.009	18.059	18.062	18.06
18	0	0	0	0	0	19.549	19.549	19.549	20.062	20.126	20.121	20.124
19	11.229	11.229	11.229	11.229	11.229	11.348	11.349	11.357	11.433	11.48	11.483	11.481
20	11.259	11.259	11.259	11.259	11.259	11.38	11.38	11.388	11.465	11.514	11.518	11.516
21	11.287	11.287	11.287	11.287	11.287	11.409	11.409	11.417	11.495	11.546	11.549	11.548
22	0	0	0	0	0	23.257	23.257	23.257	23.257	23.257	23.555	23.553
23	0	0	0	0	0	23.459	23.459	23.459	23.459	23.459	23.657	23.656
24	0	0	0	0	0	23.563	23.563	23.563	23.563	23.563	23.764	23.762
25	0	0	0	0	0	0	0	23.764	23.764	23.764	23.966	23.965
26	0	0	0	0	0	0	0	23.868	23.868	23.868	24.073	24.071
27	0	0	0	0	0	23.4	23.4	23.4	23.4	23.4	23.599	23.598
28	11.414	11.443	11.456	11.448	11.448	11.352	11.352	11.36	11.43	11.443	11.446	11.444
29	11.446	11.476	11.489	11.478	11.478	11.381	11.382	11.389	11.459	11.476	11.479	11.477
30	11.205	11.234	11.247	11.245	11.244	11.15	11.15	11.157	11.226	11.234	11.237	11.235

Table 7.10 Sale Bidding Price (Continued)

Unit \ Hour	1	2	3	4	5	6	7	8	9	10	11	12
31	11.586	11.617	11.627	11.612	11.611	11.513	11.513	11.52	11.593	11.617	11.62	11.619
32	8.763	8.731	8.702	8.681	8.68	8.545	8.545	8.555	8.654	8.731	8.725	8.729
33	8.78	8.748	8.719	8.698	8.696	8.561	8.562	8.571	8.67	8.748	8.742	8.745
34	8.796	8.762	8.734	8.712	8.711	8.575	8.576	8.586	8.685	8.763	8.757	8.76
35	8.823	8.788	8.759	8.738	8.736	8.6	8.601	8.611	8.71	8.788	8.782	8.786
36	8.916	8.88	8.851	8.829	8.828	8.691	8.691	8.701	8.801	8.88	8.874	8.877

Table 7.10 Sale Bidding Price (Continued)

Unit \ Hour	13	14	15	16	17	18	19	20	21	22	23	24
1	25.669	25.669	25.669	25.669	25.669	25.669	25.669	25.669	0	0	0	0
2-7	0	0	0	0	0	0	0	0	0	0	0	0
8	13.8	13.8	13.8	13.8	13.8	13.8	13.8	13.8	0	0	0	0
9	13.594	13.594	13.594	13.594	13.594	13.594	13.594	13.594	0	0	0	0
10	13.626	13.626	13.626	13.626	13.626	13.626	13.626	13.626	0	0	0	0
11	13.691	13.691	13.691	13.691	13.691	13.691	13.691	13.691	0	0	0	0
12	18.825	18.809	18.8	18.8	18.833	18.789	18.796	18.312	0	0	0	0
13	19.398	19.382	19.373	19.373	19.405	19.357	19.364	18.899	0	0	0	0
14	18.912	18.896	18.888	18.888	18.92	18.874	18.881	18.87	0	0	0	0
15	19.065	19.049	19.041	19.041	19.073	19.023	19.029	19.02	0	0	0	0
16	18.997	18.981	18.972	18.972	19.004	18.956	18.963	18.953	0	0	0	0
17	18.054	18.038	18.03	18.03	18.061	18.009	18.016	18.008	0	0	0	0
18	20.11	20.093	20.084	20.084	20.119	20.088	20.096	20.079	0	0	0	0
19	11.477	11.462	11.454	11.454	11.485	11.431	11.437	11.436	11.83	0	0	0
20	11.51	11.494	11.486	11.486	11.517	11.465	11.472	11.47	11.87	0	0	0
21	11.539	11.524	11.515	11.515	11.547	11.496	11.503	11.501	11.906	0	0	0
22	23.561	23.569	23.567	23.567	23.556	23.512	23.518	23.517	0	0	0	0
23	23.663	23.671	23.669	23.669	23.659	23.614	23.62	23.619	0	0	0	0
24	23.77	23.778	23.774	23.774	23.765	23.721	23.726	23.725	0	0	0	0
25	23.972	23.98	23.976	23.976	23.967	23.923	23.928	23.764	0	0	0	0
26	24.079	24.087	24.081	24.081	24.074	24.029	24.035	23.868	0	0	0	0
27	23.605	23.614	23.61	23.61	23.601	23.557	23.562	23.561	0	0	0	0
28	11.452	11.456	11.448	11.448	11.447	11.401	11.407	11.406	12.256	12.179	12.246	11.629
29	11.485	11.486	11.478	11.478	11.48	11.433	11.439	11.438	12.305	12.226	12.294	12.238
30	11.243	11.251	11.245	11.245	11.238	11.193	11.199	11.198	12.01	11.936	12	11.412
31	11.627	11.619	11.612	11.612	11.622	11.573	11.579	11.578	12.499	12.415	12.488	12.428
32	8.712	8.691	8.681	8.681	8.722	8.741	8.751	8.749	9.736	9.621	9.72	9.639
33	8.729	8.708	8.697	8.697	8.739	8.759	8.769	8.767	9.758	9.643	9.743	9.662
34	8.743	8.723	8.712	8.712	8.754	8.775	8.784	8.782	9.779	9.663	9.763	9.682
35	8.769	8.748	8.737	8.737	8.779	8.801	8.811	8.809	9.81	9.694	9.795	9.713
36	8.86	8.839	8.829	8.829	8.871	8.894	8.904	8.902	9.914	9.796	9.898	9.815

GENERATION ASSET VALUATION AND RISK ANALYSIS

Table 7.11 Purchase Bidding Price

Hour\Unit	1	2	3	4	5	6	7	8	9	10	11	12
1-12	0	0	0	0	0	0	0	0	0	0	0	0
13	0	0	0	0	0	0	0	0	0	0	0	18.899
14	0	0	0	0	0	0	0	0	0	0	18.406	18.406
15	0	0	0	0	0	0	0	0	0	18.574	18.574	18.574
16	0	0	0	0	0	0	0	0	18.499	18.499	18.499	18.499
17	0	0	0	0	0	0	0	0	17.569	17.569	17.569	17.569
18	0	0	0	0	0	0	0	0	19.549	19.549	19.549	19.549
19	0	0	0	0	0	11.212	11.211	11.201	11.094	10.999	10.999	10.999
20	0	0	0	0	0	11.242	11.241	11.232	11.123	11.025	11.025	11.025
21	0	0	0	0	0	11.27	11.27	11.26	11.151	11.05	11.051	11.05
22	0	0	0	0	0	0	0	0	0	0	23.207	23.207
23	0	0	0	0	0	0	0	0	0	0	23.208	23.208
24	0	0	0	0	0	0	0	0	0	0	23.41	23.41
25	0	0	0	0	0	0	0	0	0	0	23.611	23.611
26	0	0	0	0	0	0	0	0	0	0	23.714	23.714
27	0	0	0	0	0	0	0	0	0	0	23.249	23.249
28	11.067	11.067	11.067	11.086	11.087	11.226	11.225	11.216	11.118	11.067	11.067	11.067
29	11.091	11.091	11.091	11.113	11.115	11.254	11.254	11.245	11.146	11.091	11.091	11.091
30	10.876	10.876	10.876	10.888	10.889	11.026	11.025	11.017	10.919	10.876	10.876	10.876
31	11.206	11.206	11.211	11.239	11.24	11.383	11.382	11.373	11.272	11.206	11.206	11.206
32	7.88	7.88	7.88	7.88	7.88	7.88	7.88	7.88	7.88	7.88	7.88	7.88
33	7.893	7.893	7.893	7.893	7.893	7.893	7.893	7.893	7.893	7.893	7.893	7.893
34	7.904	7.904	7.904	7.904	7.904	7.904	7.904	7.904	7.904	7.904	7.904	7.904
35	7.926	7.926	7.926	7.926	7.926	7.926	7.926	7.926	7.926	7.926	7.926	7.926
36	8.01	8.01	8.01	8.01	8.01	8.01	8.01	8.01	8.01	8.01	8.01	8.01

Table 7.11 Purchase Bidding Price (Continued)

Hour\Unit	13	14	15	16	17	18	19	20	21	22	23	24
1-11	0	0	0	0	0	0	0	0	0	0	0	0
12	18.312	18.312	18.312	18.312	18.312	18.312	18.312	0	0	0	0	0
13	18.899	18.899	18.899	18.899	18.899	18.899	18.899	0	0	0	0	0
14	18.406	18.406	18.406	18.406	18.406	18.406	18.406	18.406	0	0	0	0
15	18.574	18.574	18.574	18.574	18.574	18.574	18.574	18.574	0	0	0	0
16	18.499	18.499	18.499	18.499	18.499	18.499	18.499	18.499	0	0	0	0
17	17.569	17.569	17.569	17.569	17.569	17.569	17.569	17.569	0	0	0	0
18	19.549	19.549	19.549	19.549	19.549	19.549	19.549	19.549	0	0	0	0
19	11.015	11.044	11.059	11.059	10.999	10.999	10.999	10.999	0	0	0	0
20	11.044	11.074	11.088	11.088	11.028	11.025	11.025	11.025	0	0	0	0
21	11.071	11.101	11.116	11.116	11.055	11.05	11.05	11.05	0	0	0	0
22	23.207	23.207	23.214	23.214	23.207	23.207	23.207	23.207	0	0	0	0
23	23.208	23.208	23.216	23.216	23.208	23.208	23.208	23.208	0	0	0	0
24	23.41	23.41	23.42	23.42	23.41	23.41	23.41	23.41	0	0	0	0
25	23.611	23.611	23.621	23.621	23.611	23.611	23.611	0	0	0	0	0
26	23.714	23.714	23.725	23.725	23.714	23.714	23.714	0	0	0	0	0
27	23.249	23.249	23.257	23.257	23.249	23.249	23.249	23.249	0	0	0	0
28	11.067	11.073	11.086	11.086	11.067	11.067	11.067	11.067	0	0	0	0
29	11.091	11.1	11.114	11.114	11.091	11.091	11.091	11.091	0	0	0	0
30	10.876	10.876	10.888	10.888	10.876	10.876	10.876	10.876	0	0	0	0
31	11.206	11.225	11.239	11.239	11.206	11.206	11.206	11.206	0	0	0	0
32	7.88	7.88	7.88	7.88	7.88	7.88	7.88	7.88	0	0	0	0
33	7.893	7.893	7.893	7.893	7.893	7.893	7.893	7.893	0	0	0	0
34	7.904	7.904	7.904	7.904	7.904	7.904	7.904	7.904	0	0	0	0
35	7.926	7.926	7.926	7.926	7.926	7.926	7.926	7.926	0	0	0	0
36	8.01	8.01	8.01	8.01	8.01	8.01	8.01	8.01	0	0	0	0

7.2.4.4 Historical Price Forecast

One historical forecast is shown in Figure 7.16, and the price difference distribution is shown in Figure 7.17. According to Figure 7.17, it is likely that the spot market price is higher than the forecasted market price (the cumulative probability of a positive price difference is larger than that of a negative price difference).

7.2.4.5 Market Price Simulation

7.2.4.5.1 Statistical Simulation Model-1. Figure 7.18 shows the probability distribution of 1000 price simulations at hour 9. The forecasted price at hour 9 is 32.6 $/MWh. The distribution is similar to the historical price difference distribution in Figure 7.17. Figure 7.19 shows a case of the simulated 24-hour market prices.

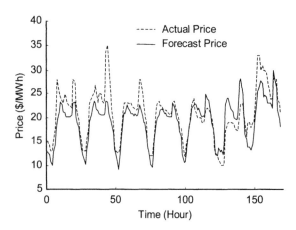

Figure 7.16 Historical Price Forecast

GENERATION ASSET VALUATION AND RISK ANALYSIS 255

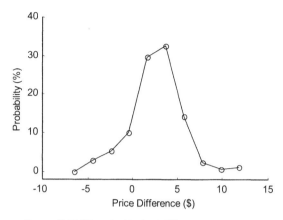

Figure 7.17 Historical Price Difference Distribution

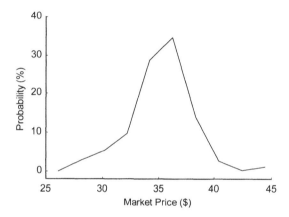

Figure 7.18 Simulated Market Price Distribution at Hour 9

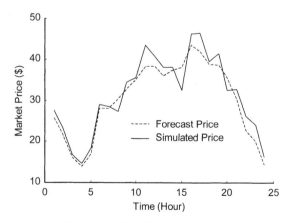

Figure 7.19 Simulated Market Prices

7.2.4.5.2 Statistical Simulation Model-2. Figures 7.20 and 7.21 show the examples of simulated market price based on the normal distribution.

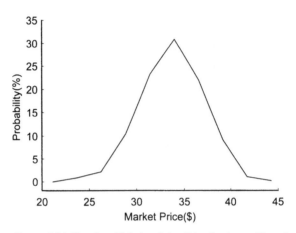

Figure 7.20 Simulated Market Price Distribution at Hour 9
(Statistical Simulation Model-2, $\sigma = 10\%\mu$)

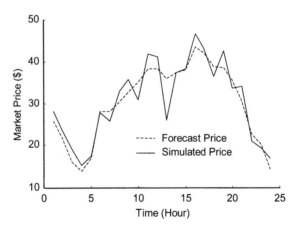

Figure 7.21 Simulated Market Prices
(Statistical Simulation Model-2, $\sigma = 10\%\mu$)

Figures 7.20 shows the probability distribution of 1000 simulations of the price at hour 9. The forecasted price at hour 9 is 32.6 \$/MWh. Figure 7.21 shows a case of the simulated market prices. In this figure, the mean μ is the forecasted market price; the standard deviation σ is 10% of the mean

GENERATION ASSET VALUATION AND RISK ANALYSIS

value. A higher standard deviation would represent a larger dispersion of the simulated market price and more volatility in the market, which translates into more risks. When the standard deviation is excessive (50%), the deviation of the actual price from the forecasted market price will be very large, and perhaps unreasonable. So, in the following simulation, we set the maximum standard deviation at 20% of the mean value.

7.2.4.6 Market Settlement, Profit Analysis and VaR Calculation

For our example of the VaR calculation, the conditions are:

- 10,000 Monte Carlo simulations
- A normal probability distribution for market price
- Standard deviation is 10% of mean
- Confidence level is 95%.

Figure 7.22 shows profits for the first 1000 simulations. The horizontal solid line indicates the expected PBUC profit. Figure 7.23 shows the CDF of 1000 simulations. The profit corresponding to (100% - 95% = 5%) is VaR, that is, $215,777.25. According to Figure 7.23, there is a 95% probability that the actual profit will be larger than $215,777.25.

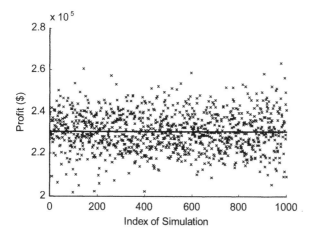

Figure 7.22 Profits Based on 1000 Simulations

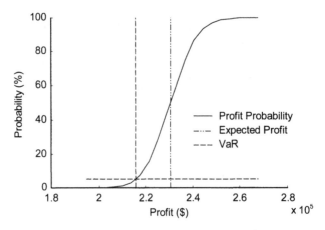

Figure 7.23 Cumulative Distribution Function of Profits

7.2.5 Sensitivity Analysis

7.2.5.1 Simulation of Bidding Strategy

We define the bidding strategy based on the bid's price and quantity as follows. For the bidding price strategy **(BPS)**:

- BPS1: bidding price is based on the forecasted price
- BPS2: bidding price is based on the marginal cost

In the case of BPS1, if the forecasted market price is **FP**, we consider bidding at 0.5FP, 0.8FP, and 0.9FP (bidding low) and bidding at 1.1FP, 1.2FP, and 1.5FP (bidding high). Similarly, in the case of BPS2, if the marginal cost is **MC**, we consider bidding at 0.5MC, 0.8MC, and 0.9MC (bidding low) and bidding at 1.1MC, 1.2MC, and 1.5MC (bidding high). To simplify the notation, we define k_p as the *bidding price coefficient* which is the ratio of bidding price to marginal cost (or forecasted market price) and set k_p at 0.5, 0.8, 0.9, 1.0, 1.1, 1.2, or 1.5 for sensitivity analysis.

The PBUC output is used as bidding quantity. A GENCO satisfies its bilateral contracts either by local generation or by purchasing from the market. Correspondingly, we define the following strategies for satisfying bilateral contracts. For the bidding quantity strategy **(BQS)**:

- BQS1: local generation is used for bilateral contracts and excess generation is offered for sale to the market.

GENERATION ASSET VALUATION AND RISK ANALYSIS

- BQS2: energy is purchased from the market (when the market price is low) for supplying bilateral contracts.

We learned in the discussion of arbitrage in Chapter 5 that the BQS2 option is more profitable. In this chapter, we study the risk associated with BQS options. If there were no bilateral contracts, then BQS1 and BQS2 would be the same.

7.2.5.2 Simulation of Market Price

In general, there are two categories of factors that would impact the market price and the associated risks. The first category is related to system demand, power exchange, and transmission network constraints. This category is relatively "objective" and can be forecasted by the usual forecasting models. Using an artificial neural network (ANN) is a good option. The second category is related to the bidding strategies of market participants. This category is relatively "subjective" and difficult to be considered by the usual forecasting models.

For the first category, the presumption is that we can forecast the market price with certain accuracy. To simulate the associated risk in this category, we assume a normal distribution for the spot market price, with the forecasted price as its mean value. By varying the standard deviation, we simulate different degrees of volatility. Suppose that the forecasted market price is μ (mean) and the volatility is σ (standard deviation). For sensitivity analysis, we consider σ to be equal to $k_\sigma \mu$, where k_σ is the *volatility coefficient* and selected as 5%, 10%, 15%, or 20%.

For the second category, we could consider gaming for analyzing the bidding strategy of the market participant and its impact on market price. In a perfect competition, no participant can change the market price by changing its bidding price. In other words, every participant is a price taker. On the contrary, in an imperfect competition, some participants may change the market price by bidding high or low. Here, we adopt a simple way to simulate the impact of the participant's bidding strategy on market price by introducing an adjustment factor [Bha00] to the forecasted market price μ. Accordingly,

$$\mu' = \begin{cases} \mu - [(\mu - \delta) - BP] & \text{if } BP < \mu - \delta \\ \mu & \text{if } \mu - \delta \leq BP \leq \mu + \delta \\ \mu + [BP - (\mu + \delta)] & \text{if } BP > \mu + \delta \end{cases} \quad (7.1)$$

where μ' is the expected spot market price considering the impact of a participant's bidding strategy, δ is a parameter that simulates the impact. If δ approaches infinity, there will be no impact. If δ is zero, the participant's bidding price **(BP)** determines the market price. The equation above assumes that BP within the range of $\mu \pm \delta$ has no impact on the market price. Beyond the range of $\mu \pm \delta$, μ will be adjusted linearly according to the difference between BP and $\mu + \delta$ or $\mu - \delta$. Simplifying (7.1), we have

$$\mu' = \begin{cases} BP + \delta & \text{if } BP < \mu - \delta \\ \mu & \text{if } \mu - \delta \leq BP \leq \mu + \delta \\ BP - \delta & \text{if } BP > \mu + \delta \end{cases} \quad (7.2)$$

Figure 7.24 shows the impact of bidding price on market price. In our study, we consider the case where δ is $k_\delta \mu$, where k_δ is denoted as the *bidding strategy coefficient* and selected as 0.0, 0.1, ..., 1.0 for sensitivity analysis.

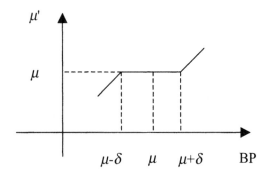

Figure 7.24 Impact of Bidding Price on Market Price

GENERATION ASSET VALUATION AND RISK ANALYSIS

7.2.5.3 Case Study

In this section, we analyze the impact of market price and bidding strategy on associated risks.

7.2.5.3.1 Perfect Competition: Market Price is Independent of Bidding Strategy. In perfect competition, no participant can change the market price by changing its bidding price (k_δ approaching infinity in (7.2)). Table 7.12 shows the VaR for different bidding strategies and different market volatilities (k_σ).

Table 7.12 VaR in Perfect Competition

Bidding Strategy	k_σ / k_p	0.05	0.10	0.15	0.20
BPS1, BQS1	0.5	598547.63	570694.20	542840.78	514994.42
	0.8	598547.63	548143.66	489476.47	453337.31
	0.9	561211.15	445033.35	385202.72	372011.18
	1.0	195304.29	208973.86	224299.98	238139.27
	1.1	-3100.00	4237.34	73435.24	117566.67
	1.2	-3100.00	-3100.00	-3100.00	6911.56
	1.5	-3100.00	-3100.00	-3100.00	-3100.00
BPS1, BQS2	0.5	607453.24	579896.76	552340.28	524920.01
	0.8	607453.24	557066.04	495575.05	462267.72
	0.9	569449.93	454816.65	396466.42	380783.47
	1.0	205808.09	219296.75	233231.86	247166.97
	1.1	7882.07	13857.22	83828.19	127460.21
	1.2	7806.86	6935.09	8020.75	17716.35
	1.5	7806.86	6744.11	5681.36	4642.47
BPS2, BQS1	0.5	598547.63	570697.74	542840.78	514987.36
	0.8	599126.68	571223.45	545844.10	523800.86
	0.9	599402.43	572382.63	550126.27	531011.27
	1.0	600374.48	574944.18	553546.12	534279.12
	1.1	599565.32	574302.55	551541.97	531185.46
	1.2	586693.35	564433.07	541120.17	522500.24
	1.5	525813.81	505789.35	487057.84	470764.26
BPS2, BQS2	0.5	607453.24	579896.76	552340.28	524783.81
	0.8	607731.91	579964.87	555174.07	532733.73
	0.9	607885.67	581377.14	558779.81	540482.97
	1.0	608687.95	584217.63	562434.97	542628.89
	1.1	608279.62	582817.55	560813.15	541075.59
	1.2	595761.65	573160.43	551561.64	531555.17
	1.5	537459.18	518583.00	498835.05	481847.68

Based on Table 7.12, we present the following analyses.

Impact of the Bidding Price Strategy. In similar conditions, VaR of BPS1 would be smaller than that of BPS2, which means that bidding at the forecasted market price would be more risky than bidding at marginal cost. This observation contradicts with the popular belief that the best strategy is to bid at the forecasted market price. This conclusion can be explained using Figure 7.24.

Figure 7.24 Bidding Price Illustration

Recall that any sale bids less than the spot market price will be accepted, and any purchase bids larger than the spot market price will be accepted. According to PBUC, the marginal cost of a sale bid will be smaller than the forecasted market price. In Figure 7.24, when the spot market price is extended from marginal cost to forecasted market price, bids at the marginal cost will have a higher chance of being accepted, thus VaR of BPS2 will be larger than that of BPS1.

At BPS1, bidding high (i.e., k_p larger than 1) is more risky than bidding low. This is because higher bids have a smaller chance of being accepted. If the accepted quantity by the market drops, the GENCO's generation cost as well as its revenue will drop because (revenue = market price × accepted quantity). Since (profit = revenue - cost), the GENCO's profit will depend on decrements in revenue and cost. At BPS1, decrement in revenue is larger than decrement in cost, which leads to smaller profit, thus lower VaR and larger risk. At BPS2, bidding high or low will introduce additional risks compared to bidding at exactly the marginal cost, since in both cases, decrement in revenue is larger than decrement in cost.

Impact of Bidding Quantity Strategy. At BQS1, a bilateral contract is guaranteed by local generation, while at BQS2 it is satisfied by purchase from the market when forecasted market price is low. In practice, BQS1 is

GENERATION ASSET VALUATION AND RISK ANALYSIS

regarded as more conservative than BQS2 and, in similar conditions, the VaR of BQS2 is regarded as larger than that of BQS1.

Since PBUC profits for BQS1 ($626,401.88) and BQS2 ($6,350,010.50) are different, to highlight the differences among VaRs, we compare in Table 7.13 the incremental differences of VaR and the PBUC profit for $k_p = 1.0$. The actual value of VaR would be the value in the table plus the corresponding PBUC profit.

In Table 7.13, BQS1 is almost as risky as BQS2, although the purchase bid for BQS2 may not be accepted (i.e., the bidding price would be smaller than the spot market price). This can be explained as follows: The purchase bid can be less profitable (i.e., the spot market price is larger than the forecasted market price but still smaller than bidding price), or the purchase bid can be more profitable (i.e., the spot market price is smaller than the forecasted market price). Since the probability of the spot market price being larger than the forecasted market price is the same as the probability of the spot market price being smaller than the forecasted market price, we expect that the presumed risk does not change for either BQS1 or BQS2.

However, if we penalize the supplier for defaulting on the bilateral contract, the BQS2 risk will increase. In Table 7.14, the VaR for BQS2 decreases, which means a larger risk. In comparison, the BQS1 risk will not change, since bilateral contract is guaranteed by local generation in BQS1. In Table 7.14, the VaR for BQS1 remains unchanged relative to the no-penalty case.

Table 7.13 Difference of VaR and PBUC Profit for $k_p = 1.0$

Bidding Strategy		k_σ			
		0.05	0.10	0.15	0.20
BPS1	BQS1	-431097.59	-417428.02	-402101.90	-388262.61
BPS1	BQS2	-429202.41	-415713.75	-401778.64	-387843.53
BPS2	BQS1	-26027.40	-51457.70	-72855.76	-92122.76
BPS2	BQS2	-26322.55	-50792.87	-72575.53	-92381.61

Table 7.14 Difference of VaR and PBUC Profit for Bidding Strategies (Penalizing Default on Bilateral Contract)

Bidding Strategy		k_σ			
		0.05	0.10	0.15	0.20
BPS1	BQS1	-431097.59	-417428.02	-402101.90	-388262.61
PS1	BQS2	-430777.07	-418366.01	-404537.17	-390708.34
BPS2	BQS1	-26027.40	-51457.70	-72855.76	-92122.76
BPS2	BQS2	-26684.12	-51049.36	-73064.95	-93124.03

Impact of Market Volatility. In a volatile market, bidding based on forecasted market price (BPS1) would increase the VaR and decrease the risk as the market's volatility increases (larger k_σ). This can be explained as follows: Since the simulated market price follows a normal distribution, the probability of a simulated market price being larger than the forecasted market price is the same as the probability of the simulated market price being smaller than the forecasted market price. For the smaller price, the simulated market price will not affect BPS1 because the bid is never accepted. For the larger price, market volatility will force the market price above the bidding price, leading to incrementally higher profit.

In a volatile market, bidding based on marginal cost (BPS2) would result in a lower VaR and larger risk as the market becomes more volatile (larger k_σ). This can be explained as follows: As the market's volatility increases, the chance of a bid not being accepted is greater, which results in smaller profits. On the other hand, a higher price will result in higher profits. For BPS2, the decrement is larger than increment, so the VaR decreases.

7.2.5.3.2 Market Price is Dependent on Bidding Strategy.

In this section, we expand the discussion of Section 7.2.5.2.2 to the case where the bidding price has an impact on the market price. We test k_δ from 0.0 to 1.0 with a step size of 0.1. Experiments show when k_δ is larger than a specific value (k_δ^*), the results would be the same as those from Section 7.5.2.2.2 where market price is independent of bidding strategy. Table 7.16 shows detailed VaR information for $k_\delta = 0.0$, Table 7.17 is for $k_\delta = 0.4$, and Table 7.12 is for k_δ approaching infinity.

When k_δ is 0.0, the bidding price determines the market price, and VaR for bidding high is larger than that for bidding low. When k_δ is larger than zero, the decision on whether to bid high or low would depend on the bidding strategy and the volatility of the market.

Table 7.15 k_δ^* of Different Bidding Strategies

Bidding Strategy		k_δ^*
BPS1	BQS1	0.5
BPS1	BQS2	0.5
BPS2	BQS1	0.8
BPS2	BQS2	>1.0

GENERATION ASSET VALUATION AND RISK ANALYSIS

Table 7.16 VaR in Imperfect Competition ($k_\delta = 0.0$)

Bidding Strategy	k_σ / k_p	0.05	0.10	0.15	0.20
BPS1, BQS1	0.5	-37704.67	-26070.54	-16691.62	-8349.16
	0.8	107865.50	119455.78	131575.55	140543.94
	0.9	152008.33	164155.62	177811.79	189435.23
	1.0	195304.29	208973.86	224299.98	238139.27
	1.1	239234.80	253732.91	271450.44	287549.19
	1.2	282675.85	298491.96	317671.47	337155.15
	1.5	412998.92	435047.48	456081.89	482726.22
BPS1, BQS2	0.5	-9889.52	925.75	9210.32	16860.24
	0.8	124245.32	134844.78	146082.92	156540.81
	0.9	164993.03	175945.97	188624.95	201241.67
	1.0	205808.09	219296.75	233231.86	247166.97
	1.1	246002.04	262186.70	277096.35	293198.83
	1.2	284837.59	303681.28	319204.08	338239.63
	1.5	407437.76	426294.79	447646.05	473361.97
BPS2, BQS1	0.5	-121565.65	-108915.58	-99557.10	-93215.59
	0.8	229668.35	224391.31	216394.47	210953.10
	0.9	344562.61	332280.83	320015.46	310238.90
	1.0	458718.23	440314.92	424021.74	409064.15
	1.1	572804.45	548237.45	527704.90	507889.40
	1.2	686173.79	656637.50	631991.22	607344.87
	1.5	1026770.11	979333.84	942721.48	904921.04
BPS2, BQS2	0.5	-103181.00	-90578.69	-82108.98	-75164.25
	0.8	338093.72	223922.18	215937.95	209760.91
	0.9	446061.99	325830.52	313888.48	302525.23
	1.0	427462.72	427462.72	410840.98	395546.21
	1.1	529837.63	529837.63	508904.72	488475.11
	1.2	630780.25	630780.25	606089.92	581982.61
	1.5	934290.86	934290.86	898359.02	862154.83

Table 7.17 VaR in Imperfect Competition ($k_\delta = 0.4$)

Bidding Strategy	k_σ / k_p	0.05	0.10	0.15	0.20
BPS1, BQS1	0.5	464306.87	439238.79	414170.71	391505.52
	0.8	598547.63	548143.66	489476.47	453337.31
	0.9	561211.15	445033.35	385202.72	372011.18
	1.0	195304.29	208973.86	224299.98	238139.27
	1.1	-3100.00	4237.34	73435.24	117566.67
	1.2	-3100.00	-3100.00	-3100.00	6911.56
	1.5	-3100.00	-3100.00	-3100.00	-3100.00
BPS1, BQS2	0.5	477319.91	452519.07	427718.24	405578.79
	0.8	607453.24	557066.04	495575.05	462267.72
	0.9	569449.93	454816.65	396466.42	380783.47
	1.0	205808.09	219296.75	233231.86	247166.97
	1.1	7882.07	13857.22	83828.19	127460.21
	1.2	7806.86	6935.09	8020.75	17716.35
	1.5	5260.88	4091.85	2965.79	2542.41
BPS2, BQS1	0.5	402007.23	378999.75	355992.26	334208.26
	0.8	603906.91	576080.86	551763.99	529073.38
	0.9	601082.48	574407.57	551609.46	533526.54
	1.0	597951.49	573301.46	550883.42	531503.26
	1.1	593439.20	567745.90	545077.11	525888.24
	1.2	576319.59	553456.75	532390.51	512164.07
	1.5	526645.69	504812.71	483510.34	463515.76
BPS2, BQS2	0.5	402007.23	378999.75	355992.26	334208.26
	0.8	603906.91	576080.86	551763.99	529073.38
	0.9	601082.48	574407.57	551609.46	533526.54
	1.0	597951.49	573301.46	550883.42	531503.26
	1.1	593439.20	567745.90	545077.11	525888.24
	1.2	576319.59	553456.75	532390.51	512164.07
	1.5	526645.69	504812.71	483510.34	463515.76

Table 7.18 lists the best values for k_p, which represent bidding high or bidding low. This table shows that the best k_p for bidding strategies (BPS1, BQS1), (BPS1, BQS2), and (BPS2, BQS1) are not responsive to market volatility (k_σ). However, the best k_p for bidding strategy (BPS2, BQS2) depends on market volatility.

For BPS1, when k_δ is small, the bidding price has a large impact on the market price. In this case, bidding close to (and slightly lower than) the forecasted market price will result in a large VaR. When k_δ is large, which means the bidding price has little impact on the market price, bidding very

GENERATION ASSET VALUATION AND RISK ANALYSIS

low will result in a large VaR. For (BPS2, BQS1), a similar conclusion can be reached. The difference is that we always need to bid high. For (BPS2, BQS2), the best k_p depends on market volatility. A general postulate can be derived as follows: when k_δ is small, bid high; as k_δ becomes larger, bid low; as k_δ becomes even larger, bid exactly at marginal cost.

Table 7.18 Best k_p Values

Bidding Strategy k_σ \ k_δ	BPS1, BQS1	BPS1, BQS2	BPS2, BQS1	BPS2, BQS2			
	All	All	All	0.05	0.1	0.15	0.2
Infinity	0.5	0.5	1.0	1.0	1.0	1.0	1.0
1.0	0.5	0.5	1.0	0.8	1.0	1.0	1.0
0.9	0.5	0.5	1.0	0.8	0.8	1.0	0.9
0.8	0.5	0.5	1.0	0.8	0.5	1.0	0.9
0.7	0.5	0.5	1.0	0.5	0.5	1.0	0.9
0.6	0.5	0.5	1.0	0.8	0.8	1.0	0.9
0.5	0.5	0.5	1.0	0.8	0.8	0.9	0.9
0.4	0.8	0.8	1.0	0.8	0.8	0.8	0.9
0.3	0.8	0.8	1.5	0.9	1.0	1.0	1.0
0.2	0.8	0.8	1.5	1.5	1.5	1.5	1.5
0.1	0.9	0.9	1.5	1.5	1.5	1.5	1.5
0.0	1.5	1.5	1.5	1.5	1.5	1.5	1.5

Note "All" corresponds to k_σ equal to 0.05, 0.1, 0.15 or 0.2.

The difference between (BPS2, BQS2) and (BPS2, BQS1) is that the bilateral contract may be satisfied by a purchase in the market. Since bidding high could increase the market price, in turn leading to a higher purchase price (sometimes even higher than the marginal cost), the probability that a purchase bid is rejected will also increase. So bidding high may not always result in higher profits. In this situation, the best strategy could only be derived from a Monte Carlo simulation.

7.3 GENERATION CAPACITY VALUATION

In generation capacity valuation, the risks are due to uncertainty in the market price and to unit availability. In this section, we first propose a framework for calculating VaR for generation capacity valuation. Then, we present some examples and examine the impact of market price simulation and physical constraints.

7.3.1 Framework of VaR Calculation

Figure 7.25 shows the framework for calculating VaR for generation capacity valuation. This framework is similar to that of Figure 7.2 for short-term generation assets valuation. Both frameworks are used for market price simulations, profit analyses and VaR calculations. However, there are some important differences:

1. In Figure 7.25, the PBUC module supersedes the bids-generating module.
2. In Figure 7.25, the market price simulation occurs prior to PBUC.
3. There is no market settling module in Figure 7.25. The PBUC profit module directly feeds into the profit analysis module.
4. Unit availability must be evaluated, for the long-term operation, in Figure 7.25. In daily practice, the availability of units is known a priori because of the short-term nature of unit availability.
5. Because of its long-term nature, the price forecasting module in Figure 7.25 provides typical price forecasts. However, in the case of generation asset valuation, short-term price forecasting is done daily.

Differences 1, 2, and 3 can be attributed to a settlement procedure. Generation capacity valuation is settled bilaterally, whereas generation asset valuation is settled through the power market. The value of the generation capacity is assessed by running PBUC extensively under different market price scenarios. However, the value of a generation asset is offered as the bid and settled as an accepted bid through the power market. Differences 4 and 5 can be attributed to the differences between long-term and short-term valuation.

7.3.2 An Example

A six-unit system, which is the same as the system used in our discussion of arbitrage in Chapter 5, is used here to study generation capacity valuation. No bilateral contracts are considered in this case. One typical energy market price scenario is shown in Table 7.19, and the profits for individual units are shown in Table 7.20. According to Table 7.20, we can make $116,573.77 profit in one day by using unit 6 to burn fuel and sell electricity. In other words, the value of unit 6 is $116,573.77.

GENERATION ASSET VALUATION AND RISK ANALYSIS

If we drop the typical market price uniformly by -5.0 $/MWh (i.e., in case 1), and increase it by 5.0 $/MWh (i.e., in case 3), or by 10.0 $/MWh (i.e., in case 4), Table 7.21 shows the profit for different cases where the original case is case 2. In Table 7.21, where the valuation of a unit depends on market price, the bidding decision may be according to an average quantity of valuation.

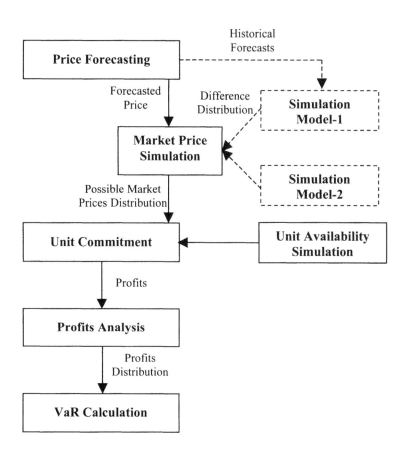

Figure 7.25 VaR Calculation for Generation Capacity Valuation

Table 7.19 Typical Energy Market Prices

Hour	1	2	3	4	5	6	7	8	9	10	11	12
Energy price ($/MWh)	16.74	12.70	6.99	5.00	8.00	19.08	18.98	21.29	23.60	26.00	29.25	29.40
Hour	13	14	15	16	17	18	19	20	21	22	23	24
Energy price ($/MWh)	27.01	28.32	29.00	34.34	33.12	29.75	29.51	26.45	21.45	13.59	11.00	5.50

Table 7.20 Unit Profits Corresponding to Typical Energy Market Prices

Unit	1	2	3	4	5	6	System
Profit ($)	274.75	1071.95	14382.56	45919.82	84180.8	116573.77	262403.66

Table 7.21 Unit Profits Corresponding to Different Typical Energy Market Prices

Unit Case	1	2	3	4	5	6	System
1	0.00	107.36	5551.41	27519.12	54418.59	77732.90	165329.38
2	274.75	1071.95	14382.56	45919.82	84180.80	116573.77	262403.66
3	1817.12	3693.68	25202.17	66395.67	121544.32	162176.39	380829.38
4	4736.74	7202.41	38247.98	90012.37	163137.48	210176.36	513513.38
Average	1707.15	3018.85	20846.03	57461.75	105820.30	141664.86	330518.95

7.3.3 Sensitivity Analysis

In this section, we use the typical example presented in the previous section to conduct additional analyses in examining the sensitivity of capacity valuation to market prices and physical constraints.

7.3.3.1 Impact of Market Price Simulation

In this section, we consider the impact of market price on generation capacity valuation. For the purpose of illustration, we assume that the market price follows a normal distribution with the typical market price as its mean (μ) and the deviation (σ) being $k_\sigma \mu$, where k_σ is selected as 5%, 10%, 15%, or 20%. Table 7.22 shows the results, where the mean value is perceived as a unit's expected profit and VaR considers the presumed risk.

GENERATION ASSET VALUATION AND RISK ANALYSIS

Table 7.22 Impact of Market Price Simulation

k_σ	Unit	1	2	3	4	5	6	System
0.05	Mean ($)	309.64	1133.39	14376.25	45922.88	84142.39	116507.18	262391.82
	VaR ($)	114.25	821.27	13172.06	44107.22	80932.45	112858.14	252274.84
0.1	Mean ($)	475.29	1337.13	14393.08	45895.19	84142.17	116442.59	262685.46
	VaR ($)	108.62	748.77	12015.56	42253.55	77716.30	109120.65	242570.75
0.15	Mean ($)	714.67	1603.13	14503.86	45892.58	84219.51	116449.84	263383.60
	VaR ($)	164.87	784.59	10885.30	40482.02	74785.09	105352.77	233136.66
0.2	Mean ($)	994.02	1909.57	14718.60	45939.84	84373.26	116481.22	264416.52
	VaR ($)	256.75	853.91	9923.80	38722.16	72116.08	101619.90	224519.58

In Table 7.22, the mean values indicate that the market price volatility has the largest impact on unit 1 and a moderate impact on unit 2, while it has not much effect on other units. This can be explained as follows: In the base case, units 1 and 2 are often shut down because of their high marginal costs. Then, in the hours when the units are OFF, a drop in market price will not impact units 1 and 2 much. However, in those hours, if the market price is high, units 1 and 2 would be utilized and their value would be increased. For other units, since market price volatility does not have much effect on the commitment, its impact on the valuation is not so significant.

In terms of VaR, its value will be smaller for units 3 to 6 in a more volatile market. However, no simple relationship exists between the volatility of the market price and the VaR for units 1 and 2. VaR for units 1 and 2 is quite different from its mean value, while VaR values are more comparable to the mean values for units 3 to 6, in particular, VaR and mean value are very close for unit 6. Consequently, the risk of units 1 and 2 due to market price volatility is greater than that of units 3 to 6.

7.3.3.2 Impact of Physical Constraints

In this section, we study the effect of physical constraints on generation capacity valuation. We assume that k_σ is 0.1.

7.3.3.2.1 Impact of Force Outage Rates (FOR). The availability of a unit affects its capacity valuation; that is to say, if a unit is unavailable, its profit will be zero. For example, Table 7.23 shows the impact of FOR on unit 6.

According to Table 7.23, with an increase in FOR, both mean and VaR will decrease. In the 95% confidence case, VaR becomes zero when

FOR is larger than 5%. This shows that VaR may be zero for a high confidence level if FOR is large (i.e., larger than 100% - confidence level). In this case, it would be essential to consider a lower confidence level such as 90% for VaR.

Note that when FOR is larger than 0.07, the VaR for 90% confidence will be larger than the mean value. This is due to the zero profit effect when the unit is unavailable. This example points to the importance of FOR in generation capacity valuation.

7.3.3.2.2 Impact of Ramp Rates. A ramp rate represents the response capability of a unit. With low ramp rates, a unit will have to be started earlier to take advantage of higher market price periods, meanwhile it may be losing revenues at low market price periods. Table 7.24 shows the impact of ramp rates on unit 6. Here both mean and VaR increase with higher ramp rates.

Table 7.23 Impact of FOR on Valuation

FOR	Mean ($)	VaR (95%) ($)	VaR (90%) ($)
0	116442.59	109120.65	110832.63
0.01	115284.07	108827.83	110551.38
0.02	114139.77	108230.48	110289.98
0.03	112992.06	107785.6	110060.49
0.04	111854.63	107200.64	109811.18
0.05	110746.49	104180.11	109288.78
0.06	109674.48	0	108932.81
0.07	108596.06	0	108426.52
0.08	107503.33	0	107903.03
0.09	106451.64	0	107376.37
0.1	105345.02	0	105725.45

Table 7.24 Impact of Ramp Rate on Valuation

Ramp Rate (MW/min)	Mean ($)	VaR (95%) ($)
3.33	116442.59	109120.65
3.0	115958.94	108636.84
2.5	115713.52	108355.19
2.0	114912.45	107568.12
1.5	114363.26	107019.03
1.0	114313.91	106967.39

7.3.3.2.3 Impact of Minimum ON/OFF Times.
The minimum ON/OFF times can affect a unit's response and thus its valuation. Table 7.25 shows the effect of minimum ON/OFF times. Since unit 6 is always running given the typical market price, unit 2 is considered in this example. According to Table 7.25, with an increase in the minimum ON/OFF times, the mean and VaR values drop.

7.3.3.2.4 Impact of Start-up Cost.
The start-up cost may affect the commitment status of a unit and its valuation. Table 7.26 shows the impact of start-up costs as unit 2 is considered. According to Table 7.26, with an increase in start-up cost, both the mean and VaR values drop.

Table 7.25 Impact of Minimum ON/OFF Times on Valuation

Min ON Time (Hour)	Min OFF Time (Hour)	Mean ($)	VaR (95%) ($)
1	-1	1337.13	748.77
2	-1	1302.30	705.45
3	-1	1266.84	676.08
4	-1	1238.58	632.20
5	-1	1213.78	601.91
3	-3	1229.92	631.59

Table 7.26 Impact of Start-up Cost on Valuation

Start-up Cost ($)	Mean ($)	VaR (95%) ($)
0	1337.13	748.77
50	1229.83	632.32
100	1143.21	547.09
150	1075.22	476.27
200	1019.02	414.54
500	716.75	109.89
1000	280.54	0.00
10000	0.00	0.00

7.4 CONCLUSIONS

In this chapter, generation asset valuation (short term) and generation capacity valuation (long term) are discussed. To consider the impact of risk, the concept of VaR is utilized.

The framework to calculate VaR in short-term generation asset valuation is described in detail, and examples are presented to illustrate the framework. The framework proposed in this chapter is general. For a real system, the market structure, market settlement rules, and market changes may be different from those described in this chapter. A framework is proposed to calculate VaR in generation capacity valuation, and the impact of market price simulation and physical constraints are discussed.

A market price simulation is the core of the proposed framework, since it reflects the market changes that are the main source of risk. Two possible ways to simulate market changes are proposed: one is based on historical information (i.e., statistical simulation model-1) and the other is based on a specific probability distribution (i.e., statistical simulation model-2). The normal distribution is extensively used in this chapter, though other distributions may also be used without any difficulty. The VaR values due to different market prices and bidding strategies are compared.

Chapter 8

Security–Constrained Unit Commitment

Impact of Reliability Constraints

8.1 INTRODUCTION

The objective of security-constrained unit commitment (SCUC) discussed in this chapter is to obtain a unit commitment schedule at minimum production cost without compromising the system reliability. The reliability of a system is interpreted as satisfying two functions: adequacy and security. An adequate amount of capacity resources must be available to meet the peak demand (adequacy), and the system must be able to withstand changes or contingencies on a daily and hourly basis (security).

In several power markets, the ISO plans the day-ahead schedule using SCUC. The traditional unit commitment algorithm determines the unit schedules to minimize the operating costs and satisfy the prevailing constraints such as load balance, system spinning reserve, ramp rate limits, fuel constraints, multiple emission requirements and minimum up and down time limits over a set of time periods. The scheduled units supply the load demands and possibly maintain transmission flows and bus voltages within their permissible limits. However, in circumstances where most of the committed units are located in one region of the system, it becomes more difficult to satisfy network constraints throughout the system. As the system becomes more congested, the ISO would consider the alternative of incorporating the network flow constraints in the unit commitment formulation (i.e., SCUC) to minimize the violation and the related costs of the normal operation of the system.

SCUC decomposes the scheduling formulation into a master problem and a subproblem based on the Benders decomposition. The master problem involves calculating unit commitment, by augmented Lagrangian relaxation, using the prevailing constraints but omitting the network constraints. Given a certain unit commitment schedule, the subproblem performs one of the following tasks:

1. Minimizes the network violations. For a linearized network, such as the one used in this study, the subproblem is decoupled into two smaller subproblems corresponding to transmission and voltage constraints. The transmission subproblem seeks to minimize transmission flow violations for the steady state and n-1 contingencies by unit generation and phase shifter adjustments. The reactive subproblem examines the voltage constraints and minimizes the violations by reactive power and tap changer adjustments.

2. Minimizes the expected unserved energy (EUE). The subproblem takes into consideration the forced outage rates of committed generating units and in-service transmission lines, and correspondingly adjusts the available control facilities to minimize EUE.

In both tasks, Benders cuts are generated if any violation is detected in the subproblems. With Benders cuts, the unit commitment in the master problem is solved iteratively to provide a minimum cost generation schedule while satisfying all constraints. Since a decomposed problem is easier to solve and requires less complicated and smaller computing capabilities, the SCUC solution is more accurate and results are obtained faster.

This chapter is organized as follows. In Section 8.2, we present the problem formulation for SCUC, where a new model on ramping constraint is discussed in detail. In Section 8.3, we discuss how to apply a Benders decomposition to solve the SCUC problem. The master problem is also presented in Section 8.3. Tasks 1 and 2 are discussed in Section 8.4 and Section 8.5, respectively. In Section 8.4, SCUC is used to minimize a network violation (Task 1), and the discussion includes transmission flow constraints and voltage constraints. In Section 8.5, SCUC is used to minimize EUE (Task 2) and this includes transmission flow constraints and EUE limits. Section 8.6 concludes the chapter.

8.2 SCUC PROBLEM FORMULATION

SCUC is treated as an optimization problem that minimizes perceived operating costs based on the incremental costs submitted by generating

SECURITY–CONSTRAINED UNIT COMMITMENT

units. The list of symbols is given in Appendix A. The objective function is given as

$$\min \sum_{i=1}^{N_g} \sum_{t=1}^{N_t} [C_i(P(i,t))I(i,t) + S(i,t)] \qquad (8.0)$$

The first term inside the bracket is the production cost $C_i(P_i(t))$, which is calculated as the product of the heat rate (MBTU/h) and the unit's fuel cost ($/MBTU). The second term represents the start-up cost of the units which depends on the length of time that the unit has been off. The start-up cost is defined as

$$S(i,t) = I(i,t)[1 - I(i,t-1)]\left[\alpha_i + \beta_i\left(1 - \exp\frac{-X^{off}(i,t)}{\tau_i}\right)\right]$$

The prevailing constraints are as follows:

System Real power Balance

$$\sum_{i=1}^{N_g} P(i,t)I(i,t) = P_D(t) \qquad t = 1,\ldots,N_t \qquad (8.1)$$

System Spinning Reserve Requirements

$$\sum_{i=1}^{N_g} r_s(i,t)I(i,t) \geq R_s(t) \qquad t = 1,\ldots,N_t \qquad (8.2)$$

where, $r_s(i,t) = \min\{10 \times MSR(i), P_{gmax}(i,t) - P(i,t)\}$.

The spinning reserve requirement, $R_s(t)$, is typically defined as a base component plus a fraction of the load requirement and a fraction of the high operating limit of the largest on-line unit.

System Operating Reserve Requirements

$$\sum_{i=1}^{N_g} r_o(i,t)I(i,t) \geq R_o(t) \qquad t = 1,\ldots,N_t \qquad (8.3)$$

where, $r_o(i,t) = \begin{cases} q(i), & \text{if unit } i \text{ is OFF} \\ r_s(i,t), & \text{if unit } i \text{ is ON} \end{cases}$

Interruptible loads are usually added to the operating reserve capacity of units (left-hand side of (8.3)). The operating reserve requirement, $R_o(t)$, is commonly defined similar to $R_s(t)$ as a function of a base component plus a fraction of the load requirement and a fraction of the high operating limit of the largest on-line unit.

Unit Generation Limits

$$P_{gmin}(i) \leq P(i,t) \leq P_{gmax}(i) \quad i=1,...,N_g \quad t=1,...,N_t \tag{8.4}$$

Thermal Unit Minimum Starting Up/Down Times

$$[X^{on}(i,t)\text{-}T^{on}(i)] \times [I(i,t\text{-}1)\text{-}I(i,t)] \geq 0 \tag{8.5}$$

$$[X^{off}(i,t\text{-}1)\text{-}T^{off}(i,t)] \times [I(i,t\text{-}1)\text{-}I(i,t)] \geq 0 \tag{8.6}$$

Ramping Constraints

$$P(i,t)\text{-}P(i,t\text{-}1) \leq UR(i) \quad \text{as unit } i \text{ ramps up} \tag{8.7}$$

$$P(i,t\text{-}1)\text{-}P(i,t) \leq DR(i) \quad \text{as unit } i \text{ ramps down} \tag{8.8}$$

A more general model on ramping constraints is discussed at the end of this section.

Fuel Constraints (for all fuel types)

$$F^{min}(FT) \leq \sum_{t=1}^{N_t} \sum_{i \in FT} C_{fi}(P(i,t)I(i,t)) + S_f(i,t) \leq F^{max}(FT) \tag{8.9}$$

where $i \in FT$ corresponds to all fuel-constrained units burning fuel FT

System Emission Limit

$$\sum_{t=1}^{N_t} \sum_{i=1}^{N_g} C_{ei}(P(i,t))I(i,t) + S_e(i,t) \leq EMS \tag{8.10}$$

SECURITY–CONSTRAINED UNIT COMMITMENT

where several emission types (e.g. SO_2, NO_X) are considered.

Regional Emission Limit

$$\sum_{t=1}^{N_t}\sum_{i\in A} C_{ei}(P(i,t))I(i,t) + S_e(i,t) \leq EMA \tag{8.11}$$

where $i \in A$ corresponds to all units in the constrained emission region A and several emission types can be considered.

Transmission Flow Limit from Bus k to Bus m

$$-P_{km}^{max} \leq P_{km}(t) = f(P(t), \varphi(t)) \leq P_{km}^{max} \qquad t = 1,...,N_t \tag{8.12}$$

where $P(t)$ is real power generation vector and $\varphi(t)$ is phase shifter control vector at time t.

Reactive Power Operating Reserve Requirement

$$\sum_{i=1}^{N_g} Q_{gmax}(i)I(i,t) \geq Q_D(t) \qquad t = 1,...,N_t \tag{8.13}$$

Reactive Power Generation Limits and Load Bus Balance

$$Q_G^{min} I(t) \leq Q_G(t) = F_1(V) \leq Q_G^{max} I(t) \qquad t = 1,...,N_t \tag{8.14}$$

$$Q_L(t) = F_2(V) \qquad t = 1,...,N_t \tag{8.15}$$

where $I(t)$ is unit commitment status vector at time t.

System Voltage and Transformer Tap Limits

$$V^{min} \leq V \leq V^{max} \tag{8.16}$$

$$T^{min} \leq T \leq T^{max} \tag{8.17}$$

Expected Unserved Energy (EUE) Limits

$$E\left\{\sum_{j=1}^{N_h} r_{jt}\right\} \leq \varepsilon_t \qquad t = 1,...,N_t \tag{8.18}$$

Formulations (8.1–8.11) are real power generation constraints. (8.12) is the transmission flow constraint. The transmission flow depends on unit generation and phase shifter controls. Equations (8.13–8.17) are reactive power generation and voltage constraints. Equation (8.18) represents the reliability constraint. It points out that EUE should be kept below a certain limit to maintain the reliability of the system. For the first task discussed before, (8.18) will not be considered.

8.2.1 Discussion on Ramping Constraints

In the restructured power market, ramping constraint deserves more attention since it relates to ancillary services (reserves) trading. Here, we study a more general ramping model by considering the ramping as a function of unit loading. The ramping constraint may consist of many segments, each segment corresponding to a unit loading level. Within each loading level, the ramp rate is assumed to be fixed. However, at different loading levels, the ramp rates could differ. If the unit is ramping-up/down from one loading level to the other, it could remain in the original loading level for some time, which is referred to as ramping delay. The traditional model for ramping is considered as a special case of the above; that is the ramp rates are fixed at all loading levels and the ramping delay is not considered.

Consider the example in Figure 8.1a for ramping. There are four segments corresponding to four loading levels. The first ramping is from 20 to 40 MW with a ramp rate of 1 MW/min. The second ramping is from 40 to 70 MW, and the ramp rate is 2 MW/min. Suppose that the unit is ramping-up from the first loading level (40 MW) to the second loading level (70 MW). After reaching 40 MW, the unit would remain at 40 MW for at least 10 minutes before ascending to the second loading level. So, in Figure 8.1a, at the first loading level, (1,10) refers to a ramp rate of 1 MW/min and a ramping delay of 10 minutes. Other loading levels in Figure 8.1a could be similarly analyzed. Note that ramping from 70 to 90 MW is not delayed, which means that it is not necessary to remain at the 90 MW loading before resuming the ramping process.

We calculate the relationship between ramping and the loading level, as shown in Figure 8.1b. Suppose the process is initiated at $t = 0$. At the 20 MW level, with a fixed ramp up rate of 1 MW/min, it takes 20 minutes to ramp up from 20 MW to 40 MW. Here, the unit remains for 10 minutes at 40 MW level before ramping to 70 MW. So, starting from 20 MW it takes 30 minutes (i.e., 20 +10) before this unit initiates the second ramping process. Then, in the second ramping process (from 40 to 70

SECURITY–CONSTRAINED UNIT COMMITMENT

MW) with a fixed ramp-up rate of 2 MW/min, it takes another 15 minutes to reach 70 MW. The rest of the process can be analyzed similarly.

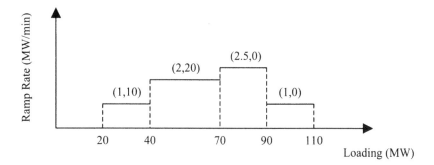

Figure 8.1a Ramp-up Rate versus Loading

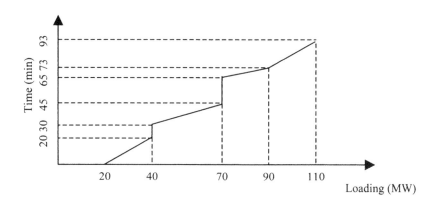

Figure 8.1b Ramping-up Process versus Loading

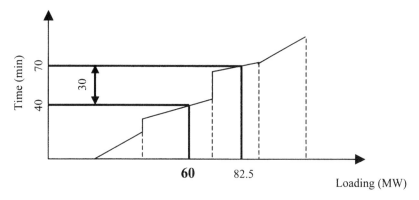

Figure 8.1c Example of Ramping-up Process

According to Figure 8.1b, we calculate the necessary time to ramp up from one loading level to another. For instance, in Figure 8.1c, suppose that the initial loading is 60 MW and we are interested in determining the maximum loading capability after 30 minutes. In this figure, the corresponding loading will be 82.5 MW.

The ramping-down process can be similarly analyzed. Figures 8.2a to 8.2c depict a ramping-down process. In Figure 8.2a, the loading levels and the corresponding ramp-down rates are the same as those in Figure 8.1a. The only difference is that the ramping delay relates to ramping-down.

According to Figure 8.2b, we calculate the necessary time to ramp down from one loading level to another. An example is shown in Figure 8.2c in which the process starts at the 60 MW loading level and takes 30 minutes to reach the 30 MW loading.

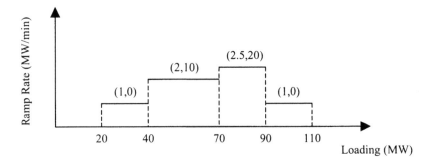

Figure 8.2a Ramp-down Rate versus Loading

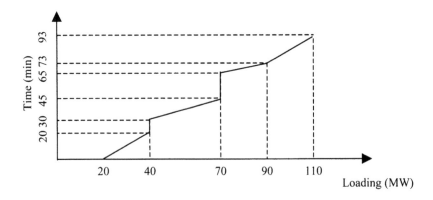

Figure 8.2b Ramping-Down Process versus Loading

SECURITY–CONSTRAINED UNIT COMMITMENT

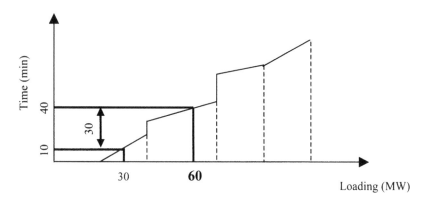

Figure 8.2c Example of Ramping-Down Process

Generally, for each unit the user can specify the number of ramping-up/down segments (loading levels). Then for each segment, the user can specify the lower limit and upper limit of the loading level, the corresponding ramp rate, and the ramping delay if necessary.

Table 8.1 shows the input data for the ramping process. Note that each unit could have different number of segments for ramping. Also, the number of ramping-up segments could be different from the number of ramping-down segments for each unit.

Table 8.1 User Input Data for Ramping

Unit Index	Ramping Process	Segment Index	Loading (MW)	Ramp Rate (MW/min)	Ramping Delay (min)
. .	. .	. .	. .	. .	. .
i	Ramping -up	1 2 . . S_{ui}	(P_{ui1}, P_{ui2}) (P_{ui2}, P_{ui3}) . . (P_{usi}, P_{usi+1})	R_{ui1} R_{ui2} . . R_{usi}	T_{ui1} T_{ui2} . . 0
	Ramping -down	1 2 . . S_{di}	(P_{di1}, P_{di2}) (P_{di2}, P_{di3}) . . (P_{dsi}, P_{dsi+1})	R_{di1} R_{di2} . . R_{dsi}	0 T_{di2} . . T_{dsi}
. .	. .	. .	. .	. .	. .

In Table 8.1, the variables are defined as follows:

S_{ui} = number of loading levels of unit i (ramping-up process)

P_{uik} = lower limit of the kth loading level of unit i (ramping-up process)

P_{uik+1} = upper limit of the kth loading level of unit i (ramping-up process)

R_{uik} = ramp-up rate of the kth loading level of unit i

T_{uik} = ramp-up delay time of unit i from the kth loading level to the $k+1^{th}$ loading level

S_{di} = number of loading levels of unit i (ramping-down process)

P_{dik} = lower limit of the kth loading level of unit i (ramping-down process)

P_{dik+1} = upper limit of the kth loading level of unit i (ramping-down process)

R_{dik} = ramp-down rate of the kth loading level of unit i

T_{dik} = ramp-down delay time of unit i from the kth loading level to the $k-1^{th}$ loading level

Corresponding to Table 8.1, we derive the ramping functions as,

$$t_u = f_u(L) \text{ for ramping-up}$$

$$t_d = f_d(L) \text{ for ramping-down}$$

Where, L is loading, and t_u and t_d are the corresponding time for ramping-up and down processes, respectively.

The inverse ramping process functions are

$$L = g_u(t_u) \text{ for ramping-up process}$$

$$L = g_d(t_d) \text{ for ramping-down process}$$

Suppose that the loading at time period t_1 is L_1 and we are interested in calculating maximum and minimum loading at the next time period t_2. First, we find the initial time point corresponding to L_1.

$$t_{u1} = f_u(L_1) \text{ for ramping-up process}$$

$$t_{d1} = f_d(L_1) \text{ for ramping-down process}$$

Next we find the time corresponding to t_2:

$$t_{u2} = t_{u1} + (t_2 - t_1) \text{ for ramping-up process}$$

$$t_{d2} = t_{d1} - (t_2 - t_1) \text{ for ramping-down process}$$

Based on the inverse ramping process functions, we calculate the maximum and minimum loading at t_2.

$$L_{u2} = g_u(t_{u2})$$

$$L_{d2} = g_d(t_{d2})$$

where, L_{u2} is the maximum loading and L_{d2} is the minimum loading.

In summary

$$L_{u2} = g_u(f_u(L_1) + (t_2 - t_1))$$

$$L_{d2} = g_d(f_d(L_1) - (t_2 - t_1))$$

The ramping constraint between t_2 and t_1 could be represented by

$$L_{d2} \leq L_2 \leq L_{u2}$$

where L_2 is the loading at t_2. Or

$$g_d(f_d(L_1) - (t_2 - t_1)) \leq L_2 \leq g_u(f_u(L_1) + (t_2 - t_1))$$

8.3 BENDERS DECOMPOSITION SOLUTION OF SCUC

In the SCUC solution process, if all transmission constraints were relaxed in the Lagrangian objective function, the problem would become vastly complicated and impossible to solve because of the large number of Lagrangian multipliers. Since the exact level of unit generation $P(i,t)$ is unknown, it is difficult to consider unit generation and phase shifter controls in the unit commitment. The Benders decomposition approach is a good way to solve the problem. The decomposition in Figure 8.3 takes into account network control actions that minimize congestion or maximize the reliability for transmission-constrained unit commitment. A brief discussion on Benders decomposition is presented next.

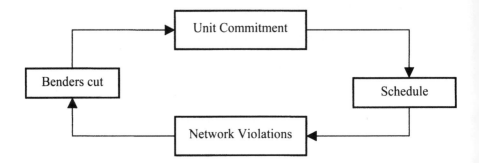

Figure 8.3 SCUC Problem with Network Constraints

8.3.1 Benders Decomposition

A standard form of Benders formulation is

$$\begin{aligned} & \min \ \boldsymbol{ux} \\ & \text{s.t.} \ \ \boldsymbol{Ax} \geq \boldsymbol{b} \\ & \ \ \ \ \ \ \boldsymbol{Ex} + \boldsymbol{Fy} \geq \boldsymbol{h} \end{aligned} \quad (8.19)$$

Using Benders decomposition, the formulation above can be decomposed into a master problem and a subproblem, which is solved as follows:

1. In the master problem, the unit commitment state x is calculated as

$$\begin{aligned} & \min \ \boldsymbol{ux} \\ & \text{s.t.} \ \ \boldsymbol{Ax} \geq \boldsymbol{b} \\ & \ \ \ \ \ \ w(x) \leq 0 \end{aligned} \quad (8.20)$$

where $w(x)$ is the cut that provides the information regarding the feasibility of the unit commitment state x in terms of transmission security and voltage constrains.

2. Given $\hat{x}$, the subproblem is formulated as

$$\begin{aligned} & \min \ w(\hat{x}) = \boldsymbol{dy} \\ & \text{s.t.} \ \ \boldsymbol{Fy} \geq \boldsymbol{h} - \boldsymbol{E}\hat{x} \end{aligned} \quad (8.21)$$

SECURITY–CONSTRAINED UNIT COMMITMENT

If the objective function $w(\hat{x})$ is larger than zero, we produce the Benders cuts $w(x) \leq 0$ once a violation is detected in the subproblem. A linear approximation of the Benders cut may be generated based on the subproblem results in which the coefficients of the linear approximation are the Simplex multipliers π_i associated with constraints in (8.19). The linear form of a Benders cut is

$$w(x) = w(\hat{x}) + \pi(x - \hat{x}) \leq 0 \qquad (8.22)$$

where $w(\hat{x})$ = optimal solution of (8.21)

$\hat{x}$ = solution for the master problem

π = simplex multiplier vector

$\pi_i = \dfrac{\partial w}{\partial x_i}$ = simplex multiplier in linear programming

8.3.2 Application of Benders Decomposition to SCUC

The SCUC problem can be rewritten as a standard Benders formulation. Corresponding to SCUC, in (8.19), x represents the unit commitment states and y represents the penalty variables for satisfying unit generation control, phase shifter, and tap transformer preventive controls. The composite formulation represents minimizing the operating cost subject to two sets of constraints. The first set of inequalities represents generating unit constraints given in (8.1–8.11) and the second set represents transmission flow, voltage and EUE constraints given in (8.12–8.18).

Applying Benders decomposition, one may solve SCUC in two stages. First, we solve the unit commitment problem for eliminating network violations or minimizing EUE without regard to the transmission flow, voltage, and EUE constraints (master problem, 8.1–8.11). Second, taking the unit commitment schedule of the first stage, we solve one of the two subproblems (Task 1 or 2) using (8.12–8.18). We create Benders cuts again for solving the master problem if any violation is detected at this stage.

8.3.3 Master Problem Formulation

The master problem is the traditional unit commitment problem without considering transmission flow, voltage, and EUE constraints. At present, one of the potential approaches for solving the unit commitment problem is the Lagrangian relaxation (LR) method. The basic idea of LR is to relax system constraints in the objective function by using Lagrangian multipliers. The relaxed problem is then decomposed into N subproblems for each unit. The dynamic programming process is used to search the optimal commitment for a single unit. The Lagrangian multipliers are updated based on violations of system constraints. The convergence criterion is satisfied if the duality gap in Lagrangian relaxation is within a given limit.

The quality of the final LR solution depends on the sensitivity of the commitment to Lagrangian multipliers. Slow and unsteady convergence of LR has always been a problem in finding the global optimum solution as reported in most unit commitment solutions. Unless a proper modification of multipliers is ensured in every iteration, unnecessary commitment of the generating units may occur, which may result in higher production costs. The difficulties are often explained by the non-convexity of the optimization problem. The non-convexity can be overcome by the augmented Lagrangian method in which quadratic penalty terms associated with power demand are added to the objective function to improve the convexity of the problem. The augmented Lagrangian function for unit commitment is given as follows:

$$\Gamma(P(i,t), I(i,t), \lambda(t), \mu_s(t), \mu_o(t), \lambda_f, \lambda_s, \lambda_a)$$

$$= \sum_{i=1}^{N_g} \sum_{t=1}^{N_t} C_i(P(i,t))I(i,t) + S(i,t)$$

$$- \sum_{t=1}^{N_t} \lambda(t) \sum_{i=1}^{N_g} P(i,t)I(i,t) - \frac{c}{2} \sum_{t=1}^{N_t} (\sum_{i=1}^{N_g} P(i,t)I(i,t) - P_D(t))^2$$

$$- \sum_{t=1}^{N_t} \mu_s(t) \sum_{i=1}^{N_g} r_s(i,t)I(i,t) - \sum_{t=1}^{N_t} \mu_o(t) \sum_{i=1}^{N_g} r_o(i,t)I(i,t)$$

$$+ \lambda_f \sum_{t=1}^{N_t} \sum_{i \in F} C_{fi}(P_i(t))I(i,t) + S_f(i,t)$$

$$+ \lambda_s \sum_{t=1}^{N_t} \sum_{i=1}^{N_g} C_{ei}(P(i,t))I(i,t) + S_e(i,t)$$

$$+ \lambda_a \sum_{t=1}^{N_t} \sum_{i \in A} C_{ei}(P(i,t))I(i,t) + S_e(i,t) \qquad (8.23)$$

SECURITY-CONSTRAINED UNIT COMMITMENT

In (8.23), the terms that are independent of the decision variable have been dropped. Obviously, the quadratic term in the equation is not separable and may disrupt the decomposability of the problem. However, by applying the decomposition and coordination technique, the non-separable quadratic penalty terms are linearized around the solution obtained from the previous iteration. Moreover, in order to improve the convergence property, we impose quadratic terms of decision variables in (8.23) in order to limit them from deviating too far from the last computed solution.

These quadratic terms are separable, convex, and differentiable with respect to $P(i,t)$ [Abd96]. It should be emphasized that these added terms do not affect the optimality since they vanish at the optimal solution. In this regard, the optimization problem in $k+1^{th}$ iteration can be decoupled into N_g subproblems, each corresponding to the optimal unit commitment of individual units over the entire study period expressed as

$$\min \ \Gamma(P(i,t), I(i,t), \hat{\lambda}(t), \hat{\mu}_s(t), \hat{\mu}_o(t), \hat{\lambda}_f, \hat{\lambda}_s, \hat{\lambda}_a)$$

$$= \sum_{t=1}^{N_t} \begin{cases} C_i(P(i,t)) - \hat{\lambda}(t)P(i,t) - \hat{\mu}_s(t)r_s(i,t) - \hat{\mu}_o(t)r_o(i,t) \\ -c \cdot P(i,t)[\sum_{i=1}^{N_g} P(i,t)^{(k)} I(i,t)^{(k)} - P_D(t) \\ + \hat{\lambda}_f C_{fi}(P(i,t)) + \hat{\lambda}_s C_{ei}(P(i,t)) + \hat{\lambda}_a C_{ei}(P(i,t)) \end{cases} I(i,t)$$

$$+ \sum_{t=1}^{N_t} \left\{ S(i,t) + \hat{\lambda}_f S_f(i,t) + \hat{\lambda}_s S_e(i,t) + \hat{\lambda}_a S_e(i,t) \right\}$$

$$+ \sum_{t=1}^{N_t} \left\{ c \cdot P_D(t)P(i,t)^{(k)} I(i,t)^{(k)} + \frac{1}{2\varepsilon}[I(i,t)P(i,t) - P(i,t)^{(k)} I(i,t)^{(k)}]^2 \right\}$$

(8.24)

subject to unit constraints (8.4–8.8).

In (8.24), the variables given with ^ are known. At this stage, $P(i,t)$ is set equal to the unit power generation corresponding to the Lagrangian multiplier $\lambda(t)$. Consequently, the integer variables $I(i,t)$, $t=1,...,N_t$ are the only unknown factors. We adopt a dynamic programming model to find the suitable values of $I(i,t)$ that minimize (8.24). The detail procedure of augmented Lagrangian relaxation for unit commitment is described in [Wan95].

Once the units are committed, economic dispatch will be solved as a continuous variable optimization process subject to a set of system (coupling) and unit constraints, such as system load balances (8.1), unit generation limits (8.4) and ramping limits (8.7–8.8). A piecewise linear fuel cost function is adopted in our economic dispatch, so we may use an LP approach to minimize the objective function. There is a possibility that we will encounter ramping violations at some hours, if we start the optimization process from $t = 1$, especially at those hours with maximum system demand or those hours with small difference between committed capacity and system demand. Therefore, we initiate the economic dispatch at the hour that corresponds to the maximum system demand. At that hour, the objective function is minimized without ramp rate constraints. Then we consider ramping and proceed forward and backward to optimize the generation schedule at other time intervals. In some cases when ramping violation is encountered, we re-initiate the scheduling process at the hour with smallest difference between the committed capacity and system demand. Then the scheduling process proceeds forward and backward to include other time intervals.

8.4 SCUC TO MINIMIZE NETWORK VIOLATION

In Task 1, the security constraints that we would consider include transmission flow constraints and voltage constraints (8.12–8.17). In the subproblem, we would try to minimize violation of these constraints. Both steady state cases and contingency cases would be considered.

8.4.1 Linearization of Network Constraints

For transmission flow constraints, since we are primarily interested in screening transmission violations in the subproblem, we could use a dc load flow to expedite the process. Accordingly, we use linear sensitivity factors (LSFs) to formulate transmission constraints. The constraints are given by:

1. Steady state constraints (no contingencies)

$$-P_{km}^{max} \leq P_{km}(t) = \sum_{i=1}^{M} A_{km}^{i} \left(P(i,t) + P_{\phi}(i,t) - P_{d}(i,t)\right) \leq P_{km}^{max} \quad (8.25)$$

2. Contingency constraints

$$-P_{km}^{max} \le P_{km}(j,t) = \sum_{i=1}^{M} E_{km}^{i}(j)\,(P(i,t) + P_{\phi}(i,t) - P_d(i,t)) \le P_{km}^{max} \quad (8.26)$$

$$j = 1, 2, ..., nc$$

M is the number of buses. $P(i,t)$, $P_{\phi}(i,t)$, and $P_d(i,t)$ are generation, load demand, and equivalent power injection from phase shifter, respectively, at bus i at time t. A_{km}^{i} and $E_{km}^{i}(j)$ are LSFs. A_{km}^{i} represents the sensitivity of the flow on line k–m to generation at bus i. $E_{km}^{i}(j)$ represents the sensitivity of the flow on line k–m to the generation at bus i due to the outage of line j. The LSFs may be computed as follows.

In the dc load flow, assuming that bus number 1 is the slack bus, we have

$$[P] = [B][\delta] \quad (8.27)$$

Therefore, the angles of the system are:

$$[\delta] = [X][P] \quad (8.28)$$

where

$$[X] = [B]^{-1}$$

For each angle, we have

$$\delta_j = \sum_{i=2}^{M} X_{ji} P_i \quad j = 2, 3, ..., M \quad (8.29)$$

For dc load flow on line k–m

$$\begin{aligned}
P_{km} &= (\delta_k - \delta_m)/x_{km} \\
&= \frac{1}{x_{km}} \left(\sum_{i=2}^{M} X_{ki} P_i - \sum_{i=2}^{M} X_{mi} P_i \right) \\
&= \sum_{i=2}^{M} \left(\frac{X_{ki} - X_{mi}}{x_{km}} \right) P_i \\
&= \sum_{i=2}^{M} A_{km}^{i} P_i
\end{aligned} \quad (8.30)$$

where $A_{km}^i = \dfrac{X_{ki} - X_{mi}}{X_{km}}$, $P_i = P(i,t) + P_\phi(i,t) - P_d(i,t)$. In the case of contingency, the corresponding E_{km}^i can be calculated from the modified $[X]$ matrix.

For voltage constraints, we assume that V^0 is the system voltage vector based on initial unit commitment state. If V^0 is infeasible, additional adjustments to the generation and/or distribution of reactive power may be necessary in order to shift the system voltage from V^0 to a desired value V^*. We neglect the effect of reactive power adjustment on bus voltage angles. Equations (8.14–8.15) are linearized in the vicinity of the initial operating point V^0 and the effect of the tap-changing transformer on voltages is considered in the linearized form as follows [Qiu87]:

1. Steady state constraints (no contingencies)

$$\Delta Q_T^{min} + Q_G^{min} I \le Q_G^0 + [J_1^{''s}]\Delta V \le Q_G^{max} I + \Delta Q_T^{max} \qquad (8.31)$$

$$[J_2^{''s}]\Delta V = 0 \qquad (8.32)$$

$$\Delta V^{min} \le \Delta V \le \Delta V^{max} \qquad (8.33)$$

2. Contingency constraints

$$\Delta Q_T^{min} + Q_G^{min} I \le Q_G^0 + [J_1^{''c}]\Delta V \le Q_G^{max} I + \Delta Q_T^{max} \qquad (8.34)$$

$$[J_2^{''c}]\Delta V = 0 \qquad (8.35)$$

$$\Delta V^{min} \le \Delta V \le \Delta V^{max} \qquad (8.36)$$

where $\Delta V = V^* - V^0$ is the incremental system voltage corresponding to ΔQ_G. $[J_1^{''s}]$ and $[J_1^{''c}]$ are the modified Jacobian matrices for buses connected to generators and transformers in steady state and contingency cases respectively. Also, $[J_2^{''s}]$ and $[J_2^{''c}]$ are the modified Jacobian matrices for load buses in steady state and contingency cases respectively.

8.4.2 Subproblem Formulation

Figure 8.4 depicts the process for SCUC problem with transmission and voltage (i.e., network) constraints. The infeasibility of network constraints can be represented by adding penalty variables to transmission and voltage constraints. Penalty variables are interpreted as the amount of transmission and voltage constraints violations associated with the base case unit commitment state $\hat{I}$. Therefore, we define the minimization of violations as the subproblem objective function $w(\hat{I})$ at each hour. Accordingly, the formulation of the subproblem is given as follows.

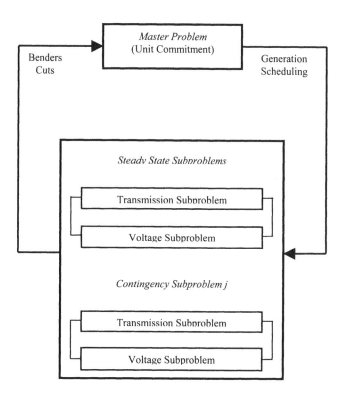

Figure 8.4 SCUC Problem with Transmission and Voltage Constraints

$$w(\hat{I}) = \min\left\{G^s + \sum_{j=1}^{nc} G^c(j) + F^s + \sum_{j=1}^{nc} F^c(j)\right\} \quad j=1,2,...,nc \quad (8.37)$$

s.t. $\quad A \cdot P + F^s \leq f^s \quad (8.38)$

$$J^{"s} \cdot \Delta V + G^s \leq g^s \quad (8.39)$$

$$E(j) \cdot P + F^c(j) \leq f^c(j) \quad j=1,2,...,nc \quad (8.40)$$

$$J^{"c}(j) \cdot \Delta V + G^c(j) \leq g^c(j) \quad j=1,2,...,nc \quad (8.41)$$

$$P^{min} \leq P \leq P^{max} \quad (8.42)$$

$$\Delta V^{min} \leq \Delta V \leq \Delta V^{max} \quad (8.43)$$

Equations (8.38–8.39) represent the steady state and (8.40–8.41) are the contingency cases. The objective function represents the minimization of penalty variables which may be multiplied by weighting factors in the objective function. P is the real power generation vector, and ΔV is the increment voltage vector. Both are functions of unit commitment vector I. The formulation (8.37–8.43) consists of several independent linear programs. It could be decomposed into steady state subproblem and contingency subproblem, each including a transmission subproblem and a voltage subproblem.

8.4.2.1 Steady State Subproblem

The steady state subproblem formulation is as follows.

8.4.2.1.1 Transmission subproblem

$$w_s(\hat{I}) = \min\{F^s\} \quad (8.44)$$

s.t. $\quad A \cdot P + F^s \leq f^s \quad (8.45)$

$$P^{min} \leq P \leq P^{max} \quad (8.46)$$

There are 24 steady state transmission subproblems that correspond to the 24-hours horizon.

SECURITY-CONSTRAINED UNIT COMMITMENT

8.4.1.1.2 Voltage subproblem

$$w_s(\hat{I}) = \min \{G^s\} \tag{8.47}$$

$$\text{s.t.} \quad J''^s \cdot \Delta V + G^s \leq g^s \tag{8.48}$$

$$\Delta V^{min} \leq \Delta V \leq \Delta V^{max} \tag{8.49}$$

There are 24 steady state voltage subproblems that correspond to the 24-hour horizon.

8.4.2.2 Contingency Subproblem

In each contingency case, the following subproblem is solved for each possible line outage j considered in that contingency case:

8.4.2.2.1 Transmission subproblem

$$w_c(\hat{I}, j) = \min \{F^c(j)\} \quad j=1,2,\ldots,nc \tag{8.50}$$

$$\text{s.t.} \quad E(j) \cdot P + F^c(j) \leq f^c(j) \quad j=1,2,\ldots,nc \tag{8.51}$$

$$P^{min} \leq P \leq P^{max} \tag{8.52}$$

There are $24 \times nc$ transmission contingency subproblems that correspond to the 24-hour horizon.

8.4.2.2.2 Voltage subproblem

$$w_c(\hat{I}, j) = \min \{G^c(j)\} \quad j=1,2,\ldots,nc \tag{8.53}$$

$$\text{s.t.} \quad J''^c(j) \cdot \Delta V + G^c(j) \leq g^c(j) \quad j=1,2,\ldots,nc \tag{8.54}$$

$$\Delta V^{min} \leq \Delta V \leq \Delta V^{max} \tag{8.55}$$

There are $24 \times nc$ contingency voltage subproblems that correspond to the 24 hours horizon.

So the total number of subproblems is $24 \times (1+1+nc+nc)$ which is $48 \times (1+nc)$. With such a large number of constraints added to the master problem, the master problem becomes bigger and more complicated which require a larger CPU time. However, the linear representation of these constraints would minimize the required CPU time.

8.4.3 Benders Cuts Formulation

After solving each subproblem, the optimal value of $w(I)$ (i.e., $w_s(\hat{I})$ or $w_c(\hat{I},j)$) would be larger than zero if violation is detected in the corresponding subproblem. Based on this we could produce a Benders cut. A linear approximation of the Benders cut is generated based on the subproblem results as

$$w(I) = w(\hat{I}) + \pi(I - \hat{I}) \leq 0 \tag{8.56}$$

where I represents the unit commitment state, $\hat{I}$ represents the base case unit commitment state, and π is the Simplex multiplier associated with the corresponding constraint in the subproblem. Specifically, the added constraints are as follows:

$$w_s^i(I) = w_s^i(\hat{I}) + \pi_s^i(I - \hat{I}) \leq 0 \quad i=1,2,\ldots,24 \tag{8.57}$$

$$w_c^i(I,j) = w_c^i(\hat{I},j) + \pi_c^i(I - \hat{I}) \leq 0 \quad i=1,2,\ldots,24 \quad j=1,2,\ldots,nc \tag{8.58}$$

8.4.4 Case Study

A 3-bus system, shown in Figure 8.5, is used to illustrate the proposed SCUC algorithm.

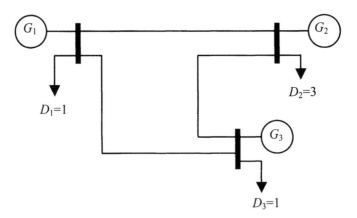

Figure 8.5 3-Bus System

SECURITY–CONSTRAINED UNIT COMMITMENT

Characteristics of generators, transmission lines data, and hourly load distribution over a 24-hour horizon are given in Tables 8.2, 8.3 and 8.4, respectively. Line flow limits are 55 MW. Lower and upper voltage limits are 0.97 and 1.06, respectively.

Table 8.2 Generators Data

Unit	Bus No.	Cost coefficients			P_{gmin} (MW)	P_{gmax} (MW)
		a	b	c		
G_3	3	118.8206	37.8896	0.01433	5.0	20.0
G_2	2	218.3350	18.1000	0.00612	10.0	150.0
G_1	1	142.7348	10.6940	0.00463	20.0	200.0

Table 8.3 Transmission Lines Parameters

Line	Resistance	Reactance (p.u.)	Flow Limit (MW)
1–2	0.0	0.20	55.0
1–3	0.0	0.40	55.0
2–3	0.0	0.25	55.0

Table 8.4 24 Hourly Load Distribution

Hr	P_D (MW)				Hr	P_D (MW)			
	D_1	D_2	D_3	Total		D_1	D_2	D_3	Total
1	132.66	44.22	44.22	221.1	13	153.06	51.02	51.02	255.1
2	122.4	40.8	40.8	204.0	14	149.64	49.88	49.88	249.4
3	115.62	38.54	38.54	192.7	15	147.96	49.32	49.32	246.6
4	112.2	37.4	37.4	187.0	16	147.96	49.32	49.32	246.6
5	108.84	36.28	36.28	181.4	17	154.74	51.58	51.58	257.9
6	110.52	36.84	36.84	184.2	18	170.04	56.68	56.68	283.4
7	112.2	37.4	37.4	187.0	19	163.26	54.42	54.42	272.1
8	119.04	39.68	39.68	198.4	20	161.52	53.84	53.84	269.2
9	136.02	45.34	45.34	226.7	21	159.84	53.28	53.28	266.4
10	149.64	49.88	49.88	249.4	22	156.42	52.14	52.14	260.7
11	153.06	51.02	51.02	255.1	23	147.96	49.32	49.32	246.6
12	154.74	51.58	51.58	257.9	24	137.76	45.92	45.92	229.6

8.4.4.1 Master Problem

In the master problem, we consider no transmission or voltage constraints. The unit commitment has a total production cost of $79,117.4 and the execution time is 1.5 seconds. The commitment schedule is shown in Table 8.5, where 1 or 0 represents ON/OFF states of units at different hours. In this table, units G_1 and G_2 are committed, while G_3 remains decommitted because of its high cost. This unit will be committed later when additional constraints for transmission and voltage are considered.

Table 8.5 Unit Commitment without Network Constraints

Unit	Hours (0–24)
G_3	0 0 0 0 0 0 0 0 0 0 0 0 0 0 0 0 0 0 0 0 0 0 0 0 0
G_2	0 1 1 0 0 0 0 0 0 1 1 1 1 1 1 1 1 1 1 1 1 1 1 1 1
G_1	0 1 1 1 1 1 1 1 1 1 1 1 1 1 1 1 1 1 1 1 1 1 1 1 1

8.4.4.2 Network Constraints

We start with the steady state constraints. Using unit commitment in Table 8.4 as $\hat{I}$, we solve the transmission subproblem (8.44–8.46) to minimize transmission flow violations at each hour. If transmission violations persist, we introduce Benders cuts for recalculating the unit commitment. However, all line flows in this case are within the limits and no Benders cuts are required. The SCUC solution is the same as that of Table 8.5, though the execution time is 3 seconds, which is higher than that in the base case for checking transmission flow violations.

Next, we consider both steady state and contingency constraints in transmission subproblems. We assume the loss of line 2–3 and, using the unit commitment schedule in Table 8.5 as $\hat{I}$, solve transmission subproblems (8.44–8.46) and (8.50–8.52) to minimize flow violations at each hour. Each contingency subproblem represents a line outage possibility. The maximum number of Benders cuts resulting from each subproblem is 24, since we check the violations hourly over the 24-hour horizon. Since transmission violations persist, we introduce Benders cuts as constraints in the unit commitment and solve the revised unit commitment. The results given in Table 8.6 satisfy all constraints with a total production cost of $79,628.9, which is higher than that of the base case due to the commitment of unit G_3 at hour 18. In Figure 8.5, as we lose line 2–3, load D_3 is larger than the line 1–3 flow limit, so G_3 is committed

SECURITY–CONSTRAINED UNIT COMMITMENT

to supply D_3. The execution time is 4.5 seconds, which is higher than that in previous cases due to the inclusion of line 2–3 contingency and a Benders cut added at hour 18.

Next we consider the impact of outages of individual lines (2–3, 1–3, 1–2) on SCUC. Using the unit commitment schedule in Table 8.5 as $\hat{I}$, we solve (8.44–8.46) and (8.50–8.52) to minimize transmission flow violations at each hour. The results are given in Table 8.7. Hour 18 represents the peak load, which creates most of transmission violations as shown in Table 8.7. At hours 3-8, Table 8.7 indicates that there will be flow violations when line 1-3 is on outage. The reason is given in Table 8.6 as G_1 is the only committed unit at hours 3–8 and the remaining lines in each case are not strong enough to carry D_2 and D_3 requirements. In order to check the SCUC results, we run a security-constrained economic dispatch. Power generation and line flows are given in Table 8.8 which satisfy the corresponding limits.

Table 8.6 SCUC Line 2-3 contingency

Unit	Hours (0–24)
G_3	0 0 0 0 0 0 0 0 0 0 0 0 0 0 0 0 0 0 1 0 0 0 0 0 0
G_2	0 1 1 0 0 0 0 0 0 1 1 1 1 1 1 1 1 1 1 1 1 1 1 1 1
G_1	0 1 1 1 1 1 1 1 1 1 1 1 1 1 1 1 1 1 1 1 1 1 1 1 1

Table 8.7 Cost and Transmission Violations

Outaged Line	Cost ($)	Number of Cuts	Hours with Flow Violations
2–3	79628.9	1	18
1–3	81236.3	7	3–8,18
1–2	80724.7	6	3–8

Table 8.8 Generation and Line Flows at Hour 18

Generators and Lines	Steady State (MW)	Contingency (MW)		
		Line 2–3 out	Line 1–3 out	Line 1–2 out
G_1	200.0	200.00	200.00	200.00
G_2	83.4	78.40	78.40	78.40
G_3	OFF	5.00	5.00	5.00
1–2	6.24	-21.72	29.96	0.00
2–3	32.96	0.00	51.68	21.72
1–3	23.72	51.68	0.00	29.96

Next, we consider the following transmission security cases:

- Case I ($nc = 1$): line 2–3 possibly on outage (same as the previous case)
- Case II ($nc = 2$): line 2–3 or 1–3 possibly on outage
- Case III ($nc = 3$): line 2–3 or 1–3 or 1–2 possibly on outage

For instance, in Case III with $nc = 3$, we consider independent outages of three lines (three single line outages) in SCUC. This case will require 3 × 24 transmission subproblems for contingencies in addition to 24 subproblems for steady state transmission flows in 24 hours. We show the results in Table 8.9. The total production cost is increased due to the commitment of more expensive units to cover more possible contingencies. However, the cost in Case III is the same as that in Case II since the same number of units are committed. As the number of possible outages increases, the number of Benders cuts may increase, which could further complicate the optimization problem.

Table 8.10 shows the SCUC results for Cases II and III with execution times of 6.5 and 8 seconds, respectively. Note that unit G_2 is committed for additional hours as shown with bold numbers.

Table 8.9 Transmission Flows in SCUC

Case	Cost ($)	Number of Cuts	Hours with Flow Violations
I	79628.9	1	18
II	81236.3	8	3–8, 18
III	81236.3	14	3–8, 18

Table 8.10 SCUC with Transmission Constraints (Cases II and III)

Unit	Hours (0–24)
G_3	0 0 0 0 0 0 0 0 0 0 0 0 0 0 0 0 0 1 0 0 0 0 0 0
G_2	0 1 1 **1 1 1 1 1** 1 1 1 1 1 1 1 1 1 1 1 1 1 1 1 1
G_1	0 1 1 1 1 1 1 1 1 1 1 1 1 1 1 1 1 1 1 1 1 1 1 1

To study the effect of line flow limits, we increase line flow limits in SCUC from 55 MW to 75 MW for all three lines. The SCUC results with transmission constraints for Case I are similar to the base case since the subproblem encounters no flow violations. The SCUC results with transmission constraints for Cases II and III are given in Table 8.11 for the 75 MW line flow limit. The production cost in this case is $79,668.5 which is lower than that of the case with the 55 MW line flow limit shown in Table 8.10.

Table 8.11 SCUC with 75 MW Transmission Constraints (Cases II and III)

Unit	Hours (0–24)
G_3	0 0 0 0 0 0 0 0 0 0 0 0 0 0 0 0 0 0 0 0 0 0 0 0 0
G_2	0 1 1 1 0 0 0 0 1 1 1 1 1 1 1 1 1 1 1 1 1 1 1 1 1
G_1	0 1 1 1 1 1 1 1 1 1 1 1 1 1 1 1 1 1 1 1 1 1 1 1 1

8.4.4.3 Voltage Constraints

First, we consider the steady state case. Using the unit commitment in Table 8.4 as $\hat{I}$, we solve the steady state transmission and voltage subproblems (8.44–8.49) at each hour. There are 4 voltage Benders cuts at hours 18–21 that are added in the revised master problem. The SCUC results given in Table 8.12 satisfy all constraints with a total production cost of $80,267.1 and an execution time of 10 seconds. The production cost is increased as compared with the base case because of the commitment of unit G_3 for additional hours to satisfy voltage constraints.

Table 8.12 SCUC with Steady State Constraints

Unit	Hours (0–24)
G_3	0 0 0 0 0 0 0 0 0 0 0 0 0 0 0 0 0 1 1 1 1 0 0 0
G_2	0 1 1 0 0 0 0 0 0 1 1 1 1 1 1 1 1 1 1 1 1 1 1 1 1
G_1	0 1 1 1 1 1 1 1 1 1 1 1 1 1 1 1 1 1 1 1 1 1 1 1 1

Next, we consider voltage constraints in steady state as well as transmission and voltage contingency subproblems (8.44–8.55) for losing line 2–3 at each hour. There are 14 voltage Benders cuts at hours 10–23 that are added to the revised master problem, in addition to 4 voltage and 1 transmission Bender cuts discussed earlier. The results given in Table 8.13 satisfy all constraints and represent the possible loss of line 2–3 with a total production cost of $82,407.7. The production cost increases due to the commitment of unit G_3 for additional hours to accommodate voltage and transmission security constraints. The execution time is 16 seconds, which is higher than that in the previous case.

Table 8.14 shows the hours with voltage violations and the corresponding number of voltage Benders cuts in each possible line outage. Most of the Benders cuts are produced at high load hours when the load is more than 87% of the peak load (at hour 18), for the first two cases and more than 94% of the peak load in the third case. Each contingency subproblem represents a line outage possibility.

The same contingency cases, discussed in section 8.4.4.2, will be adopted in this section to show the effect of voltage constraints on SCUC. In Case III with $nc = 3$ we consider 3 x 24 contingency transmission subproblems and 3 x 24 contingency voltage subproblems in addition to 24 steady state transmission subproblems and 24 steady state voltage subproblems. For each contingency case, the revised production cost for SCUC is given in Table 8.15. As the number of Benders cuts increases in proportion to possible outages, the optimization problem becomes bigger and more complicated with more CPU time.

Table 8.13 SCUC with Line 2–3 Outage and Voltage Constraints

Unit	Hours (0–24)
G_3	0 0 0 0 0 0 0 0 0 1 1 1 1 1 1 1 1 1 1 1 1 1 1 1 0
G_2	0 1 1 0 0 0 0 0 0 1 1 1 1 1 1 1 1 1 1 1 1 1 1 1 1
G_1	0 1 1 1 1 1 1 1 1 1 1 1 1 1 1 1 1 1 1 1 1 1 1 1 1

Table 8.14 Cost and Voltage Violations

Outaged line	Cost ($)	Number of Cuts	Hours with Voltage Violations
2–3	82407.7	14	10–23
1–3	84015.0	14	10–23
1–2	81874.4	4	18–21

Table 8.15 Costs and Benders Cuts for Voltage Constraints

Contingency	Cost ($)	Number of Trans. Cuts	Number of Voltage Cuts	Number of Cuts
I	82407.7	1	18	19
II	84015.0	8	32	40
III	84015.0	14	36	50

Table 8.16 shows the same scheduling results for Cases II and III while the execution times are 24 and 32 seconds respectively. The production cost is $84,015.0 as shown in Table 8.15. In comparing Tables 8.16 and 8.10, we learn that the most expensive unit G_3 is committed for additional hours to satisfy voltage constraints.

Table 8.16 SCUC with Transmission and Voltage Constraints (Cases II and III)

Unit	Hours (0–24)
G_3	0 0 0 0 0 0 0 0 0 1 1 1 1 1 1 1 1 1 1 1 1 1 1 1 0
G_2	0 1 1 1 1 1 1 1 1 1 1 1 1 1 1 1 1 1 1 1 1 1 1 1 1
G_1	0 1 1 1 1 1 1 1 1 1 1 1 1 1 1 1 1 1 1 1 1 1 1 1 1

8.5 SCUC APPLICATION TO MINIMIZE EUE – IMPACT OF RELIABILITY

The need to represent operating conditions uncertainties in SCUC is widely recognized. In this section, uncertainties related to equipment outages and load variations (composite reliability evaluation) are taken into account. Corresponding to the problem formulation described in Section 8.1, we consider the security constraints discussed in Task 2 which include transmission flow constraints and EUE limits (8.12 and 8.18). Here, we disregard voltage constraints for simplicity, and use a transportation model to represent the transmission network, though we could also use the dc transmission network model as in Task 1. In the subproblem, we try to minimize the violation of EUE caused by equipment outages and load variations. EUE is defined as the expected unserved energy by generating units due to a capacity deficiency. In this chapter, we use EUE as the reliability index.

8.5.1 Subproblem Formulation and Solution

Again, we use a Benders decomposition in which the master problem encompasses the unit commitment without any regards to the transmission network model or equipment outages. The master problem here is the same as that for Task 1, which was discussed in Section 8.3. The difference between this solution and that for Task 1 is that here we consider forced outage rates (FOR) of transmission lines and generating units and our objective is to minimize the expected value (EUE) of the penalty variables in the formulation.

The formulation of subproblem for the set of committed units is given as (for a specific time period t):

$$\min\ E\left\{\sum_{j=1}^{N_b} r_{jt}\right\}$$

$$\text{s.t.}\quad Sf + P + r = P_D(\varphi) \qquad (8.59)$$
$$r \leq P_D(\varphi)$$
$$P \leq \overline{P}(\varphi)$$
$$|f| \leq \overline{f}(\varphi)$$

The objective again represents the penalty term (unserved load) in supplying the load, except that we are now considering the expected value since the solution of (8.59) involves a reliability evaluation of the composite system. The constraints in (8.59) represent the transportation network model for the transmission network, including load balance equation, system operation limits, and generation and line flow limits. The procedure for solving (8.59) is as follows:

1. Select a system state φ according to the given FOR; that is to define load levels, equipment availability, operating conditions, and so on.
2. Calculate (8.59) for the selected state, namely verify whether the specific configuration of generators and transmission lines is able to supply the specific load without violating system limits.
3. Update the master problem solution if the subproblem is infeasible (EUE is larger than the required value).
4. Repeat step 1.

The subproblem is infeasible if the EUE cannot be kept below the desired level, in which case an infeasibility cut will be generated. For each infeasible subproblem resulting from the nth trial solution of the master problem, the deterministic infeasibility cut (Benders cut) is given as

$$E\left\{\sum_{j=1}^{N_b} r_{jt}\right\} + \sum_{i=1}^{N_g} \lambda_{it}^n \overline{P}_i (I_{it}^n - I_{it}) \leq \varepsilon_t \qquad (8.60)$$

SECURITY–CONSTRAINED UNIT COMMITMENT

where λ is the dual variable for P in the third constraint of (8.59) and ε is the required EUE. The multiplier λ_{ti}^n may be interpreted as the marginal decrease in unserved energy with a 1 p.u. increase in generation, given the nth trial commitment schedule. The infeasibility cuts (8.60) will be added to the master problem and will eliminate those commitment schedules which can not satisfy the reliability requirement.

However, if there are not enough units available to meet the minimum reliability requirement (EUE), we would consider purchasing energy from other energy providers. The energy purchased for period t, denoted as F_t, will depend on the utilization of available generating units to satisfy load constraints in each time period subject to maintaining the reliability above a certain level. Thus, the minimization of energy purchased from outside in period t can be expressed as

$$F_t = \min \chi_t \int_{\sum_{i \leq N_g} \overline{P_i} I_{it}^n}^{\infty} ELDC_{N_g}(P) dP \qquad (8.61)$$

where χ_t is cost per MWh of energy purchased from outside at time t, N_g is number of units, and $ELDC_{N_g}$ is equivalent load duration curve for N_g units.

The solution of (8.61) for all t yields a set of dual multipliers from which another cut is constructed. The new cut to be added to the master problem is of the form

$$z \geq \sum_t \left\{ F_t^n + \pi_{it}^n \, \overline{P_i}(I_{it}^n - I_{it}) \right\} \qquad (8.62)$$

where F_t^n is the expected purchased energy in period t associated with the nth trial solution. The optimal dual multipliers are given as

$$\pi_{it} = \frac{\partial F_t}{\partial (P_{it} I_{it})} \qquad (8.63)$$

where π_{it}^n may be interpreted as expected marginal costs associated with 1 p.u. decrease in power generation, given the nth trial solution. The new cut (8.62) tends to increase the lower bounds obtained from successive master problem solutions.

Figure 8.6 demonstrates the method to find dual multipliers for generating unit i as ELDC is derived based on the cumulant method [Str80]. Cumulants exhibit highly desirable characteristics for our analysis:

first, individual unit cumulants are calculated and used repeatedly throughout the analysis; second, the derivative term is determined using the Gram-Charlier approximation. ELDC is evaluated at L_i after convolving all available units (which includes unit i operating at its installed capacity). Then, unit i is subtracted from the system load and ELDC is evaluated at L_{i-1}. If we assume that ELDC is linear in the region of interest, the dual multiplier of the ith unit is approximated as

$$\pi_{it} = \chi_t \frac{(ELDC_{i-1}(L_{i-1})) + ELDC_i(L_i))}{2} \qquad (8.64)$$

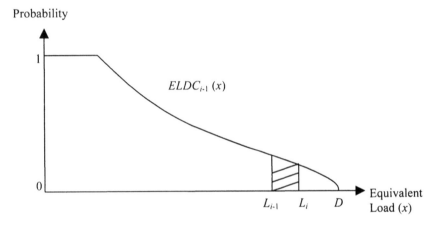

Figure 8.6 Equivalent Load Curve

8.5.2 Case Study

We use the same three-bus system in Figure 8.5 as an example. The maximum EUE (ε) is 0.5 p.u. and generator and line data are given in Tables 8.17 and 8.18. Load data are depicted in Figure 8.5. We assume that lines have negligible FOR, the study period has only one time interval, and loads are constant during the study period.

First, we solve the initial master problem. The solution is: $I_1 = 1$, $I_2 = 1$, and $I_3 = 0$.

SECURITY–CONSTRAINED UNIT COMMITMENT

Table 8.17 Generator Data for 3-bus System

Unit	Min Cap. (p.u.)	Max Cap. (p.u.)	Cost ($)	FOR
1	0.5	2.5	10 P_1	0.05
2	0.6	2.5	20 P_2	0.1
3	0.6	3.0	30 P_3	0.02

Table 8.18 Line Data for 3-bus System

Line	Number of lines	Capacity/line (p.u.)
1-2	2	0.25
2-3	2	0.5
1-3	2	0.25

Then, we check the feasibility of the sub-problem given the first trial of commitment schedule. The formulation of the subproblem is as follows:

min $r_1+r_2+r_3$

s.t. $-f_{12}-f_{13}+P_1+r_1=1$ Load balance at bus 1

$-f_{23}+f_{12}+P_2+r_2=3$ Load balance at bus 2

$f_{13}+f_{23}+P_3+r_3=1$ Load balance at bus 3

$0.5 \leq P_1 \leq 2.5$ Generator 1 limit

$0.6 \leq P_2 \leq 2.5$ Generator 2 limit

$0.0 \leq P_3 \leq 0.0$ Generator 3 limit

$-2 \times 0.25 \leq f_{12} \leq 2 \times 0.25$ Line 1–2 flow limit

$-2 \times 0.25 \leq f_{13} \leq 2 \times 0.25$ Line 1–3 flow limit

$-2 \times 0.5 \leq f_{23} \leq 2 \times 0.5$ Line 2–3 flow limit

The primal solutions for all state spaces in the feasibility check are given in Table 8.19. The subproblem solution is infeasible, since $E\{r_1+r_2+r_3\} = 0.85 > 0.5$. To compute the infeasible cut, we first compute λ_{P1}, λ_{P2}, and λ_{P3}.

To find λ_{P1}, we increase the capacity of P_1 by 1 p.u. and the primal solutions are given in Table 8.20. To find λ_{P2}, we increase the capacity of P_2 by 1 p.u. and the primal solutions are given in Table 8.21. To find λ_{P3}, we increase the capacity of P_3 by 1 p.u. and the primal solutions are given in Table 8.22.

Table 8.19 Feasibility Check State Spaces

P_1	φ P_2	Probability	Σr_i	Probability x Σr_i
2.5	2.5	0.855	0.5	0.4275
0	2.5	0.045	2.5	0.1125
2.5	0	0.095	3	0.2850
0	0	0.005	5	0.0250

$E\{\Sigma r_i\} = 0.85$

Table 8.20 Feasibility Check State Spaces

P_1	φ P_2	Probability	Σr_i	Probability x Σr_i
3.5	2.5	0.855	0.5	0.4275
0	2.5	0.045	2.5	0.1125
3.5	0	0.095	3	0.2850
0	0	0.005	5	0.0250

$E\{\Sigma r_i\} = 0.85$
$\lambda_{P1} = 0.85 - 0.85 = 0.00$

Table 8.21 Feasibility Check State Spaces

P_1	φ P_2	Probability	Σr_i	Probability x Σr_i
2.5	3.5	0.855	0	0.0000
0	3.5	0.045	1.5	0.0675
2.5	0	0.095	3	0.285
0	0	0.005	5	0.025

$E\{\Sigma r_i\} = 0.3775$
$\lambda_{P2} = 0.85 - 0.3775 = 0.4725$

SECURITY-CONSTRAINED UNIT COMMITMENT

Table 8.22 Feasibility Check State Spaces

P_3	φ P_1	P_2	Probability	Σr_i	Probability x Σr_i
1	2.5	2.5	0.8379	0	0.00000
1	0	2.5	0.0441	1.5	0.06615
1	2.5	0	0.0931	2	0.18620
1	0	0	0.0049	4	0.01960
0	2.5	2.5	0.0171	0.5	0.00855
0	0	2.5	0.0009	2.5	0.00225
0	2.5	0	0.0019	3	0.00570
0	0	0	0.0001	5	0.00050

$E\{\Sigma r_i\} = 0.28895$
$\lambda_{P3} = 0.85 - 0.28895 = 0.56105$

From the preceding calculation, we obtain the dual prices of the subproblem as: $\lambda_{P1} = 0$, $\lambda_{P2} = 0.4725$, and $\lambda_{P3} = 0.56105$. With the dual prices, the infeasibility cut is as:

$0.85 + 0 \times 2.5 \times (1-I_1) + 0.4725 \times 2.5 \times (1-I_2) + 0.56105 \times 3 \times (0 - I_3) < 0.5$

With the infeasibility cut, we resolve the master problem. The solution is: $I_1 = 0$, $I_2 = 1$, and $I_3 = 1$. We then perform the feasibility check in the subproblem as follows:

min $r_1 + r_2 + r_3$

s.t.
$-f_{12} - f_{13} + P_1 + r_1 = 1$ Load balance at bus 1
$-f_{23} + f_{12} + P_2 + r_2 = 3$ Load balance at bus 2
$f_{13} + f_{23} + P_3 + r_3 = 1$ Load balance at bus 3
$0.0 \leq P_1 \leq 0.0$ Generator 1 limit
$0.6 \leq P_2 \leq 2.5$ Generator 2 limit
$0.6 \leq P_3 \leq 3.0$ Generator 3 limit
$-2 \times 0.25 \leq f_{12} \leq 2 \times 0.25$ Line 1-2 flow limit
$-2 \times 0.25 \leq f_{13} \leq 2 \times 0.25$ Line 1-3 flow limit
$-2 \times 0.5 \leq f_{23} \leq 2 \times 0.5$ Line 2-3 flow limit

The primal solutions for the feasibility check are given in Table 8.23.

Table 8.23 Feasibility Check State Spaces

P_2	P_3	Probability	Σr_i	Probability x Σr_i
2.5	2.5	0.882	0	0
0	2.5	0.098	2.5	0.245
2.5	0	0.018	2.5	0.045
0	0	0.002	5	0.010

$$E\{\Sigma r_i\}=0.3$$

The subproblem solution is feasible, since $E\{r_1 + r_2 + r_3\} = 0.23 < 0.5$. So, the final solution is: $P_1 = 0$, $P_2 = 2.5$, $P_3 = 2.5$, $f_{21} = 0.5$, $f_{31} = 0.5$, $f_{32} = 1$. This solution satisfies the reliability requirement and is different from the initial solution.

8.6 CONCLUSIONS

This chapter presents a new approach to solving the transmission and voltage security-constrained unit commitment problem in a restructured market structure using linear sensitivity factors and Benders decomposition. The inclusion of n-1 contingency constraints in unit commitment ensures the security of the system. The tests on a 3-bus system show the effectiveness of the proposed approach. This chapter also presents formulation considering operating conditions uncertainties (including equipment outages and load variations) and an efficient algorithm. An example of a 3-bus system is used to illustrate the solution procedure.

Chapter 9

Ancillary Services Auction Market Design

9.1 INTRODUCTION

As the electric power industry moves toward the full competition, various services previously provided by electric utilities are being unbundled. Much of the attention given to the ISO development has focused on the structure of markets for energy and power transmission, and the market for ancillary services which is getting to be more critical. Ancillary services are generally referred to as those services other than energy that are essential for ensuring the reliable operation of the electrical grid. The reliable operation of a power system requires generation reserves to be available in order to cover generation and transmission contingencies. As restructuring evolves, determining the cost of supplying ancillary services and finding out how these costs would change with respect to operating decisions is becoming a major issue [Wil96].

According to FERC, ancillary services are necessary to support the transmission of power from sellers to buyers given the obligation of control areas and transmission utilities to maintain a reliable operation of the interconnected transmission system. The 1995 FERC rule defined six ancillary services and developed *pro forma* tariffs for these services. These services account for 5% to 25% of total generation and transmission costs, with an average of 10%. Based on the U.S. energy production of 2,900,000 GWh in 1994, ancillary services would cost almost $12 billion a year [Kir96]. Operating reserves would account for the bulk of ancillary services costs; reliability (spinning reserve) averages 16% of the total cost and supplemental (non-spinning reserve) averages 18%. Real power losses are 30% and voltage control represents 12% of total costs. Energy imbalances, assuming that 1% of customer loads are subject to this penalty, are 11% of the total cost. Load following averages 9% and scheduling and dispatch accounts for 4% of the total cost.

To facilitate an efficient trading of energy and ancillary services, a reasonable market structure is of great importance. In practice, market structures could differ according to their timing, the amount of information individual suppliers provide to the ISO, and the role of ISO in facilitating or controlling these markets. In general, there are forward and real-time markets for electricity. The day-ahead forward market is for scheduling resources at each hour of the following day, and the hour-ahead forward market is for adjusting deviations from the day-ahead schedule. The real-time market is for balancing the production and the consumption in real time. Energy trading is usually operated in forward markets, while ancillary services trading is operated in both forward and real-time markets.

In forward markets, there are two different approaches to energy and ancillary services auctions: sequential and simultaneous, depending on the amount of control delegated to the ISO. The sequential approach involves sequential computations in energy and ancillary services markets in which the results of one market would represent the starting point for the next market. The ISO plays an important role in balancing supply, demand, and prices in a sequential auction market structure. The simultaneous approach involves the simultaneous computation of supply, demand, and prices in all auction markets. In a simultaneous auction market, the ISO would not redispatch the generation in an already closed market to adjust the second auction market. The simultaneous approach would simplify auction market processes and reduce auction market prices due to the integration of energy and ancillary services markets.

In real-time markets, an important responsibility of the ISO is to maintain the real-time balance of energy and supply. One of the indispensable tools for the task is the automatic generation control (AGC). AGC is offered in ancillary services markets for minimizing frequency deviations, which would lead to a balance of energy and supply, and for regulating tie-line flows, that would facilitate bilateral contracts spanning over several control areas.

The chapter is organized as follows: Section 9.2 discusses definitions and requirements of ancillary services. Section 9.3 describes the sequential approach to the markets computations and an alternative ancillary service auction design where we introduce a weighting factor. Section 9.4 describes the simultaneous approach for the markets computations. Results for the application of the proposed designs to different auctions are presented and discussed for the sequential and simultaneous approaches in Sections 9.3 and 9.4 respectively. Section 9.5 discusses AGC operation and its pricing with examples. Section 9.6 concludes this chapter.

9.2 ANCILLARY SERVICES FOR RESTRUCTURING

FERC has determined that the following six ancillary services must be included in an open access transmission tariff:

- Scheduling, control, and dispatch, in which transmitting utilities would schedule and coordinate transactions with other entities and confirm the power exchange in and out of their control areas.

- Reactive supply and voltage control, where generation sources help maintain a proper transmission line voltage. This service would supply reactive power and voltage control, which is unbundled from basic transmission rates.

- Regulation and frequency response, for following moment-to-moment variations in customer demand or scheduled generation delivery, in order to maintain the 60 Hz frequency.

- Energy imbalance, to correct the hourly mismatch between a transmission customer's (TC's) energy supply and the load being served in a control area.

- Operating reserves, where spinning reserve and non-spinning reserve are defined as extra energy for supplying the load in the case of unplanned events such as the outage of a major generation facility.

  - Spinning reserve should be on-line and operate at less than maximum output, and be ready to immediately serve load.

  - Non-spinning reserve should generate capacity for emergency conditions but not be available immediately. Non-spinning reserve capacity should be started up very quickly (usually within 10 minutes).

Although a transmission provider (TP) must be equipped to offer all six services to TCs, FERC clarified that only the first two ancillary services must be offered to all TCs. In addition, FERC ruled that TCs must buy the first two ancillary services from a TP because services are local by nature and the TP is best suited to provide these services. For the other four ancillary services, FERC allows TCs to obtain the service in any of the following three ways: from the TP, from another source, or by self-provision.

The FERC requirement that the six services be included in an open access transmission tariff does not preclude TPs from offering other ancillary services voluntarily to TCs along with the supply of basic transmission service.

FERC discussed other ancillary services such as real power loss replacement, dynamic scheduling, backup supply, and black start capability. However, it didn't require TPs to unbundle and offer these services as separate services because these services are either very inexpensive or highly location-specific. Real power loss replacement is the use of generation to compensate for transmission system losses. Dynamic scheduling provides metering, telemetering, computer software, hardware, communications, engineering, and administration required to electronically transfer some or all of a generator's output or a customer's load from one control area to another. Backup supply is a generating capacity that can be made available within one hour. It is used to back up operating reserves and for commercial purposes. Black start capability is the ability of a generating unit to go from a shutdown condition to an operating condition without any assistance from the electrical grid, and to energize the grid to help other units start after a blackout occurs.

Other entities such as the Oak Ridge National Laboratory (ORNL) [Orn01] and the North American Electric Reliability Council (NERC) [Ner01] have also developed comprehensive lists and definitions of ancillary services. The Interconnected Operations Services Working Group, which was established by NERC to develop an industry consensus on definitions, requirements, obligations, and management for ancillary services, defined two other ancillary services: load following and network stability service. Load following is the use of generation to meet hourly and daily variations in system load. Network stability refers to the use of special equipment for maintaining the transmission system stability and reliability. Special equipment could include stabilizers, dynamic braking resistors, and FACTS devices.

Since operating reserves account for the bulk of ancillary services costs, special attention is directed toward operating reserve requirements in different reliability councils. NERC's criteria for operating reserves point out that each control area should maintain a certain level of operating reserve sufficient to account for factors such as errors in forecasting, generation and transmission equipment unavailability, number and size of generating units, system equipment forced outage rates, maintenance schedules, regulation requirements, and regional and system load diversity.

The minimum operating reserve requirements could differ from region to region. The 10-minute requirement for a complete response is set by NERC and is therefore consistent throughout the North America. However, the required amount of time for these reserves to remain in service after a complete response could differ among regions and is sometimes unspecified. Some tariffs could require operating reserves to be maintained for two hours. Others require operating reserves to be available until the end of the clock hour after the contingency has occurred. Some regions require operating reserves to be restored to the stated minimum levels as soon as practicable, without providing any specific meaning to "practicable."

It is possible, but not easy, to establish competitive markets for the provision, acquisition, and pricing of ancillary services. The difficulty stems from the complexity of these services, the relationship among these services, and their relationship with the energy service for supplying loads.

9.3 FORWARD ANCILLARY SERVICES AUCTION – SEQUENTIAL APPROACH

In the PoolCo model, Figure 9.1, GENCOs would interact with the ISO by providing bids for supplying the system load and ancillary services. The ISO is responsible for trading energy to supply the load in the forward energy markets and for trading ancillary services in forward and real-time energy markets. TRANSCOs post their information regarding the availability and the capability of transmission lines via the Open Access Same Time Information System (OASIS). DISCOs submit their demand bids to the ISO to be matched with GENCO's bids while satisfying the FERC's regulations and utilizing TRANSCOs' transmission information.

Since ancillary services auctions are operated by the ISO, the ISO is the single buyer party to meet the reliability obligations. The ISO's objective is to minimize ancillary services payments to GENCOs while encouraging GENCOs to provide sufficient ancillary services. GENCOs would anticipate submitting a bid that would maximize their profits as allocations are made. The ancillary services bids should include financial information for capacity reservation and energy, as well as operational information such as location, ramp-rate, and quantity blocks. Based on the ISO's requirements for ancillary services and participants' bids submitted to the ISO, the price and quantity of each service is determined, and payments are calculated by the ISO.

Let us consider an example of ancillary services. The California ISO (CAISO) procures four types of reserves: regulation for AGC, spinning reserves that are synchronized and available within 10 minutes, non-spinning reserves that are not synchronized but can be made available within 10 minutes, and replacement reserves that can be made available within 60 minutes. In its initial market design, the auction for the four reserve products was conducted sequentially, starting from the regulation, as shown in Figure 9.2. In each round, GENCOs would be allowed to rebid their uncommitted resources at new prices.

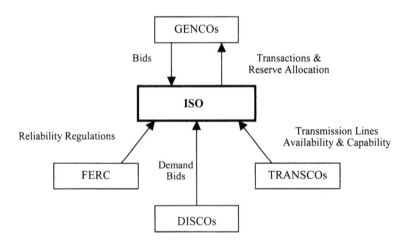

Figure 9.1 Restructured Market Participants

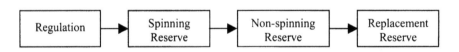

Figure 9.2 Sequential Auction of Ancillary Services

A GENCO submits two bids to each ancillary service auction: a capacity reservation bid ($/MW) and an energy bid ($/MWh) [Cai01]. However, the auction is cleared solely based on capacity bids. In this section, we propose an alternative that would clear an ancillary service

ANCILLARY SERVICES AUCTION

auction based on both capacity bids and energy bids. We demonstrate that our alternative is superior to the original scheme, as it encourages GENCOs to supply more ancillary services which could lower the ISO's total payment. The proposed alternative for auction design is discussed here in a very general sense, which can be implemented for any of the four ancillary services.

9.3.1 Two Alternatives in Sequential Ancillary Services Auction

We consider the following two alternatives for auctioning a specific ancillary service in the sequential auction process.

Alternative I. This is our proposed scheme. In this alternative, bids are ranked on the basis of both capacity reservation bids and energy bids. The ISO will select bids with the lowest combined price for capacity and energy, based on hypotheses concerning the use of each service. To minimize the expected energy payment and consequently the total payment for reserves, the ISO would use a weighting factor x as in $[C(i,t) + x \cdot E(i,t)]$ to rank bids, where $C(i,t)$ and $E(i,t)$ are capacity bid and energy bid for the specific ancillary service of generating unit i at time t respectively. We propose the use of energy bids in our Alternative I to hedge the ISO's risk against extremely high energy bids at the time of utilization. In this chapter, we will show how to design an optimal value for the weighting factor.

Alternative II. This is the traditional alternative for the ancillary services auction market design. Alternative II is a special case of Alternative I, in which x is set to zero and bids are ranked solely on the basis of capacity reservation $[C(i,t)]$. The ISO would select bids with the lowest capacity reservation price. The energy bid is not considered in the ranking process but considered in calculating ISO's payments.

In the real-time energy market, which is operated by the ISO, the energy component of ancillary services bid is used for energy balancing and ex-post pricing systems. Resources available in the energy balancing system include regulation, spinning, non-spinning, and replacement reserves, as well as resources that submitted supplemental bids for real-time imbalances. These resources are pooled in the energy balancing system and arranged in the merit order based on their energy bid prices. Only those bids that are considered for the ancillary services market-clearing price (MCP) calculation will be used in the balancing system. The MCP calculation will be presented in the following sections.

9.3.2 Ancillary Services Scheduling

If resources are ranked solely on the basis of capacity bids, as in the current sequential markets, then resources with low capacity bids and extremely high energy bids will appear in the real-time balancing stack for utilization. These resources will dramatically affect the ISO's energy payment and consequently the ISO's total payment. The reserve capacity payment in Alternative I includes an "opportunity cost"[1] payment, which is expressed as a function of capacity bids, energy bids, and weighting factor x. This expression increases the reserve MCP and reduces the ISO's total payment, which is the market's main objective. We will discuss this opportunity cost further in our case studies.

Choosing a proper value for x in Alternative I is the ISO's responsibility. The value of x will affect MCP and the ISO's capacity payment directly (as shown later). The value of x will affect the ISO's energy payment indirectly. In other words, after the resources are ranked and the bidders accepted, only the energy bids part will be considered for the real-time utilization. Since x affects ranking, it will automatically affect energy bids, the ISO's energy payment, and the ISO total payment. This will be shown later in the case studies.

Figure 9.3 shows the procedure of scheduling ancillary services in Alternative I. As can be seen in the figure, the ISO determines the reserve requirements in each category based on FERC regulations. Accordingly, GENCOs apply PBUC to define and submit their ancillary services bids to the ISO. GENCOs should submit two bids for ancillary services, one for capacity reservation and another for energy. The ISO sets the weighting factor x so as to minimize the payment for ancillary services while encouraging GENCOs to participate in this auction. The ISO then applies the matching process, using the submitted bids in the ancillary services

---

[1] The opportunity cost is the cost that a decision-maker would sacrifice in making a choice [Buc69]. In the case of ancillary services, suppose that a GENCO's energy bid is conveniently lower than the MCP for a forward energy market. However, the GENCO opts to bid in the ancillary services market instead of bidding in the forward energy market. The GENCO's bidding in ancillary services market would make it lose the opportunity of making profit in the forward energy market.

ANCILLARY SERVICES AUCTION

market, to calculate MCP. If reliability requirements are met at this stage, capacity reservation payments will be calculated and the final schedule will be published. Otherwise, the ISO will send signals to the GENCOs to modify their bids. Energy payments will be calculated at the time of utilization. If congestion exists, the ISO will hold ancillary services auctions on a zonal basis. For Alternative II, since the value of x is zero, it is not necessary for the ISO to set the weighting factor.

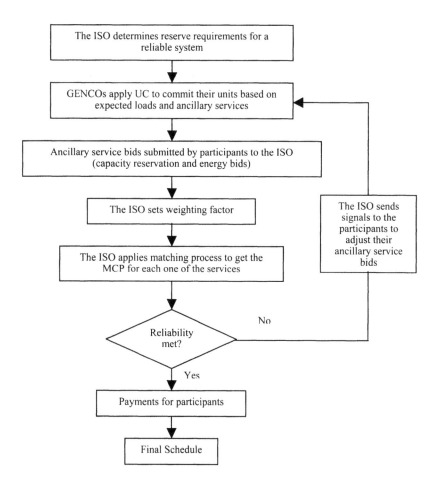

Figure 9.3 Ancillary Services Scheduling

9.3.3 Design of the Ancillary Services Auction Market

Formulation (9.1–9.5) shows how to auction a specific type of ancillary service at a specific time period t using Alternative I.[2] The objective (9.1) is to minimize the ISO's total payment for an ancillary service. The ISO's payment is formulated as a combined payment, which is the sum of the capacity reservation payment plus the energy payment assuming different values of real-time utilization factor y. The value of y is between 0% and 100%, which corresponds to the reserved capacity not utilized at all or utilized fully, respectively. Note that y only affects the ISO's energy payment and has no effect on the MCP for the ancillary services or on the ISO's capacity payment.

Here, the ISO is responsible for setting a reasonable value for x to minimize the capacity reservation and energy payments while encouraging participants to supply enough ancillary services. The energy payment is not known until the time of utilization. Hence, we calculate the combined payment for different values of y to study the effect of x. The value of x is zero for Alternative II.

The ISO's energy payment is affected indirectly by capacity bids and the choice of x, since in real time, the energy bids utilized are those of successful bidders in the ranking process where the combined bid prices including capacity and energy bids are considered.

$$\min \sum_i \{[MCP - x.E(i,t)] + [y.E(i,t)]\}.Q(i,t) \quad (9.1)$$

$$\text{s.t.} \quad \sum_i Q(i,t) \geq Q^{req}(t) \quad (9.2)$$

$$Q(i,t) \leq Q^{max}(i,t) \quad (9.3)$$

$$0.0 \leq x \leq 1.0 \quad (9.4)$$

$$0.0 \leq y \leq 1.0 \quad (9.5)$$

where MCP is the market-clearing price for the ancillary service, which will be discussed later, $E(i,t)$ is the energy bid ($/MWh) component for the ancillary service offered by unit i at time t, $Q(i,t)$ is the quantity of the ancillary service (MW) of unit i that is accepted at time t, $Q^{req}(t)$ is the

---

[2] Note that this formulation could be applied to any ancillary services auction in the sequential auction process.

ancillary service requirement at time t, and $Q^{max}(i,t)$ is the maximum quantity of the ancillary service that unit i can offer based on its ramp rate at time t.

Formulation (9.1-9.5) is a nonlinear optimization problem with three variables, x, y, and $Q(i,t)$. The value of y is not known until the time of utilization. The procedure to solve the formulation is as follows:

Step 1: Start with $x = 0.0$.

Step 2: Determine the MCP for the ancillary service.

The MCP for the ancillary service is calculated by the ISO through a matching process. MCP is the highest price among the accepted bids at that the ancillary service requirement is met. In Alternative I, the matching process is defined as minimizing the combined payment, whereas in Alternative II, the matching process is only to minimize the capacity payment. This is a major difference between the two alternatives. Mathematically, the matching process for the ancillary service at time t is expressed as

$$\min \sum_i \{[C(i,t) + \hat{x}.E(i,t)] \cdot Q(i,t)\} \tag{9.6}$$

s.t.
$$\sum_i Q(i,t) \geq Q^{req}(t) \tag{9.7}$$

$$Q(i,t) \leq Q^{max}(i,t) \tag{9.8}$$

where $C(i,t)$ is the capacity bid ($/MW) component for the ancillary service offered by unit i at time t, and $\hat{x}$ is the specified value of x. Formulation (9.6–9.8) is a linear optimization problem since $Q(i,t)$ is the only variable. It would be solved for all values of x.

In Alternative I, the ISO sorts the bids based on combined bid prices $C(i,t) + \hat{x}.E(i,t)$ for all participants. The MCP for the ancillary service is calculated as $[\max\{C(i,t) + \hat{x}.E(i,t)\}]$, representing the highest combined price among the accepted bids at which the ancillary services requirements are met. Successful bidders are paid at $[\text{MCP-}\hat{x}.E(i,t)]$ as capacity payment, which includes implicitly the lost opportunity cost payment. This is in addition to their energy price $E(i,t)$, if they are called upon to deliver in real time. If a tie happens in the ranking process, the participants will equally share the quantity. In Alternative II, the capacity payment is only [max$\{C(i,t)\}$] for accepted bids.

Step 3: The ISO's total payment is formulated as a function of y:

$$\sum_i \{[\text{MCP} - \hat{x}.E(i,t)] + [y.E(i,t)]\}.\hat{Q}(i,t) \qquad (9.9)$$

where $\hat{Q}(i,t)$ is the accepted quantity of the ancillary service of unit i at time t, which is the result of (9.6) – (9.8).

The difference between Alternatives I and II is the consideration of the opportunity cost, which could encourage GENCOs to participate in the ancillary services auction. The lost opportunity cost is embedded implicitly in the capacity payment formula (i.e., the first term in equation 9.9) for Alternative I. The GENCOs could lose the chance of bidding in energy market if they participated in the ancillary services auction. So, as formulated in Alternative I for a lost opportunity, GENCOs will receive the difference between the MCP and their energy bid price, whether or not they are utilized in real time, to encourage them to bid in the ancillary services auction.

Step 4: Increment x by a step size (Δx) (e.g., 0.01) to define the new $\hat{x}$ as long as $\hat{x} \leq 1.0$ and go to step 2.

Step 5: List the ISO's total payment as a function of x and y to choose the minimum total payment.

9.3.4 Case Study

In this section, we apply several examples to discuss the following issues:

1. Is there any direct relationship between the MCP of ancillary services and the ISO's total payment?
2. What is the role of x? What is the optimal value of x? Does the value of x depend on y?
3. What are the effects of transmission congestion on the ISO's total payment, ancillary services MCP, and on GENCOs revenues?
4. Which of the two alternatives is better and why?

ANCILLARY SERVICES AUCTION

Two different auctions are studied. In the first auction, there are four bids from two GENCOs and no transmission congestion is assumed. In the second auction, there are ten bids from three GENCOs and the auction is analyzed with and without transmission congestions in the forward energy market. However, in the following studies, we do not concern the reader with a specific type of ancillary services auction. We use a design that may be applied to any ancillary service auction in the sequential auction process.

9.3.4.1 4-Bid Auction with No Congestion

In this auction, there are two GENCOs A and B. Each has 2 generating units with a total reserve capacity of 40 MW. The bids submitted by these GENCOs are given in Table 9.1. The ISO has a reserve requirement of 40 MW at a certain hour.

Table 9.2 shows the ranking process and accepted quantities with the corresponding MCP for certain values of x. In this case, for all values of x, GENCO A's bids are always accepted and GENCO B's bids are rejected since the combined bid price of GENCO A is always lower. MCP is the highest accepted combined price as shown in Table 9.2 by bold numbers. Note the combined bid price is defined as $C(i,t) + x \cdot E(i,t)$.

Table 9.1 Ancillary Services Bids

GENCO		Capacity Bid ($/MW)	Energy Bid ($/MWh)	Quantity (MW)
A	G_1	10.0	50.0	20.0
	G_2	15.0	60.0	20.0
B	G_3	20.0	70.0	20.0
	G_4	25.0	80.0	20.0

Table 9.2 Ranking Process, Accepted Quantities, and MCP

GENCO		Combined Price at $x = 0.0$	Accepted Reserved Quantity	Combined Price at $x = 0.5$	Accepted Reserved Quantity	Combined Price at $x = 1.0$	Accepted Reserved Quantity
A	G_1	10	20	35	20	60	20
	G_2	**15**	20	**45**	20	**75**	20
B	G_3	20	0	55	0	90	0
	G_4	25	0	65	0	105	0

The energy payment is calculated by ranking the accepted bids again based on the energy component of the bid price as shown in Table 9.3. The results for the ISO's total payments are shown in Table 9.4. The optimal value for the weighting factor x is zero for all values of y.

To show how the optimal weighting factor will be affected by the participants' energy bid, we change the bids in Table 9.1 to those in Table 9.5. Note that we only changed GENCO B's energy bid.

Table 9.3 Energy Bid Part Ranking and Utilized Quantities

GENCO		Accepted Reserved Quantity (MW)	Energy Bid ($/MWh)	Utilized Reserve Quantity (MW)				
				$y=0.0$	$y=0.25$	$y=0.5$	$y=0.75$	$y=1.0$
A	G1	20	50	0	10	20	20	20
	G2	20	60	0	0	0	10	20

Table 9.4 MCP and the ISO's Total Payment

Real-time Utilization Factor (y)	Weighting Factor (x)				
	0.00	0.125	0.25	0.5	1.0
0.000	600	625	650	700	800
0.125	850	875	900	950	1050
0.250	1100	1125	1150	1200	1300
0.375	1350	1375	1400	1450	1550
0.500	1600	1625	1650	1700	1800
0.625	1900	1925	1950	2000	2100
0.750	2200	2225	2250	2300	2400
0.875	2500	2525	2550	2600	2700
1.000	2800	2825	2850	2900	3000
MCP ($/MW)	15.0	22.5	30.0	45.0	75.0

Table 9.5 Ancillary Service Bids

GENCO		Capacity Bid ($/MW)	Energy Bid ($/MWh)	Quantity (MW)
A	G_1	10.0	50.0	20.0
	G_2	15.0	60.0	20.0
B	G_3	20.0	10.0	20.0
	G_4	25.0	10.0	20.0

ANCILLARY SERVICES AUCTION

Table 9.6 shows the ranking process and accepted quantities with the corresponding MCPs for different values of x. In this case, the accepted bids are different, and the situation is changed to the advantage of GENCO B. The MCP is the highest accepted combined price as shown by bold numbers in Table 9.6.

Table 9.7 shows the MCP and the ISO's total payment for different values of y. Also, it shows the effect of x on the ISO's total payment for different values of y. The results support the idea that x has an important role in minimizing the ISO's total payment. The reserve MCP is increasing with increasing x while the ISO's total payment is fluctuating. The optimal value for x is greater than zero for most values of y. According to Table 9.7, if y is 0.0, the optimal values of x are 0.0 or 0.25. For y between 0.0 and 0.75, the optimal value of x is 0.25; for y larger than 0.75, the optimal value of x is larger than 0.375.

Table 9.6 Ranking Process, Accepted Quantities and MCP

GENCO		Combined Price at $x = 0.0$	Accepted Quantity	Combined Price at $x = 0.5$	Accepted Quantity	Combined Price at $x = 1.0$	Accepted Quantity
A	G_1	10	20	35	0	60	0
A	G_2	15	20	45	0	75	0
B	G_3	20	0	25	20	30	20
B	G_4	25	0	**30**	20	**35**	20

Table 9.7 MCP and the ISO's Total Payment

Real-Time Utilization Factor (y)	Weighting Factor (x)					
	0.00	0.125	0.25	0.375	0.5	0.625-1.0
0.000	**600**	700	**600**	700	1000	1000
0.125	850	750	**650**	750	1050	1050
0.250	1100	800	**700**	800	1100	1100
0.375	1350	850	**750**	850	1150	1150
0.500	1600	900	**800**	900	1200	1200
0.625	1900	1150	**1050**	1150	1250	1250
0.750	2200	1400	**1300**	1400	**1300**	**1300**
0.875	2500	1650	1550	1650	**1350**	**1350**
1.000	2800	1900	1800	1900	**1400**	**1400**
MCP ($/MW)	15.0	21.25	22.5	28.75	30.0	31.25-35.0

9.3.4.2 10-Bid Auction with No Congestion

A restructured 30-bus system with 10 generating units and 42 transmission lines is used to illustrate the proposed ancillary service auction. The system consists of three GENCOs: A, B and C. GENCOs A and B each has 3 generating units with a reserve capacity of 30 MW, and GENCO C has 4 units with a reserve capacity of 50 MW. Bids submitted by GENCOs are given in Table 9.8. The ISO has a reserve requirement of 65 MW at the given hour. The system is depicted in Figure 9.4.

Table 9.8 Ancillary Service Bids

GENCO		Capacity Bid ($/MW)	Energy Bid ($/MWh)	Quantity (MW)
A	G_1	10.0	15.0	5.0
	G_2	12.0	10.0	10.0
	G_3	13.0	11.0	15.0
B	G_4	12.2	13.0	5.0
	G_5	9.5	29.0	10.0
	G_6	11.0	33.0	15.0
C	G_7	11.0	19.0	5.0
	G_8	11.5	36.0	10.0
	G_9	12.0	14.0	15.0
	G_{10}	9.0	56.0	20.0

Table 9.9 shows the results for $x = 0.2$ in which MCP is 15.3 \$/MW. Let us take a closer look at the accepted bidders and the corresponding capacity and energy payments shown in Table 9.9. The ISO pays \$794.5 as capacity payment, which includes \$391 to GENCO A, \$158.5 to GENCO B, and \$245 to GENCO C. In Table 9.9, capacity payment for each unit is different since it accounts for the lost opportunity cost. Differences in the lost opportunity cost originate from differences in the energy part of bids as shown in Table 9.8.

Note that the capacity payment is different from MCP and the capacity bid, as explained earlier. If the energy bid is high, then the capacity payment will be low, as shown in Tables 9.8 and 9.9, for G_2 and G_4. The capacity payment for G_2 is higher than that of G_4, but the capacity bid of G_2 is lower than that of G_4. This is due to the fact that the energy bid of G_2 is lower than that of G_4, which contributes to the lost opportunity cost. So, accounting for the lost opportunity cost in the capacity payment

ANCILLARY SERVICES AUCTION

formula will encourage GENCOs to lower their energy bid, which will lower the ISO's total payment accordingly.

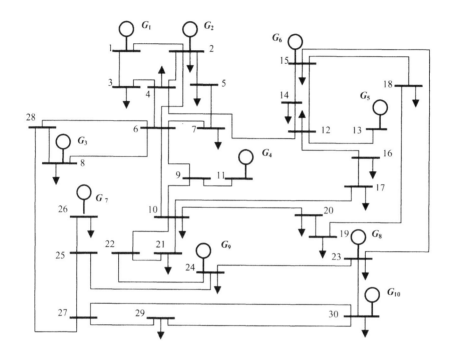

Figure 9.4 30-Bus System (no congestion)

Table 9.9 Accepted Bidders and Payments by the ISO ($x = 0.2$)

GENCO		Combined Price ($/MW)	Capacity Payment				Energy Payment ($/MWh)[3]
			Price ($/MW)	Quantity (MW)	Payment ($)	Sub-total ($)	
A	G_1	13.0	12.3	5.0	61.5	391.0	15.0
	G_2	14.0	13.3	10.0	133.0		10.0
	G_3	15.2	13.1	15.0	196.5		11.0
B	G_4	14.8	12.7	5.0	63.5	158.5	13.0
	G_5	15.3	9.5	10.0	95.0		29.0
C	G_7	14.8	11.5	5.0	57.5	245.0	19.0
	G_9	14.8	12.5	15.0	187.5		14.0
		MCP=15.3		Total: 65.0		Total: 794.5	

---

[3] We assume energy payment is the same as energy bid price.

For $x = 0.05$ in Table 9.10, MCP is 12.65 $/MW, which is lower than that of the previous case. The ISO pays $713.5 as capacity payment, which includes $181 to GENCO A, $277 to GENCO B and $255.5 to GENCO C. Note that the ISO's capacity reservation payment decreases as energy payment increases by accepting G_6 and G_{10} instead of G_3, G_4, and G_9.

Table 9.11 shows the results for $x = 0.0$ (based on capacity reservation bids only), which is Alternative II. MCP is 11.5 $/MW, which is lower than that in Alternative I. However, the energy payment is higher for accepting the G_8 bid in addition to those of G_6 and G_{10}. The ISO pays $747.5 as capacity payment, which is higher than that of $x = 0.05$. This includes $57.5 to GENCO A, $287.5 to GENCO B and $402.5 to GENCO C. The ISO's total payment in this case would be higher than that for $x = 0.05$ which means that $x = 0.0$ is not the optimal weighting factor for minimizing the ISO's payment.

Table 9.10 Accepted Bidders and Payments by the ISO ($x = 0.05$)

GENCO		Combined Price ($/MW)	Capacity Payment				Energy Payment ($/MWh)
			Price ($/MW)	Quantity (MW)	Payment ($)	Sub-total ($)	
A	G_1	10.75	11.90	5.0	59.5	181.0	15.0
	G_2	12.50	12.15	10.0	121.5		10.0
B	G_5	10.95	11.20	10.0	112.0	277.0	29.0
	G_6	12.65	11.00	15.0	165.0		23.0
C	G_7	11.95	11.70	5.0	58.5	255.5	19.0
	G_{10}	11.80	9.85	20.0	197.0		56.0
		MCP=12.65		Total: 65.0		Total: 713.5	

Table 9.11 Accepted Bidders and Payments by the ISO ($x=0.0$)

GENCO		Combined Price ($/MW)	Capacity Payment				Energy Payment ($/MWh)
			Price ($/MW)	Quantity (MW)	Payment ($)	Sub-total ($)	
A	G_1	10.0	11.5	5.0	57.5	57.5	15.0
B	G_5	9.5	11.5	10.0	115.0	287.5	29.0
	G_6	11.0	11.5	15.0	172.5		33.0
C	G_7	11.0	11.5	5.0	57.5	402.5	19.0
	G_8	11.5	11.5	10.0	115.0		36.0
	G_{10}	9.0	11.5	20.0	230.0		56.0
		MCP = 11.5		Total: 65.0		Total: 747.5	

ANCILLARY SERVICES AUCTION

Table 9.12 shows the effect of x on the ISO's total payment for different values of y. In Table 9.12, MCP is increased by incrementing x while the ISO's total payment is fluctuating. The optimal value for x is 0.07 if y is expected to be less than 50%. Otherwise, the optimal value of x is 0.19 as shown by bold numbers in Table 9.12.

Note that when the ISO uses the optimal value for x, the ISO's total payment is minimized and consequently DISCOs' payments are minimized. Also, MCP for the optimal x is higher than that of $x = 0.0$ as shown in Table 9.12, which presumably encourages GENCOs to participate in this market.

If the ISO uses the optimal value of $x = 0.07$ for y less than 50%, then the calculated MCP is 12.98 $/MW, with results shown in Table 9.13. The ISO pays $711.4 as capacity payment which includes $182.45 to GENCO A, $109.5 to GENCO B and $419.45 to GENCO C.

Table 9.12 MCP and the ISO's Total Payment for Different Values of x

Real-Time Utilization Factor (y)	Weighting Factor (x)					
	0.00	0.05	0.07	0.1	0.19	0.2
0.000	747.500	713.5	**711.400**	816.5	790.850	794.5
0.125	881.875	794.75	**792.650**	897.75	872.100	875.75
0.250	1098.750	912.25	**898.90**	985.25	959.600	963.25
0.375	1351.875	1110.375	**1012.65**	1074.625	1048.975	1052.625
0.500	1620.000	1356.0	1168.90	1181.5	**1155.850**	1159.5
0.625	1905.000	1624.125	1404.525	1295.25	**1269.600**	1273.25
0.750	2272.500	1978.5	1640.150	1412.75	**1387.100**	1390.75
0.875	2727.500	2433.5	2061.400	1630.875	**1605.225**	1608.875
1.000	3182.500	2888.5	2516.400	1866.5	**1840.850**	1844.5
MCP ($/MW)	11.50	12.65	12.98	14.10	15.09	15.30

Table 9.13 Accepted Bidders and Payments by the ISO ($x = 0.07$)

GENCO		Combined Price ($/MW)	Capacity Payment				Energy Payment ($/MWh)
			Price ($/MW)	Quantity (MW)	Payment ($)	Sub-total ($)	
A	G_1	11.05	11.93	5	59.65	182.45	15.0
	G_2	12.70	12.28	10	122.80		10.0
B	G_5	11.53	10.95	10	109.50	109.50	29.0
C	G_7	12.33	11.65	5	58.25	419.45	19.0
	G_9	12.98	12.00	15	180.00		14.0
	G_{10}	12.92	9.06	20	181.20		56.0
		MCP=12.98		Total: 65.0		Total: 711.4	

Similarly, if the ISO uses the optimal value of $x = 0.19$ for y greater than or equal to 50%, the calculated MCP is 15.09 $/MW, with results shown in Table 9.14. In this case, the ISO pays $790.85 as capacity payment which includes $388.1 to GENCO A, $158.9 to GENCO B and $243.85 to GENCO C.

Figure 9.5 shows the effect of x on the ISO's total payment for different values of y. It is apparent that in the case of Alternative II, the ISO could face higher energy payments in real time. However, GENCOs' total revenue at $x=0.19$ and $y<0.05$ is higher than that of $x=0.0$ which points to the design of an optimal value for x by the ISO.

Table 9.14 Accepted Bidders and Payments by the ISO ($x=0.19$)

GENCO		Combined Price ($/MW)	Capacity Payment				Energy Payment ($/MWh)
			Price ($/MW)	Quantity (MW)	Payment ($)	Sub-total ($)	
A	G_1	12.85	12.24	5.0	61.20	388.10	15.0
	G_2	13.90	13.19	10.0	131.90		10.0
	G_3	15.09	13.00	15.0	195.00		11.0
B	G_4	14.67	12.62	5.0	63.10	158.90	13.0
	G_5	15.01	9.58	10.0	95.80		29.0
C	G_7	14.61	11.48	5.0	57.40	243.85	19.0
	G_9	14.66	12.43	15.0	186.45		14.0
		MCP=15.09		Total: 65.0		Total: 790.85	

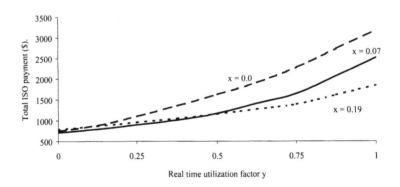

Figure 9.5 ISO's Total Payment versus Real-time Utilization Factor

9.3.4.3 10-Bid Auction with Congestion

The same system as the preceding one is used here. For the congested case, the ISO would establish two zones as shown in Figure 9.6. Bids in Table 9.8 are used again based on zones as shown in Table 9.15. Hence, for each ancillary service, we will have an auction in each zone. Here, we assume that the ISO's ancillary service requirements are 20 and 45 MW in zones 1 and 2, respectively. Also we assume that each unit participates in the ancillary service auction in its own zone.

Table 9.15 Zonal Ancillary Service Bids

Zone	GENCO		Capacity Bid ($/MW)	Energy Bid ($/MWh)	Quantity (MW)
1	A	G_3	13.0	11.0	15.0
	C	G_7	11.0	19.0	5.0
		G_9	12.0	14.0	15.0
2	A	G_1	10.0	15.0	5.0
		G_2	12.0	10.0	10.0
	B	G_4	12.2	13.0	5.0
		G_5	9.5	29.0	10.0
		G_6	11.0	33.0	15.0
		G_8	11.5	36.0	10.0
	C	G_{10}	9.0	56.0	20.0

Zonal ancillary service auctions for $x = 0.19$ are given in Table 9.16 in which zonal MCPs are different from those without congestion. However, the most important issue is the effect of congestion on the ISO's total payment. The ISO pays $817.65 as capacity payment, which is higher than that without congestion. The payment represents $225.8 to GENCO A, $356.6 to GENCO B, and $235.25 to GENCO C. Note that the energy payment is higher than that without congestion due to the ISO's accepting G_6 instead of G_3. Here, if we assume that $y = 1.0$ for the energy payment calculations, the ISO's total payment would be $2147.65, which is higher than that without congestion where the total payment is $1840.85.

GENCO B has a higher capacity reservation payment than it had without congestion and a higher energy payment because G_6 was accepted instead of G_3. By contrast, GENCO A has a lower capacity reservation payment than that without congestion, and it has lost the energy payment for G_3. GENCO C was not affected much compared to GENCOs A and B.

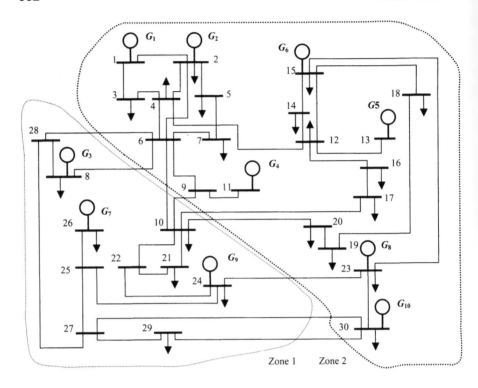

Figure 9.6 30-Bus System (Case with Congestion)

Table 9.16 Zonal Ancillary Services Auction ($x = 0.19$)

Zone	GENCO		Combined Price ($/MW)	Capacity Payment				Energy Payment ($/MWh)
				Price ($/MW)	Quantity (MW)	Payment ($)	Sub-total ($)	
1	C	G_7	14.61	11.05	5.0	55.25	235.25	19.0
		G_9	14.66	12.00	15.0	180.00		14.0
			MCP: 14.66		Total: 20.0			
2	A	G_1	12.85	14.42	5.0	72.10	225.8	15.0
		G_2	13.90	15.37	10.0	153.70		10.0
	B	G_4	14.67	14.80	5.0	74.00	356.0	13.0
		G_5	15.01	11.76	10.0	117.60		29.0
		G_6	17.27	11.00	15.0	165.00		33.0
			MCP: 17.27		Total: 45.0			
							Total: 817.65	

ANCILLARY SERVICES AUCTION

Table 9.17 shows the results for Alternative II, $x = 0$. From Table 9.17, MCP in zone 1 is 12.0 $/MW, which is higher than that without congestion where MCP was 11.5 $/MW, and the MCP in zone 2 is 11.0 $/MW, which is lower than that without congestion. So, the zonal MCP could be higher or lower in the case of congestion.

The ISO pays $735.0 as capacity payment which includes $55.0 to GENCO A, $220.0 to GENCO B, and $460.0 to GENCO C. GENCO A receives a lower capacity payment since MCP in zone 2 is lower than that of no-congestion case. Also, GENCO B receives a lower capacity payment in the case of congestion. GENCO C has a higher capacity payment since MCP in zone 1 is higher than that of no-congestion case with a lower energy payment. This is because the G_8 bid is rejected in the case of congestion.

In the congestion case, if we assume $y = 1.0$ for the energy payment calculations, the ISO's total payment is $2855.0, which is lower than that of no-congestion case where the total payment was $3182.5. Note that the energy payment is lower than that of no-congestion case due to the ISO's accepting the G_9 bid instead of the G_8 bid. This shows that congestion can be advantageous to DISCOs.

Table 9.17 Zonal Ancillary Services Auction ($x = 0$)

Zone	GENCO		Combined Price ($/MW)	Capacity Payment				Energy Payment ($/MWh)
				Price ($/MW)	Quantity (MW)	Payment ($)	Sub-total ($)	
1	C	G_7	11.0	12.0	5.0	60.0	240.0	19.0
		G_9	12.0	12.0	15.0	180.0		14.0
			MCP: 12.0		Total: 20.0			
2	A	G_1	10.0	11.0	5.0	55.0	55.0	15.0
	B	G_5	9.5	11.0	10.0	110.0	220.0	29.0
		G_6*	11.0	11.0	10.0	110.0		33.0
	C	G_{10}	9.0	11.0	20.0	220.0	220.0	56.0
			MCP: 11.0		Total: 45.0			
							Total: 735.0	

* Only 10 MW out of 15 MW is accepted.

9.3.5 Discussions

Competitive markets for energy supply require competitive markets for ancillary services. This section presents two alternative designs that can be used for any ancillary services auction in the sequential auction process. The first alternative, which is the proposed ancillary services auction design, allows bids to be ranked on the basis of both the capacity reservation and the energy. The second alternative, which follows the initial market design of the California ISO, allows ranking that is based on only capacity reservation bids.

Through case studies, we find the first alternative design to be superior to the second design. First, it shows higher MCPs of ancillary services, which should encourage GENCOs to participate in the ancillary services market. Second, it accounts for the lost opportunity cost in the capacity payment, which should encourage GENCOs to bid low in the energy part of the auction. Third, it lowers the ISO's total payment, which benefits the DISCOs. Last, it hedges the ISO and DISCOs payments against the GENCOs' extremely high energy bids. A difficulty of the first alternative is how to set an optimal weighting factor as it depends on the real-time utilization of the ancillary services, addressed in this section.

The second alternative may result in higher systemwide revenues for GENCOs. The disadvantage of this alternative for DISCOs is the high total payment and the possibility of having extremely high energy payments. The low MCPs in the second alternative do not mean lower payments by DISCOs. On the contrary, their energy payment may be very high due to the fact that the energy bids are neglected in the ranking process. In this alternative, MCP does not depend on the energy bids.

The designs of ancillary services auctions reveal an interesting dilemma between cost minimization by the ISO and profit maximization by the bidders. Congestion has a great effect on the decision played out in the ancillary services auctions. In times of congestion, some GENCOs have worse bidding chances while intelligent bidders with alternative design have higher chances of success.

9.4 FORWARD ANCILLARY SERVICES AUCTION–SIMULTANEOUS APPROACH

In the initial market design for ancillary services, CAISO adopted the sequential approach. Each market was operated separately and cleared sequentially, proceeding from regulation, to spinning reserve, to non-spinning reserve, and last to replacement reserve. This approach was easy to implement. However, some unexpected phenomena emerged after the implementation. For instance, lower cost bids with higher value were not utilized, and prices for lower quality services (e.g., replacement reserve) were higher than those for higher quality services (e.g., spinning reserve)[4]. Similar problems also existed in the New England ISO and the New York ISO.

The reason why lower cost/higher value bids are left unused is that substitutability of ancillary services is not well utilized. An important aspect of ancillary services is their hierarchical nature that allows the substitution of a higher quality service for a lower quality one. Both the social efficiency and the rational procurement behavior dictate that such substitution should be allowed. In a sequential auction, it is possible that some of higher value bids are not accepted in the initial auction round. If those bids are not considered in a later auction round, these ancillary services are not utilized. To solve this problem, CAISO redesigned the auction rules and proposed a "rational buyer" auction that allows a simultaneous auction of ancillary services and the substitution of a higher quality ancillary service for a lower quality ancillary service[5]. Specifically, CAISO would need to procure four types of reserves: regulation, spinning, non-spinning, and replacement reserves. These products are hierarchically substitutable. Regulation resource could be used for any other services. Spinning reserve could be used for non-spinning reserve and replacement reserve. Non-spinning reserve could provide replacement reserve.

The phenomenon that prices for lower quality services would be higher than prices for higher quality services is called "price reversal." Ideally, in a competitive market, higher quality services should possess higher prices. However, a sequential auction with independent uniform MCPs in each round and without a substitution of a higher quality service for a lower quality service could result in a price reversal. Price reversals could pose serious incentive compatibility problems, since the price-taking generators that could anticipate such a reversal might be inclined to understate their capability and wait for a later round of the sequential auction that could fetch a higher MCP. With market power, the situation could be exacerbated as losing players in the early rounds raise their bids in subsequent rounds when they perceive a potential scarcity of services. The

---

[4] For reserves, faster response reserves are graded as higher quality or higher value.

[5] Another purpose of the rational buyer auction is to decrease the total cost of the procurement of ancillary services.

rational buyer auction could reduce the price reversal mechanism to some extent but could not avoid it. It has been observed that a simultaneous auction with marginal pricing and substitution can avoid price reversals[6].

In this section, we introduce a few features of the simultaneous auction design of ancillary services. Two simultaneous approaches are discussed and examples are presented. For simplicity and clarity, we consider four ancillary services as in CAISO: regulation, spinning reserve, non-spinning reserve, and replacement reserve.

9.4.1 Design Options for Simultaneous Auction of Ancillary Services

Within the framework of a simultaneous auction, we could have various design options depending on the *objective of the auction* and the *settlement procedures* (i.e., settlement rule or pricing rule) in the auction. These two principles regarding the design options would enable the substitution of high quality services for low quality services to reduce (or prevent) price reversals.

The objective of our auction market could be to minimize the social cost or minimize the procurement cost. The social cost refers to the actual cost of the required services. In the restructuring paradigm, it is not easy to measure the social cost since services are rendered as bids, which may not represent the true cost of services. If we assume that the bids are close to their true costs, we could minimize the social costs for the accepted bids. The procurement cost refers to the cost to procure the required services.

If the accepted bids are paid using a uniform price, the social cost might be different from the procurement cost. For example, if accepted bids for regulation are 100 MW @ 5 $/MW and 50 MW @ 10 $/MW, the MCP would be 10 $/MW. The social cost would be 100 x 5 + 50 x 10 = $1000, while the procurement cost would be (100 + 50) x 10 = $1500.

In general, there are two market settlement rules: *uniform pricing* and *pay-as-bid pricing*. In uniform pricing, all market participants with accepted bids would be paid a uniform price regardless of their bids. In pay-as-bid pricing, the market participants with accepted bids are paid according to their bids. In uniform pricing, there are different ways of determining the uniform price, namely through marginal pricing and substitution pricing (including demand substitution and supply substitution).

---

[6] The problem with marginal pricing is, however, that it may significantly increase the procurement costs.

ANCILLARY SERVICES AUCTION

Marginal pricing means that the marginal cost of a service is its price. In demand substitution pricing, the demand for a lower quality service can be substituted by a higher quality service and the price is set to the highest accepted bid of that service. In a supply substitution pricing, the supply for a higher quality service can be used for a lower quality service and the price is set to the highest accepted bid of that service.

There is a slight distinction between the demand substitution pricing and the supply substitution pricing. Assume that bids submitted to the ISO are 100 MW @ 5 $/MW and 50 MW @ 10 $/MW for regulation, and 200 MW @ 8 $/MW for spinning reserve. The demand for regulation is 50 MW and for spinning reserve is 150 MW.

- By a demand substitution, to meet the demand for regulation, the ISO would procure 50 MW @ 5 $/MW regulation, thus the procurement price for regulation would be 5 $/MW with a procurement cost of 50 x 5 = $250. To meet the demand for spinning reserve, the ISO would procure 50 MW @ 5 $/MW regulation but would use it for spinning reserve and another 100 MW @ 8 $/MW spinning reserve, thus the procurement price for spinning reserve would be 8 $/MW with a procurement cost of 150 x 8 = $1200. So, by a demand substitution, the total procurement cost would be 250 +1200 = $1450.

- By a supply substitution, the ISO would procure 50 MW @ 5 $/MW regulation. Next, the ISO would procure another 50 MW @ 5 $/MW regulation but would use it as spinning reserve. Thus, the total procured regulation would be 100 MW, the procurement price would be 5 $/MW with a procurement cost of 100 x 5 = $500. Then, the ISO would procure 100 MW @ 8 $/MW spinning reserve, thus the procurement price for spinning reserve would be 8 $/MW with a procurement cost of 100 x 8 = $800. So, by a supply substitution, the total procurement cost would be 500 + 800 = $1300.

Comparing the two options, we conclude that the procured services (100 MW regulation and 100 MW spinning reserve), and the social costs (100 x 5 + 100 x 8 = $1300) are the same. However, the procurement costs are different: the supply substitution saves the procurement cost by (1450 - 1300 = $150).

Figure 9.7 summarizes the different options for ancillary services auction design. In general, we could select any of the objectives and any of the settlement rules to design the auction market. Some exceptions are as follows: In pay-as-bid pricing, the procurement cost is the same as the social cost, so there is no difference between the two objectives. In marginal pricing, we could not use the objective of procurement cost minimization, since the procurement price (marginal cost) would depend on the objective function and could not appear at the objective function.

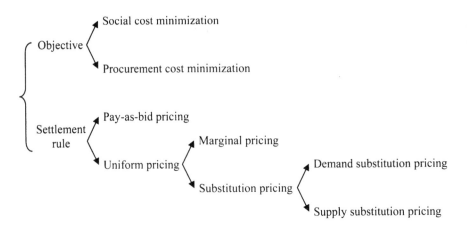

Figure 9.7 Simultaneous Auction Design Options for Ancillary Services

9.4.2 Rational Buyer Auction

The rational buyer auction is the redesigned auction mechanism for ancillary services in California. The impetus behind the rational buyer auction is CAISO's intend to adopt a common sense rule of substituting higher quality lower cost services for lower quality higher cost services, when it would reduce the total procurement cost. The objective of the rational buyer is to minimize the total procurement cost. The rational buyer auction would utilize a supply substitution pricing in which all market participants would be paid a uniform price for each ancillary service, and the price for an ancillary service would be set as the highest accepted bid of that ancillary service. According to this pricing rule, the buyers of an ancillary service, say a spinning reserve, might end up paying more if the spinning reserve is also used as a non-spinning reserve, thus raising the MCP for the spinning reserve.

In the rational buyer auction, each generator would submit a single bid for each ancillary service simultaneously, which would specify the type, the price, and the quantity of the ancillary service. After receiving the bids, the ISO could use any of the procured ancillary services to meet the

ANCILLARY SERVICES AUCTION

demand. Hence, the ISO would enable the substitution of a high quality service for a low quality service. In summary, there are two substantial differences between the rational buyer auction design and the original auction design in California: it shifts from a sequential design to a simultaneous design, and it enables the ISO to exploit the substitutability among different ancillary services.

9.4.2.1 Formulation of the Rational Buyer Auction

Mathematically, the substitutability can be expressed as follows:

$$s_{RG} \geq d_{RG} \tag{9.10a}$$

$$s_{RG} + s_{SR} \geq d_{RG} + d_{SR} \tag{9.10b}$$

$$s_{RG} + s_{SR} + s_{NR} \geq d_{RG} + d_{SR} + d_{NR} \tag{9.10c}$$

$$s_{RG} + s_{SR} + s_{NR} + s_{RR} \geq d_{RG} + d_{SR} + d_{NR} + d_{RR} \tag{9.10d}$$

where s refers to supply, d refers to demand, RG refers to regulation, SR refers to spinning reserve, NR refers to non-spinning reserve and RR refers to replacement reserve.

In comparison, without a substitution, we could meet the demand for each ancillary service as follows:

$$s_{RG} \geq d_{RG} \tag{9.11a}$$

$$s_{SR} \geq d_{SR} \tag{9.11b}$$

$$s_{NR} \geq d_{NR} \tag{9.11c}$$

$$s_{RR} \geq d_{RR} \tag{9.11d}$$

Mathematically, (9.11) is more viable than (9.10). In essence, if (9.11) could hold, (9.10) would also hold. However, the reverse of this axiom may not be true.

An ancillary service bid would represent a price and a quantity. So, for each ancillary service, it is possible to rank bids according to bidding prices and get a stepwise supply curve, as shown in Figure 9.8. The MCP for an ancillary service is the price on the supply curve corresponding to the total procured service, which is the price of the highest accepted bid for that ancillary service.

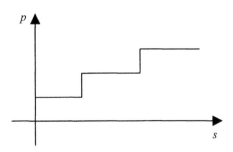

Figure 9.8 Supply Curve for Ancillary Service

The mathematical formulation of the rational buyer auction implemented in California is as follows:

$$\min \left\{ \begin{array}{l} s_{RG} \times p_{RG}(s_{RG}) + s_{SR} \times p_{SR}(s_{SR}) \\ + s_{NR} \times p_{NR}(s_{NR}) + s_{RR} \times p_{RR}(s_{RR}) \end{array} \right\} \quad (9.12)$$

subject to (9.10), where $p_{RG}(s_{RG})$ is the regulation supply curve, $p_{SR}(s_{SR})$ is the spinning reserve supply curve, $p_{NR}(s_{NR})$ is the non-spinning reserve supply curve, and $p_{RR}(s_{RR})$ is the replacement reserve supply curve.

9.4.2.2 Solution of the Rational Buyer Auction

Analyzing the objective function, we notice that the MCP for an ancillary service is a function of its procured quantity, which is a variable in the formulation. So the problem cannot be solved using the standard linear programming technique. One option is to use an exhaustive search that would evaluate all possible price combinations, since the total number of possible MCP combinations is a finite number [Liu00]. However, this finite number may be too large to make the method practical for a reasonably large system. Another option is to formulate the problem as a simple dynamic programming problem as follows [Ore01]:

ANCILLARY SERVICES AUCTION

$$\min_{D_{RR} \geq S_{NR} \geq D_{NR}} \left[C_{RR}(D_{RR} - S_{NR}) + \min_{D_{NR} \geq S_{SR} \geq D_{SR}} \left[C_{NR}(S_{NR} - S_{SR}) + \min_{D_{SR} \geq S_{RG} \geq D_{RG}} \left[C_{SR}(S_{SR} - S_{RG}) + C_{RG}(S_{RG}) \right] \right] \right] \quad (9.13)$$

In the formulation above, for simplicity, we define

$S_{RG} = s_{RG} \quad D_{RG} = d_{RG}$ (9.14a)

$S_{SR} = s_{RG} + s_{SR} \quad D_{SR} = s_{RG} + s_{SR}$ (9.14b)

$S_{NR} = s_{RG} + s_{SR} + s_{NR} \quad D_{NR} = d_{RG} + d_{SR} + d_{NR}$ (9.14c)

$S_{RR} = s_{RG} + s_{SR} + s_{NR} + s_{RR} \quad D_{RR} = d_{RG} + d_{SR} + d_{NR} + d_{RR}$ (9.14d)

$C_{RG}(x) = x \times p_{RG}(x)$ (9.15a)

$C_{SR}(x) = x \times p_{SR}(x)$ (9.15b)

$C_{NR}(x) = x \times p_{NR}(x)$ (9.15c)

$C_{RR}(x) = x \times p_{RR}(x)$ (9.15d)

The physical meaning of S_{RR} (or D_{RR}) is the cumulative supply (or demand) for replacement reserve and better quality ancillary services. Others can be similarly explained. $C_{RG}(.)$ is the total payment for regulation, $C_{SR}(.)$ for spinning reserve, $C_{NR}(.)$ for non-spinning reserve, and $C_{RG}(.)$ for replacement reserve.

Equation (9.13) is a four-stage dynamic programming problem. The solution involves one forward and one backward pass. In the forward pass, for each ancillary service, we first compute the cost of acquiring any feasible quantity if the demand for an ancillary service is only satisfied by the same type of ancillary service (no substitution). Then, we start with the most inner minimization and compute the cost of acquiring any feasible quantity of regulation capacity in the range between the demand for regulation and the combined demand for all services, which is trivial with the previous step. Next, we compute the least cost feasible mix of regulation and spinning reserve for any total amount of the two in the range between the combined demand for the two and the demand for all services combined, subject to the constraint that the regulation capacity exceeds the demand for regulation. We then compute the least cost feasible mix of the first two services and non-spinning reserve for each possible total amount of the three, and so on. In the backward pass we start with the total amount of the four ancillary services and trace back the least cost path from which we can extract the optimal procured quantity of each resource type. Once we have these quantities, we can use the supply functions to determine the corresponding market-clearing prices.

In practice, we could discretize the supply and demand quantities using an appropriate increment for the required precision. The computation time will depend on that increment.

9.4.2.3 Numerical Example

Consider the following example. There are two bidders, each bidding for the four ancillary services markets. The bids are shown in Table 9.18 and the requirements for all ancillary services are shown in Table 9.19. For this example, we use an increment of 2 MW.

Step 0: For each ancillary service, compute the cost of acquiring any feasible quantity if the demand for an ancillary service could only be satisfied by the same type of ancillary service (no substitution). It is easy to obtain Table 9.20. Consider a regulation example. From the bidding information, if the demand is less than or equal to 16MW, the only accepted bid would be from *B* with an MCP of 6 $/MW. If the demand is larger than 16MW, part of the bid from *A* would be accepted and *A* would set the MCP at 7 $/MW.

Table 9.18 Ancillary Services Bids

Bidder	Regulation		Spinning Reserve		Non-Spinning Reserve		Replacement Reserve	
	Price ($/MW)	Quantity (MW)	Price ($/MW)	Quantity (MW)	Price ($/MW)	Quantity (MW)	Price ($/MW)	Quantity (MW)
A	7	10	10	10	8	6	12	14
B	6	16	5	4	9	10	3	10

Table 9.19 Ancillary Services Demand

Type	Regulation	Spinning Reserve	Non-Spinning Reserve	Replacement Reserve
Demand (MW)	12	12	12	12

ANCILLARY SERVICES AUCTION

Step 1: Compute the cost of acquiring any feasible quantity of the regulation capacity within a range between the demand for regulation (12 MW) and the combined demand for all services (12 + 12 + 12 + 12 = 48 MW). Since the maximum regulation that A and B can provide is (10 + 16 = 26 MW), the actual range would be between 12 MW and 26 MW. From Table 9.20, we can easily get Table 9.21.

Step 2: compute the least cost feasible mix of regulation and spinning reserves within a range that is between the combined demand for the two services (12 + 12 = 24 MW) and the combined demand for all services (48 MW), subject to the constraint that the regulation capacity would exceed the demand for regulation (12 MW). Since the maximum amount of regulation and spinning reserve that A and B would be able to provide is (10 + 16 + 10 + 4 = 40 MW), the range would be between 24 MW and 40 MW.

Table 9.20 Step 0 of Example on Rational Buyer Auction

Demand (MW)	For Regulation			For Spinning Reserve			For Non-spinning Reserve			For Replacement Reserve		
	A	B	Cost	A	B	Cost	A	B	Cost	A	B	Cost
2	0	2	12	0	2	10	2	0	16	0	2	6
4	0	4	24	0	4	20	4	0	32	0	4	12
6	0	6	36	2	4	60	6	0	48	0	6	18
8	0	8	48	4	4	80	6	2	72	0	8	24
10	0	10	60	6	4	100	6	4	90	0	10	30
12	0	12	72	8	4	120	6	6	108	0	12	36
14	0	14	84	10	4	140	6	8	126	0	14	42
16	0	16	96	-	-	-	6	10	144	2	14	192
18	2	16	126	-	-	-	-	-	-	4	14	216
20	4	16	140	-	-	-	-	-	-	6	14	240
22	6	16	154	-	-	-	-	-	-	8	14	264
24	8	16	168	-	-	-	-	-	-	10	14	288
26	10	16	182	-	-	-	-	-	-	-	-	-

Table 9.21 Step 1 of Example on Rational Buyer Auction

Demand (MW)	Regulation (MW)		Total Cost ($)
	A	B	
12	0	12	72
14	0	14	84
16	0	16	96
18	2	16	126
20	4	16	140
22	6	16	154
24	8	16	168
26	10	16	182

To understand Table 9.22, consider the column for spinning reserve in Table 9.20. For a total demand of 32 MW, the possible combinations of regulation and spinning reserve are shown in Table 9.23. From the table, the optimal combination would be to procure 26 MW regulation and 6 MW spinning reserve (2 MW from A and 4 MW from B) and with a total procurement cost of $242. Other demand scenarios would be handled similarly.

Table 9.22 Step 2 of the Example on Rational Buyer Auction

Total Demand (MW)	Spinning Reserve			Regulation of A and B		Total Cost ($)
	A (MW)	B (MW)	Cost ($)	MW	Cost ($)	
24	0	4	20	20	140	160
26	0	4	20	22	154	174
28	0	4	20	24	168	188
30	0	4	20	26	182	202
32	2	4	60	26	182	242
34	4	4	80	26	182	262
36	6	4	100	26	182	282
38	8	4	120	26	182	302
40	10	4	140	26	182	322

Table 9.23 Explanation for Step 2 of Example on Rational Buyer Auction

Regulation		Spinning Reserve		Total Cost ($)
MW	Cost ($)	MW	Cost ($)	
18	126	14	140	266
20	140	12	120	260
22	154	10	100	254
24	168	8	80	248
26	182	6	60	242

It is important to note that in Step 2, the regulation capacity should exceed the demand for regulation. In our example, this constraint is naturally satisfied. Consider the bids in Table 9.24, which are similar to those in Table 9.18 but the price of spinning reserve is lower than that of regulation.

If we minimize the procurement cost while observing the above constraint, the result would be shown in Table 9.25. Table 9.25 also shows the results without considering the constraints (in parentheses), which are close except that the regulation is only 10 MW when the demand is 24 MW, and this violates the constraint.

ANCILLARY SERVICES AUCTION

Step 3: Compute the least cost feasible mix of the first two resources and non-spinning reserves for a total amount, which is within a range between the combined demand for the three services (12 + 12 + 12 = 36 MW) and the combined demand for all services (48 MW). This computation is subject to the constraint that the sum of the first two capacities would exceed the corresponding demand for the first two (24 MW). Note in Table 9.26 that the maximum regulation, spinning reserve, and non-spinning reserve is (10 + 16 + 10 + 4 + 6 + 10 = 56 MW), which is larger than the demand for all services combined. If the demand is 44 MW or 46 MW, the two combinations would lead to the same total procurement cost.

Table 9.24 Ancillary Services Bids

Bidder	Regulation		Spinning Reserve		Non-Spinning Reserve		Replacement Reserve	
	Price ($/MW)	Quantity (MW)	Price ($/MW)	Quantity (MW)	Price ($/MW)	Quantity (MW)	Price ($/MW)	Quantity (MW)
A	7	10	5	10	8	6	12	14
B	6	16	4	4	9	10	3	10

Table 9.25 Step 2 of Example on Rational Buyer Auction

Total Demand (MW)	Spinning Reserve			Regulation of A and B		Total Cost ($)
	A (MW)	B (MW)	Cost ($)	MW	Cost ($)	
24	8 (*10*)	4	60 (*70*)	12 (*10*)	72 (*60*)	132 (*130*)
26	10	4	70	12	72	142
28	10	4	70	14	84	154
30	10	4	70	16	96	166
32	10	4	70	18	126	196
34	10	4	70	20	140	210
36	10	4	70	22	154	224
38	10	4	70	24	168	238
40	10	4	70	26	182	252

Table 9.26 Step 3 of Example on Rational Buyer Auction

Total Demand (MW)	Non-Spinning Reserve			Regulation and Spinning Reserve of A and B		Total Cost ($)
	A (MW)	B (MW)	Cost ($)	MW	Cost ($)	
36	6	0	48	30	202	250
38	6	2	72	30	202	274
40	6	4	90	30	202	292
42	6	6	108	30	202	310
44	6	6 (*0*)	108 (*48*)	32 (*38*)	242 (*302*)	350
46	6	6 (*0*)	108 (*48*)	34 (*40*)	262 (*322*)	370
48	6	6	108	36	282	390

Step 4: Compute the least cost feasible mix of the first three resources and replacement reserves for a combined demand of all services (48 MW), subject to the constraint that the sum of the first three capacities would exceed the demand for the first three (36 MW). See Table 9.27.

Table 9.27 Step 4 of Example on Rational Buyer Auction

Total Demand (MW)	Replacement Reserve (MW)			Regulation, Spinning and Non-spinning Reserve of A and B		Total Cost ($)
	A (MW)	B (MW)	Cost ($)	MW	Cost ($)	
48	0	10	30	38	274	304

Step 5: Start with the total amount for the four ancillary services and trace back the least cost path to extract the optimal procured quantity of each resource type.

According to Step 4, the optimal replacement reserve is 10 MW from B and the sum of regulation, spinning and non-spinning reserve services is 38 MW. According to Step 3, if the sum of regulation, spinning and non-spinning reserve services is 38 MW, the optimal non-spinning reserve would be 6 MW from A, 2 MW from B, and the sum of regulation and spinning reserve services would be 30 MW. According to Step 2, if the sum of regulation and spinning reserve service is 30 MW, the optimal spinning reserve would be 4 MW from B and the sum of regulation would be 26 MW. According to Step 1, if the regulation service is 26 MW, the optimal regulation procurement would be 10 MW from A and 16 MW from B. The final procurement schedule for all ancillary services is shown in Table 9.28. Note that although the demand for regulation is only 12 MW, we procure 26 MW regulation. The difference (16 - 12 = 14 MW) is for spinning reserve (12 - 4 = 8 MW), non-spinning reserve (12 - 8 = 4 MW) and replacement reserve (12 - 10 = 2 MW).

Step 6: Use the supply functions to determine the corresponding MCPs. The MCP for a specific ancillary service is the bidding price of the highest accepted bid for that ancillary service. The results are shown in Table 9.29. Table 9.30 compares the rational buyer auction with the auction without substitution. The rational buyer auction saves a procurement cost of (444 - 304 = $140).

ANCILLARY SERVICES AUCTION

Table 9.28 Procurement Schedule

Bidder	Regulation (MW)	Spinning Reserve (MW)	Non-spinning Reserve (MW)	Replacement Reserve (MW)
A	10	0	6	0
B	16	4	2	10
Total	26	4	8	10

Table 9.29 Market Clearing Prices

Regulation ($/MW)	Spinning Reserve ($/MW)	Non-spinning Reserve ($/MW)	Replacement Reserve ($/MW)
7	5	9	3

Table 9.30 Comparison of Rational Buyer Auction and Auction without Substitution

Service	Demand (MW)	No Substitution			Rational Buyer		
		Purchase (MW)	MCP ($/MW)	Cost ($)	Purchase (MW)	MCP ($/MW)	Cost ($)
Regulation	12	12	6	72	26	7	182
Spinning reserve	12	12	10	120	4	5	20
Non-spinning reserve	12	12	9	108	8	9	72
Replacement reserve	12	12	12	144	10	3	30
Total	48	48	–	444	48	–	304

9.4.3 Marginal Pricing Auction

According to the previous section, a rational buyer auction can reduce the total procurement cost because it would enable the substitution of a higher quality lower cost service for a lower quality higher cost service. However, the problem of price reversal would still exist. For example, the MCP for non-spinning reserve (9 $/MW) is larger than the MCP for regulation (7 $/MW) and the MCP for spinning reserve (5 $/MW). In this regard, marginal pricing is an option to avoid price reversal.

In marginal pricing auction, the objective is to minimize the total social cost and the substitution of a high quality service for a low quality service is enabled. All market participants are paid a uniform price for each ancillary service, and the price for an ancillary service is determined as the marginal cost of that ancillary service.

Note that in marginal pricing, the objective cannot be the minimization of procurement cost since the procurement price is the marginal cost, which is not given prior to optimization.

9.4.3.1 Formulation of the Marginal Pricing Auction

Assume that bids would represent their true costs. According to the Figure 9.8, the social costs would be given as $\int_0^{S_{RG}} p_{RG}(s)ds$, $\int_0^{S_{SR}} p_{SR}(s)ds$, $\int_0^{S_{NR}} p_{NR}(s)ds$, and $\int_0^{S_{RR}} p_{RR}(s)ds$ for regulation, spinning reserve, non-spinning reserve, and replacement reserve, respectively. The formulation of marginal pricing auction is presented as follows.

$$\min \left\{ \begin{array}{l} \int_0^{S_{RG}} p_{RG}(s)ds + \int_0^{S_{SR}} p_{SR}(s)ds \\ + \int_0^{S_{NR}} p_{NR}(s)ds + \int_0^{S_{RR}} p_{RR}(s)ds \end{array} \right\} \quad (9.16)$$

subject to (9.10).

9.4.3.2 Solution of the Marginal Pricing Auction

Formulation (9.16) is similar to the formulation of the rational buyer auction (9.12), except that the objective here is to minimize the social cost while the objective of the rational buyer auction is to minimize the procurement cost. So it is possible to solve this problem using the same technique as we presented for the rational buyer auction. A simpler solution method is presented as follows.

Considering the stepwise characteristics of the supply curve, we solve the problem by a simple "greedy algorithm." This algorithm would successively fill the demand for each service from the highest to the lowest quality using bids in their ascending merit order of price and pushing any unused bids to the next level. For example, regulation bids are selected in the ascending merit order of bid prices until the cumulative quantity could satisfy the demand for regulation. The remaining regulation bids are pushed forward and mixed with spinning reserve bids, and the selection out

ANCILLARY SERVICES AUCTION

of the combined pool is again done in the ascending order of bid prices until the cumulative quantity could satisfy the spinning reserve demand. Similar procedures would be applied to non-spinning and replacement reserves.

The major difference between marginal pricing and rational buyer auction is that the MCP for an ancillary service is what it marginally costs to serve the next MW for that ancillary service. It could be shown that marginal pricing would avoid a price reversal. The conclusion is [Kam01]

$$\text{MCP}_{RG} = \mu_{RG} + \mu_{SR} + \mu_{NR} + \mu_{RR} \tag{9.17a}$$

$$\text{MCP}_{SR} = \mu_{SR} + \mu_{NR} + \mu_{RR} \tag{9.17b}$$

$$\text{MCP}_{NR} = \mu_{NR} + \mu_{RR} \tag{9.17c}$$

$$\text{MCP}_{RR} = \mu_{RR} \tag{9.17d}$$

Where, μ_{RG}, μ_{SR}, μ_{NR}, and μ_{RR} are the Lagrangian multipliers for constraints (9.10a)–(9.10d) respectively, which are all positive. From (9.17), we reach the following conclusion, which points out that there will be no price reversal:

$$\text{MCP}_{RG} \geq \text{MCP}_{SR} \geq \text{MCP}_{NR} \geq \text{MCP}_{RR} \tag{9.18}$$

9.4.3.3 Example

Consider the simple example given in Table 9.31. In this example, there is only one bidder and only regulation and a spinning reserve are considered. In Cases 1 and 2, the bidding price for the spinning reserve is higher than that for regulation, while in Cases 3 and 4, the bidding price for spinning reserve is lower than that of regulation.

The schedule, MCP, and cost are shown in Table 9.32. Consider Case 2 as an example. To compute the MCP for regulation, which is the cost of providing the next MW regulation, we suppose that the demand for regulation is 401MW. The schedule would be 500 MW of regulation and 101 MW of spinning reserve with a total social cost of 500 x 5 + 101 x 6 = $3106. The increased cost is 3106 - 3100 = $6. Thus the MCP for regulation is 6 $/MW.

Similarly, to compute the MCP for spinning reserve, which is the cost of providing the next MW spinning reserve, we suppose the demand for spinning reserve is 201 MW. The schedule would be 500 MW of regulation and 101 MW of spinning reserve with a total social cost of 500 × 5 + 101 × 6 = $3106. The increased cost is 3106 - 3100 = $6. Thus the MCP for spinning reserve is 6 $/MW. In all four cases, the MCP for regulation is larger than or equal to the MCP for spinning reserve, which verifies our conclusion that no price reversal would happen in a marginal pricing auction.

In comparison, the result for the same example using a rational buyer auction is shown in Table 9.33. The schedule and social cost are the same as that in the marginal pricing auction. However, in Case 2, the MCP for regulation is less than the MCP for spinning reserve, which is a price reversal. The procurement cost for Case 2 is $3100, which is the same as the social cost. However, in the marginal pricing auction, the procurement cost is $3600, which is $500 higher than the social cost.

Table 9.31 Ancillary Services Bids

Case Index	Demand (MW)		Bids for Regulation		Bids for Spinning Reserve	
	Regulation	Spinning Reserve	Quantity (MW)	Price ($/MW)	Quantity (MW)	Price ($/MW)
1	200	200	500	5	300	6
2	400	200	500	5	300	6
3	200	200	500	5	300	4
4	400	200	500	5	300	4

Table 9.32 Schedule, Market Clearing Price and Cost

Case Index	Schedule (MW)		MCP ($/MW)		Total Cost ($)	
	Regulation	Spinning Reserve	Regulation	Spinning Reserve	Social Cost	Procurement Cost
1	400	0	5	5	2000	2000
2	500	100	6	6	3100	3600
3	200	200	5	4	1800	1800
4	400	200	5	4	2800	2800

Table 9.33 Schedule, Market Clearing Price and Cost

Case Index	Schedule (MW)		MCP ($/MW)		Total Cost ($)	
	Regulation	Spinning Reserve	Regulation	Spinning Reserve	Social Cost	Procurement Cost
1	400	0	5	5	2000	2000
2	500	100	5	6	3100	*3100*
3	200	200	5	4	1800	1800
4	400	200	5	4	2800	2800

ANCILLARY SERVICES AUCTION

In general, if the bidding price for a lower quality service is higher than that for a higher quality service, a price reversal is possible in the rational buyer auction. In both Cases 1 and 2, the bidding price for spinning reserve is higher than that for regulation, while a price reversal would only occur in Case 2. That is because in Case 1, no spinning reserve is scheduled, while in Case 2, some spinning reserve is scheduled. In comparison, if the bidding price for a lower quality service is lower than that for a higher quality service, a price reversal is not possible in the rational buyer auction. Cases 3 and 4 point to these conditions. In the case of a marginal pricing auction, both theory and experiment show that a price reversal would not occur.

Consider the same example in the rational buyer auction, which is more complex. The next set of steps describes the scheduling procedure in detail.

Step 1: Schedule for regulation. The bids are,

A: 10 MW @ 7 \$/MW, B: 16 MW @ 6 \$/MW.

To satisfy the demand (12 MW) at the minimum social cost, the schedule would be A: 0 MW, B: 12 MW.

The unused bids are A: 10 MW @ 7 \$/MW, B: 4 MW @ 6 \$/MW.

Step 2: Schedule for spinning reserve. The bids are

Regulation: A: 10 MW @ 7 \$/MW, B: 4 MW @ 6 \$/MW

Spinning reserve: A: 10MW @ 10 \$/MW, B: 4MW @ 5 \$/MW.

To satisfy the demand (12MW) at the minimum social cost and include substitution, the schedule would be

Regulation: A: 4 MW, B: 4 MW

Spinning reserve: A: 0 MW, B: 4 MW

The unused bids are

Regulation: A: 6 MW @ 7 \$/MW

Spinning reserve: A: 10 MW @ 10 \$/MW

Step 3: Schedule for non-spinning reserve. The bids are

Regulation: A: 6 MW @ 7 \$/MW

Spinning reserve: A: 10MW @ 10 \$/MW

Non-spinning reserve: A: 6 MW @ 8 \$/MW, B: 10 MW @ 9 \$/MW

To satisfy the demand (12 MW) at the minimum social cost and include substitution, the schedule would be

Regulation: A: 6 MW

Non-spinning reserve: A: 6 MW, B: 0 MW

The unused bids are

Spinning reserve: A: 10 MW @ 10 \$/MW

Non-spinning reserve: B: 10 MW @ 9 \$/MW

Step 4: Schedule for replacement reserve. The bids are

Spinning reserve: A: 10 MW @ 10 \$/MW

Non-spinning reserve: B: 10 MW @ 9 \$/MW

Replacement reserve: A: 14MW @ 12\$/MW, B: 10MW @ 3\$/MW

To satisfy the demand (12 MW) at minimum social cost considering substitution, the schedule would be

Non-spinning reserve: B: 2 MW

Replacement reserve: A: 0 MW, B: 10MW

The unused bids are

Spinning reserve: A: 10 MW @ 10 \$/MW

Non-spinning reserve: B: 8 MW @ 9 \$/MW

Replacement reserve: A: 14 MW @ 12 \$/MW

The preceding procurement procedure is summarized in Table 9.34. Of course we could obtain the results very quickly by way of a computer program. The final procurement schedule is shown in Table 9.35, which happens to be the same as the schedule of the rational buyer auction (Table 9.28). The social cost is \$252.

We use the definition of marginal pricing to compute MCPs. Consider the regulation example. The marginal cost would be the cost it takes to serve the next MW regulation. Since the increment in this example is 2 MW, we consider the cost it would take to serve the next 2 MW regulation. Thus the demand for regulation would be 14 MW. Repeating the procedure above, which would be very fast, we find the schedule shown in Table 9.36. The social cost is $270 and the increased cost is (270 - 252 = $18). So the MCP for regulation would be (18/2 = 9 $/MW). MCPs for other services can be calculated similarly. The result is shown in Table 9.37.

Table 9.34 Procurement Schedule

	Regulation (MW)				Spinning Reserve (MW)				Non-spinning Reserve (MW)				Replacement Reserve (MW)			
Step	1	2	3	4	1	2	3	4	1	2	3	4	1	2	3	4
Bidder A	0	4	6	-	-	0	0	-	-	-	0	0	-	-	-	0
Bidder B	12	4	-	-	-	4	0	-	-	-	6	2	-	-	-	10

Table 9.35 Procurement Schedule

Bidder	Regulation (MW)	Spinning Reserve (MW)	Non-spinning Reserve (MW)	Replacement Reserve (MW)
A	10	0	6	0
B	16	4	2	10

Table 9.36 Procurement Schedule

Bidder	Regulation (MW)	Spinning Reserve (MW)	Non-spinning Reserve (MW)	Replacement Reserve (MW)
A	10	0	6	0
B	16	4	4	10

Table 9.37 Market Clearing Prices

Regulation ($/MW)	Spinning Reserve ($/MW)	Non-spinning Reserve ($/MW)	Replacement Reserve ($/MW)
9	9	9	9

In this example, the MCPs for all ancillary services are the same without any price reversal. The total procurement cost is 9 × 48 = $432, which is much larger than that of the rational buyer auction ($304). This high procurement cost is a big disadvantage for a marginal pricing auction, and it could render the auction impractical in industry. In general, a marginal pricing auction could avoid a price reversal but it could increase the total procurement cost.

9.4.4 Discussions

In industry applications, the sequential auction design for ancillary services has created a few problems. Higher quality and lower cost services have remained unused and there are price reversals. To overcome these problems, various simultaneous approaches enabling the substitution of a higher quality service for a lower quality service have been proposed and implemented. In this section, we discussed in detail two design options: rational buyer and marginal pricing auctions. Both auctions have advantages and disadvantages. The rational buyer auction, which has been implemented in California, would reduce the procurement cost but cannot avoid a price reversal. The marginal pricing auction could avoid a price reversal but might increase the procurement cost. More theoretical and experimental analyses are needed to predict the performances of these options more accurately. It should be noted that even a good market design's performance might be discounted by market power, which may lead to noncompetitive pricing and inefficient allocation of resources.

9.5 AUTOMATIC GENERATION CONTROL (AGC)

AGC would provide an effective method for adjusting the generation to minimize frequency deviations and regulate tie-line flows. This crucial role would continue in restructured markets with some modifications to account for bilateral contracts that span over several control areas.

9.5.1 AGC Functions

In a vertically integrated structure, utilities were responsible for supplying loads and maintaining reliability. Utilities have been providing these services for about six decades to match the generation with the load and maintain the frequency within the 60 Hz range. In general, a system load could consist of three components:

- Constant base load during the hour
- Hourly load trend
- Fluctuations around the underlying trend.

ANCILLARY SERVICES AUCTION

Load uncertainty could complicate the decision on how to meet the system load as control area[7] operators are obliged to provide sufficient on-line generation to respond to uncertainties.

In the restructured market structure, FERC Order 888 required electric utilities to unbundle generation and transmission services and defined six ancillary services of which regulation and frequency response would track moment-to-moment fluctuations in the system load. Regulation and frequency response signifies the use of generators to help meet the NERC control area performance criteria. These criteria require the control areas to maintain its area control error (ACE) within tight limits. ACE is measured in MW and defined as the instantaneous difference between the actual and scheduled interchanges plus a frequency bias.

Originally, there were two control area performance criteria. The first criterion was called A1, which required the control area to be balanced with the rest of the interconnection at least once every 10 minutes. The second criterion was called A2, which required the control area's energy imbalance net of frequency bias to be within a certain limit every 10 minutes. Utilities would deploy AGC to manage ACE to meet both NERC criteria.

In 1997, NERC adopted two new criteria, Control Performance Standard 1 (CPS1) and CPS2, in place of the old A1 and A2 criteria, for the evaluation of load frequency control (LFC) in each control area [Ner01]. The new criteria are more sophisticated and would require more measurement and data collection. All control areas in North America are required to report CPS1 and CPS2 performances to the NERC monthly.

CPS1 is the measure of short-term error between load and generation. CPS1's performance will be good if a control area matches generation with load exactly, or if the mismatch causes system frequency to be driven closer to 60 Hz. CPS1's performance would be degraded if the system frequency is driven away from 60 Hz. CPS2 would place boundaries on CPS1 to limit net unscheduled power flows that are unacceptably large. CPS2 would prevent excessive generation/load mismatches even if a mismatch is in the proper direction. Large mismatches could cause excessive power flows and potential transmission overloads between areas with over-generation and those with insufficient generation. CPS1 and

---

[7] The definition of a control area is somewhat determined by pooling arrangements of utilities. Sometimes the physical boundaries of a vertically integrated utility define a control area. All such control areas are interconnected by tie lines.

CPS2 are measurable and normal functions of each control area's energy management system (EMS). Measurements are taken continuously with data recorded at each minute of operation. To obtain a "pass" control compliance rating, a control area must demonstrate that CPS1 is greater than or equal to 100%, and CPS2 is greater than or equal to 90%. Otherwise, the control compliance rating would be "fail." Perfect control results are 200% CPS1 and 100% CPS2.

9.5.2 AGC Response

The AGC's main objective is to control tie-line flows at scheduled values defined by utilities' contracts. The tie-line flow control will match the generation and the local load in order to maintain the frequency within control areas as close to the nominal value as possible. In the classical AGC system, the balance between generation and load was achieved by detecting frequency and tie-line flow deviations which were used in generating and ACE signal. The ACE signal was used for an integral feedback control strategy.

In general, generators would respond to fast load fluctuations (i.e., 1–2 seconds) depending on the droop characteristics of governors. Generators would respond to slower fluctuations (2–6 seconds) based on signals received from the control area's AGC system based on measuring ACE. Generators respond to longer-term load changes (several minutes) based on manual directions that would utilize the economics of the AGC system to minimize operating costs. The AGC system would realize generation changes by sending signals to units under its control. The performance of an AGC system is very much dependent on how those units would respond to the signals. The generating unit response characteristics are dependent on many factors such as type of unit, type of fuel, plant type, type of plant control, operating point, and operator actions.

NERC separated generator actions into two parts: those associated with large frequency deviations where generators would respond through governor action and then in response to AGC signals, and those associated with a continuous regulation process in response to AGC signals only. Large frequency deviations would be due to generation or transmission outages, which might occur rarely. During a sudden area load change, the area frequency experiences a transient drop. At the transient state, there would be flows of power from other areas to supply the excess load in this area. Certain generators within each area would be on regulation to meet this load change. At steady state, the generation would be matched exactly

ANCILLARY SERVICES AUCTION

with the load, causing tie-line power and frequency deviations to drop to zero.

A DISCO has the freedom to contract with any GENCO in its own area or sign bilateral contracts with a GENCO in another area that would be cleared by the ISO. If a bilateral contract exists between DISCOs in one control area and GENCOs in other control areas, the scheduled flow on a tie line between two control areas must exactly match the net sum of the contracts that exist between market participants on opposite sides of the tie line. If the bilateral contract is adjusted, the scheduled tie-line flow must be adjusted accordingly. In general, using bilateral contracts, DISCOs would correspond demands to GENCOs, which would introduce new signals that did not exist in the vertically integrated environment. These signals would give information as to "which GENCO ought to follow which DISCO". Moreover, these signals would provide information on scheduled tie-line flows adjustments and ACEs for control areas.

9.5.3 AGC Units Revenue Adequacy

Resources for the real-time energy imbalances include ancillary services (e.g., regulation, spinning, non-spinning, and replacement reserves) and supplemental energy. Ancillary services bids should include the financial information for capacity reservation and energy, as well as operational information such as location, ramp rate, and quantity of services. So ancillary services are procured competitively using a two-part bid format. Suppliers who have committed capacity to one of the ancillary services markets except regulation, and who also produce energy, receive the imbalance energy price in addition to their respective ancillary services capacity payment. Supplemental energy bids would only include an energy bid with no capacity reservation bid. Suppliers who provide energy through supplemental energy bids receive the imbalance energy payment only.

There are two approaches to ensure a revenue adequacy for AGC units. The first is to compensate AGC units on the basis of their energy price bids instead of the real-time MCP for that 10-minute interval. This would ensure that the units obtain adequate revenue even if the real-time MCP in that interval is lower than their offered energy price. The other approach is to let AGC bidders internalize the risk of revenue adequacy within their capacity reservation price. The current mechanism in some markets is to pay suppliers of regulation energy an amount based the total (up and down) adjustable capacity that they provide during an hour.

9.5.4 AGC Pricing

Load and supply schedule deviations could be instructed or uninstructed. Instructed deviations occur because of planned line and unit outages by the ISO. Moreover, instructed deviations are procured by the ISO in response to energy imbalances caused by uninstructed deviations. Uninstructed deviations occur because of load forecasting errors, normal variations of load and generation from scheduled levels, and unplanned line and unit outages.

The interactions among different resources in a real-time market are shown in Figure 9.9. Generators with AGC would respond to uninstructed deviations from schedules within few seconds. Then, to return AGC units to their set points, the ISO will utilize other resources, which have submitted energy price/quantity bids for the real-time energy imbalances, by means of instructed deviations.

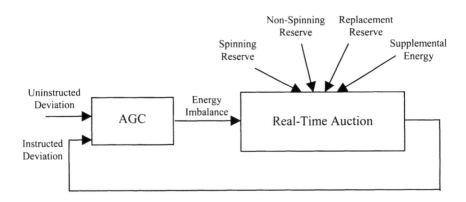

Figure 9.9 The Real-Time Market

Suppose that the real-time auction takes place every 10 minutes, which means 6 times every hour, for an instructed deviations settlement. The ISO would conduct a single-sided auction every 10 minutes for trading energy. If the ISO is a buyer, the resource with the highest energy price is used in the 10-minute auction to set the price for the instructed deviations. If the ISO is a seller, the resource with the lowest energy price is used in the 10-minute auction to set the price for the instructed deviations. A weighted average of the 10-minute prices is calculated at the end of the hour to settle all uninstructed deviations. The weighted average is called

ANCILLARY SERVICES AUCTION

the hourly ex-post price. At any dispatch interval, an energy imbalance is either positive (supply exceeds load) or negative (load exceeds supply). In the following, we discuss cases with negative energy imbalance, when the ISO is a buyer. Other cases can be similarly handled.

9.5.4.1 Formulation and Solution

If the ISO is a buyer, the real-time MCP for instructed deviations is calculated every 10 minutes by the ISO through a matching process as the highest energy price among accepted bids at which energy imbalances are met. The matching process can be represented mathematically as follows.

$$\min \sum_i \{E(i,t) \cdot Q(i,t)\} \quad (9.19)$$

$$\text{s.t.} \quad \sum_i Q(i,t) \geq Q^{proc}(t) \quad (9.20)$$

$$Q(i,t) \leq Q^{max}(i,t) \quad (9.21)$$

where $E(i,t)$ is the energy bid \$/MWh offered by unit i in interval t. $Q(i,t)$ is the quantity of imbalance energy procured by the ISO from unit i in interval t, and $Q^{proc}(t)$ is the required quantity of imbalance energy to be procured by the ISO in interval t. $Q^{max}(i,t)$ is the maximum quantity that the ISO can utilize from unit i in interval t based on its ramp rate.

The preceding problem can be solved as a linear optimization problem since $Q(i,t)$ is the only variable. The ISO sorts the bids based on the energy bid price $E(i,t)$ for all participants to calculate the real-time MCP and to decide which bids to accept. The reserve MCP is calculated as [$Max\{E(i,t)\}$] for those accepted bids to satisfy energy imbalance requirements. This MCP is the highest energy price among accepted bids at which the energy imbalance requirements are met.

9.5.4.2 Case Studies

Three case studies are presented. In the first case, there is no congestion in any dispatch intervals of the given hour. In the second case, there are congestions in all dispatch intervals. In the third case, congestion exists in some of the dispatch intervals but not in all intervals.

9.5.4.2.1 Case without Congestion. The same system used in Section 9.3.4.2 (Figure 9.4) is used here to illustrate the proposed real-time auction. The system consists of three GENCOs: A, B and C, which compete for providing imbalance energy. GENCOs A and B each has 3 generating units with a reserve capacity of 30 MW, and GENCO C has 4 units with a reserve capacity of 50 MW. The ISO has a reserve requirement of 5.0/6, 10.0/6, 15.0/6, 27.5/6, 45.0/6, and 55.0/6 MW at intervals 1, 2, 3, 4, 5, and 6 respectively.

Energy bids submitted by accepted GENCOs in the ancillary services market are given in Table 9.38. Note that G_5, G_6, and G_{10} have ramping limitations of 50% of their capacities and only 10 MW is accepted from G_6 as a result of the ancillary service auction. Table 9.39 shows the dispatch instructions at each of the 10-minute intervals within the hour. The real-time MCP is set to the highest accepted bid in each 10-minute interval.

The calculated real-time MCPs are used to calculate the instructed deviation settlements as follows.

Payment from the ISO to the GENCOs for the instructed deviations are

GENCO A: 5/6x15 + 5/6x19 + 5/6x29 + 5/6x33 + 5/6x36 + 5/6x56 = $156.67

GENCO B: 5/6x29 + 17.5/6x33 + 25/6x36 + 25/6x56 = $503.75

GENCO C: 5/6x19 + 5/6x29 + 5/6x33 + 15/6x36 + 25/6x56 = $390.83

Total ISO payment = 156.67 + 503.75 + 390.83 = $1051.25

The average hourly ex-post price used for uninstructed deviation settlements is (5/6x15 + 10/6x19 + 15/6x29 + 27.5/6x33 + 45/6x36 +55/6x56) / (157.5/6) = 40.048 $/MWh

Based on Table 9.39, the total uninstructed deviation is 157.5/6 = 26.25 MWh. So, the ISO would collect from the uninstructed deviation settlement: 40.048 x 26.25 = $1051.25.

ANCILLARY SERVICES AUCTION

Table 9.38 Accepted Bidders, Capacity and Energy Bids

GENCO		Capacity (MW)	Energy Bid ($/MWh)
A	G_1	5.0	15.0
B	G_5	10.0	29.0
	G_6	15.0	33.0
C	G_7	5.0	19.0
	G_8	10.0	36.0
	G_{10}	20.0	56.0

Table 9.39 Dispatch Instructions

GENCO		Capacity MW	Energy Bid $/MWh	Interval 1	2	3	4	5	6	Total
A	G_1	5	15	5	5	5	5	5	5	5 MW
				5/6	5/6	5/6	5/6	5/6	5/6	5 MWh
B	G_5	10	29			5	10	10	10	10 MW
						5/6	10/6	10/6	10/6	35/6 MWh
	G_6	15	33				7.5	15	15	15 MW
							7.5/6	15/6	15/6	37.5/6 MWh
C	G_7	5	19		5	5	5	5	5	5 MW
					5/6	5/6	5/6	5/6	5/6	25/6 MWh
	G_8	10	36					10	10	10
								10/6	10/6	20/6 MWh
	G_{10}	20	56						10	10
									10/6	10/6 MWh
Real-time MCP ($/MWh)				15	19	29	33	36	56	
Total Available MW				5	10	15	27.5	45	55	55 MW
Total Required MWh				5/6	10/6	15/6	27.5/6	45/6	55/6	157.5/6 MWh

9.5.4.2.2 Case with Congestion–Impact of Transmission Congestion. In the case of congestion in any 10-minute settlement, the ISO would set separate real-time auctions for each zone in that 10-minute interval. So imbalance energy would be procured by the ISO separately in each zone to meet the uninstructed deviations. Consequently, different instructed MCP would be calculated for each zone in the same 10-minute settlement. Since congestion could occur in one settlement or more, one should be careful in calculating the uninstructed deviation price for that hour, for then the price would be the weighted average of the six instructed MCPs. Therefore, if congestion exists in any 10-minute settlement of that hour, there would be different hourly uninstructed deviation prices for each zone.

The same system discussed above is used in this section for illustration. For the congested case, the ISO would decide to establish two zones as shown in Figure 9.6. Here, we assume that congestion exists in all six 10-minute intervals of the hour. Hence, for each real-time settlement, we would have an auction in each zone. The ISO has a reserve requirement of 2.5/6, 5.0/6, 10.0/6, 17.5/6, 20.0/6, and 20.0/6 MW at intervals 1, 2, 3, 4, 5, and 6 respectively in zone 1. Also, the ISO has a reserve requirement of 2.5/6, 5.0/6, 5.0/6, 10.0/6, 25.0/6, and 35.0/6 MW at intervals 1, 2, 3, 4, 5, and 6, respectively, in zone 2.

The accepted bids in each zone are shown in Table 9.40. Note that G_5, G_6, and G_{10} have ramping limitations of 50% of their capacities and only 10 MW is accepted from G_6 as a result of the ancillary service auction. Tables 9.41 and 9.42 show the dispatch instructions at each of the 10-minute intervals within the hour for zones 1 and 2, respectively. The real-time MCP is set to the highest accepted bid in each 10-minute interval in each zone.

The calculated real-time MCPs in zone 1 are used for calculating the instructed deviation settlements in zone 1 as follows.

Payments from the ISO to the GENCOs for the instructed deviations are

GENCO A: $0.0

GENCO B: $0.0

GENCO C: 2.5/6x14 + 5/6x14 + 10/6 x14 + 17.5/6x19 + 20/6x19 + 20/6x19 = $222.92

Total ISO payment = 0.0 + 0.0 + 222.92 = $222.92

The average hourly ex-post price used for the uninstructed deviation settlements in zone 1 is calculated as: (2.5/6x14 + 5/6x14 + 10/6x14 + 17.5/6x19 + 20/6x19 + 20/6x19) / (75/6) = 17.833 $/MWh. Note it is lower than that without congestion:

Table 9.40 Accepted Bidders, Capacity and Energy Bids

Zones	GENCO		Capacity (MW)	Energy Bid ($/MWh)
1	C	G_7	5.0	19.0
		G_9	15.0	14.0
2	A	G_1	5.0	15.0
	B	G_5	10.0	29.0
		G_6	10.0	33.0
	C	G_{10}	20.0	56.0

ANCILLARY SERVICES AUCTION

Table 9.41 Dispatch Instructions in Zone 1

GENCO		Capacity MW	Energy Bid $/MWh	Interval						Total
				1	2	3	4	5	6	
C	G_7	5	19				5	5	5	5 MW
							2.5/6	5/6	5/6	12.5/6 MWh
	G_9	15	14	15	15	15	15	15	15	15 MW
				2.5/6	5/6	10/6	15/6	15/6	15/6	62.5/6 MWh
Real-time MCP $/MWh				14	14	14	19	19	19	
Total available MW				15	15	15	20	20	20	20 MW
Total required MWh				2.5/6	5/6	10/6	17.5/6	20/6	20/6	75/6 MWh

Table 9.42 Dispatch Instructions in Zone 2

GENCO		Capacity MW	Energy Bid $/MWh	Interval						Total
				1	2	3	4	5	6	
A	G_1	5	15	5	5	5	5	5	5	5 MW
				2.5/6	5/6	5/6	5/6	5/6	5/6	27.5/6 MWh
B	G_5	10	29				5	10	10	10 MW
							5/6	10/6	10/6	25/6 MWh
	G_6	10	33					7.5	10	10 MW
								7.5/6	10/6	17.5/6 MWh
C	G_{10}	20	56					10	10	10 MW
								2.5/6	10/6	12.5/6 MWh
Real-time MCP ($/MWh)				15	15	15	29	56	56	
Total Available MW				5	5	5	10	32.5	35	35 MW
Total Required MWh				2.5/6	5/6	5/6	10/6	25/6	35/6	82.5/6 MWh

From Table 9.41, the total uninstructed deviation is 75/6 = 12.50 MWh. So, the ISO would collect from the uninstructed deviation settlement: 17.833 × 12.50 = $222.92.

The calculated real-time MCPs in zone 2 are used for calculating the instructed deviation settlements in zone 2 as follows:

Payments from the ISO to the GENCOs for the instructed deviations are

GENCO *A*: 2.5/6×15 + 5/6×15 + 5/6×15 + 5/6×29 + 5/6×56 + 5/6×56
= $148.75

GENCO *B*: 5/6 × 29 + 17.5/6 × 56 + 20/6 × 56 = $374.17

GENCO *C*: 2.5/6 × 56 + 10/6 × 56 = $116.67

Total ISO payment = 148.75 + 374.17 + 116.67 = $639.58

The average hourly ex-post price used for the uninstructed deviation settlements in zone 2 is calculated as: (2.5/6 × 15 + 5/6 × 15 + 5/6 × 15 + 10/6 × 29 + 25/6 × 56 + 35/6 × 56) / (82.5/6) = 46.515 $/MWh. Note it is higher than that without congestion.

From Table 9.28, the total uninstructed deviation is 82.5/6 = 13.75 MWh. So the ISO will collect from the uninstructed deviation settlement: 46.515 × 13.75 = $639.58.

So, in the event of congestion, the total ISO payments in zones 1 and 2 to the GENCOs for the instructed deviations is calculated as follows:

GENCO A: 0.0 + 148.75 = $148.75

GENCO B: 0.0 + 374.17 = $374.17

GENCO C: 222.92 + 116.67 = $339.58

The total ISO payment = 148.75 + 374.17 + 339.58 = $862.50

Note that both the ISO's total payment and the ISO's payment to any GENCO in case of congestion are lower than those without congestion. This shows that congestion will affect the ISO payment to the GENCOs.

9.5.4.2.3 Some Intervals with Congestion. Unlike the analysis of the previous section, we assume here that congestion exists only in the last two 10-minute intervals of the hour. Accordingly, the ISO reserve requirements are 5.0/6, 10.0/6, 15.0/6, and 27.5/6 MW at intervals 1, 2, 3, and 4 respectively. The reserve requirements for the last two intervals are 20.0/6 and 20.0/6 MW at intervals 5 and 6, respectively, in zone 1 and 25.0/6 and 35.0/6 MW at intervals 5 and 6, respectively, in zone 2.

The accepted bids for both zones are the same as those in Table 9.38 for the first four intervals, and the same as those in Table 9.40 for the last two intervals. Note that G_5, G_6, and G_{10} have ramping limitations of 50% of their capacities, and only 10 MW is accepted from G_6 as a result from the ancillary service auction. Tables 9.43, 9.44, and 9.45 show the dispatch instructions at each of the 10-minute intervals within the hour. The real-time MCP is set to the highest accepted bid in each 10-minute interval in each zone.

ANCILLARY SERVICES AUCTION

Table 9.43 First Four Intervals Dispatch Instructions

GENCO		Capacity MW	Energy Bid $/MWh	Interval						Total
				1	2	3	4	5	6	
A	G_1	5	15	5	5	5	5			5 MW
				5/6	5/6	5/6	5/6			20/6 MWh
B	G_5	10	29			5	10			10 MW
						5/6	10/6			15/6 MWh
	G_6	15	33				7.5			15 MW
							7.5/6			7.5/6 MWh
C	G_7	5	19		5	5	5			5 MW
					5/6	5/6	5/6			15/6 MWh
	G_8	10	36							
	G_{10}	20	56							
Real-time MCP ($/MWh)				15	19	29	33			
Total available MW				5	10	15	27.5			27.5 MW
Total required MWh				5/6	10/6	15/6	27.5/6			57.5/6 MWh

Table 9.44 Last Two Intervals Dispatch Instructions in Zone 1

GENCO		Capacity MW	Energy Bid $/MWh	Interval						Total
				1	2	3	4	5	6	
C	G_7	5	19					5	5	5 MW
								5/6	5/6	10/6 MWh
	G_9	15	14					15	15	15 MW
								15/6	15/6	30/6 MWh
Real-time MCP ($/MWh)				15	19	29	33	19	19	
Total available MW								20	20	20 MW
Total required MWh								20/6	20/6	40/6 MWh

Table 9.45 Last Two Intervals Dispatch Instructions in Zone 2

GENCO		Capacity MW	Energy Bid $/MWh	Interval						Total
				1	2	3	4	5	6	
A	G_1	5	15					5	5	5 MW
								5/6	5/6	10/6 MWh
B	G_5	10	29					10	10	10 MW
								10/6	10/6	20/6 MWh
	G_6	10	33					7.5	10	10 MW
								7.5/6	10/6	17.5/6 MWh
C	G_{10}	20	56					10	10	10 MW
								2.5/6	10/6	12.5/6 MWh
Real-time MCP ($/MWh)				15	19	29	33	56	56	
Total available MW								32.5	35	35 MW
Total required MWh								25/6	35/6	60/6 MWh

The calculated real-time MCPs are used to calculate the instructed deviation settlements in as follows.

Payments from the ISO to the GENCOs for the instructed deviations are

GENCO A: 5/6x15 + 5/6x19 + 5/6x29 + 5/6x33 + 5/6x56 + 5/6x56 = $173.33

GENCO B: 5/6x29 + 17.5/6x33 + 17.5/6x56 + 20/6x56 = $470.42

GENCO C: 5/6x19 + 5/6x29 + 5/6x33 + 20/6x19 + 20/6x19 + 2.5/6x56 + 10/6x56 = $310.83

Total ISO payment = 173.33 + 470.42 + 310.83 = $954.58

The average hourly ex-post price used for the uninstructed deviation settlements in zone 1 is calculated as follows: (5/6x19 + 5/6x29 + 5/6x33 + 20/6x19 + 20/6x19) / (15/6 + 40/6) = 21.182 $/MWh. Note it is lower than that without congestion.

From Tables 9.43 and 9.44, the total uninstructed deviation is 15/6 + 40/6 = 55/6 = 9.17 MWh. So, the ISO would collect from the uninstructed deviation settlement: 21.182 x 9.17 = $194.17.

The average hourly ex-post price used for the uninstructed deviation settlements in zone 2 is calculated as follows: (5/6x15 + 5/6x19 + 10/6x29 + 22.5/6x33 + 25/6x56 + 35/6x56) / (42.5/6 + 60/6) = 44.512 $/MWh. Note it is higher than that without congestion.

From Tables 9.43 and 9.45, the total uninstructed deviation is 42.5/6 + 60/6 = 102.5/6 = 17.08 MWh. So, the ISO would collect from the uninstructed deviation settlement: 44.512 x 17.08 = $760.42.

From the preceding analyses, the ISO's total payment in times of partial congestion is higher than that when all intervals experience congestion, but lower than when no congestion occurs at any interval.

9.5.5 Discussions

We learned in this section that AGC would provide an effective method of adjusting generation to minimize frequency deviations and regulate tie-line flows. This crucial role would continue in restructured markets with some modifications to account for issues such as bilateral contracts that span control areas.

This section described pricing of AGC for the real-time energy imbalance market that is managed by the ISO. The presented case study demonstrated the situation of negative imbalances when the actual load exceeds the scheduled supply and the ISO is a buyer. The analysis was extended to the case of congestion in the forward energy market. The results show that congestion has an important effect on decision making in a real-time auction. Theoretically, GENCOs have more of a chance to get better bidding situations (e.g., by applying gaming techniques) in times of congestion.

9.6 CONCLUSIONS

The chapter discusses an auction market design for ancillary services. The chapter also presents definitions and requirements for ancillary services. Two different approaches for the auction design of ancillary services are presented, sequential and simultaneous. In the sequential approach, an alternative option is introduced with a weighting factor that is superior to a regular design option. The proposed alternative would encourage GENCOs to participate in the ancillary services market with higher ancillary service MCPs, encourage GENCOs to bid low in the energy part of the bid by accounting for the lost opportunity cost in the capacity payment, and enhance the benefits of the ISO and DISCOs by lowering the ISO's total payment and hedging the ISO and DISCOs against extremely high energy bids. The simultaneous approach is proposed to overcome the problems incurred by the sequential approach concerning unused higher quality lower cost services and price reversal. A well-known characteristic of simultaneous approaches is their substitution capability of a higher quality service for a lower quality service. Of all simultaneous approach auction designs, the rational buyer auction can reduce procurement cost but cannot avoid price reversal, whereas the marginal pricing auction can avoid price reversal but may increase the procurement cost. In the real-time market, AGC provides an effective method of adjusting generation to minimize frequency deviations and regulate tie-line flows. This crucial role will continue in restructured markets with some modifications to account for issues such as bilateral contracts that span control areas. The chapter ends with a discussion of AGC pricing for the real-time energy imbalance market.

Chapter 10

Transmission Congestion Management and Pricing

10.1 INTRODUCTION

Despite the fact that transmission charges represent a small percentage of operating expenses in utilities, the transmission network is a vital mechanism in competitive electricity markets. In a restructured power system, the transmission network is where generators compete to supply large users and distribution companies. Thus, transmission pricing should be a reasonable economic indicator used by the market to make decisions on resource allocation, system expansion, and reinforcement.

The competitive environment of electricity markets necessitates wide access to transmission and distribution networks that connect dispersed customers and suppliers. Moreover, as power flows influence transmission charges, transmission pricing may not only determine the right of entry but also encourage efficiencies in power markets. For example, transmission constraints could prevent an efficient generating unit from being utilized. A proper transmission pricing scheme that considers transmission constraints or congestion could motivate investors to build new transmission and/or generating capacity for improving the efficiency. In a competitive environment, proper transmission pricing could meet revenue expectations, promote an efficient operation of electricity markets, encourage investment in optimal locations of generation and transmission lines, and adequately reimburse owners of transmission assets. Most important, the pricing scheme should implement fairness and be practical.

However, it is difficult to achieve an efficient transmission pricing scheme that could fit all market structures in different locations. The ongoing research on transmission pricing indicates that there is no generalized agreement on pricing methodology. In practice, each country or each restructuring model has chosen a method that is based on the particular characteristics of its network. Measuring whether or not a certain transmission pricing scheme is technically and economically adequate would require additional standards.

During the last few years, different transmission pricing schemes have been proposed and implemented in various markets [Shi96, Lim96]. The most common and unsophisticated approach to transmission pricing is the postage-stamp method. In this method, regardless of the distance that the energy travels, an entity pays a rate equal to a fixed charge per unit of the energy transmitted within a particular utility system. Postage-stamp rates are based on average system costs. In addition, the rates often include separate charges for peak and off-peak periods, which are functions of season, day, and holiday usage. Under this approach, when energy is transmitted across several utility systems, it can suffer from a pancaking problem[1]. Another commonly used method is the contract path method, which is proposed for minimizing transmission charges and overcoming the pancaking problem. However, this pricing method does not reflect actual flows through the transmission grid that include loop and parallel path flows[2]. As an alternative to the contract path method, the MW-mile method is introduced as a flow-based pricing scheme. In this scheme, power flow and the distance between injection and withdrawal locations reflect transmission charges.

The main drawback of the aforementioned approaches is that they do not consider transmission congestion. In the new environment, it is essential to involve transmission tariffs in transmission pricing according to flow-based pricing and congestion-based pricing. Congestion pricing would allocate each limited transmission resource to customers who value it the most [Alo99a, Alo99b, Alo00a, Alo00b]. A proper pricing scheme should allocate congestion charges to participants who cause congestion, and should reward participants whose schedules tend to relieve congestion.

When the transmission becomes congested, meaning that no additional power can be transferred from a point of injection to a point of extraction, more expensive generating units may have to be brought on-line

---

1 Pancaking occurs when energy transmitted across several utility systems accumulates utility access charges.
2 As a transaction occurs in the system, electrons flow through all available transmission paths between generators and points of extraction according to the laws of physics. The contractual paths do not represent the actual use of the system.

TRANSMISSION CONGESTION MANAGEMENT AND PRICING 371

on one side of the transmission system. In a competitive market, such an occurrence would cause different locational marginal prices (LMPs) between the two locations. If transmission losses are ignored, a difference in LMPs would appear when lines are congested. Conversely, if flows are within limits (no congestion), LMPs will be the same at all buses and no congestion charges would apply. The difference in LMPs between the two ends of a congested line is related to the extent of congestion and MW losses on this line [Alo99a, Alo99b, Alo00a]. Since LMP acts as a price indicator for both losses and congestion, it should be an elementary part of transmission pricing.

Firm transmission rights (FTRs) are proposed as purchased rights that can hedge congestion charges on constrained transmission paths. By holding FTR, a transmission customer has a mechanism to offset congestion charges when transmission lines are congested [Alo99a, Alo99b, Alo00a, Pjm]. Besides providing financial certainty, FTR could maximize the efficient use of the system and make users pay for the actual use of congested paths.

In this chapter, we will discuss some of the existing approaches for transmission pricing. First, we will present an overview of recent techniques used for designing equitable access fees to recover fixed transmission charges. Among these methods, we will show some recent methods that determine a generator's (load's) contribution to a line power flow and a consumer load. The so-called distribution factors used to determine transmission usage would also be shown. Numerical examples for using different methods will be presented.

As LMPs and FTRs have been widely utilized in many power restructuring models to resolve problems associated with transmission congestion and pricing, we will detail these key subjects in a comprehensive manner. Then, a transmission congestion management scheme that incorporates LMPs and FTRs will be discussed.

In this chapter, we will also introduce a comprehensive scheme for the ISO to modify preferred schedules, trace participants' contributions, and allocate transmission usage and congestion charges. The ISO would adjust preferred schedules on a nondiscriminatory basis to keep the system within its limits and applies curtailment priority according to the participants' willingness to avoid curtailing transactions. We assume a general restructuring model where pool, bilateral, and multilateral contracts exist concurrently. In this scheme, transmission congestion and losses are calculated based on LMPs. A flow-based tracing method is utilized to allocate transmission charges. FTR holders' credits are calculated based on line flow calculations and LMPs. Examples based on a testing system will be presented to support the approach.

10.2 TRANSMISSION COST ALLOCATION METHODS

An efficient transmission pricing mechanism should recover transmission costs by allocating the costs to transmission network users in a proper way. The transmission costs may include:

- Running costs, such as costs for operation, maintenance, and ancillary services.
- Past capital investment.
- Ongoing investment for future expansion and reinforcement associated with load growth and additional transactions.

Running costs are small compared with the capital investment (or embedded transmission costs). Consequently, transmission charges for embedded cost recovery would largely exceed running costs over the investment recovery period.

The study objectives and market structures are main factors for choosing algorithms in the evaluation of transmission pricing. Regardless of the market structure, it is important to accurately determine transmission usage in order to implement usage-based cost allocation methods. However, determining an accurate transmission usage could be difficult due to the nonlinear nature of power flow. This fact necessitates using approximate models, sensitivity indices, or tracing algorithms to determine the contributions to the network flows from individual users or transactions.

In the following, we discuss major transmission cost allocation methods. Some of these methods are used widely by electric utilities, while others are still in developmental stages.

10.2.1 Postage-Stamp Rate Method

Postage-stamp rate method is traditionally used by electric utilities to allocate the fixed transmission cost among the users of firm transmission service [Hap94]. This method is an embedded cost method, which is also called the rolled-in embedded method. This method does not require power flow calculations and is independent of the transmission distance and network configuration. In other words, the charges associated with the use of transmission system determined by postage-stamp method are independent of the transmission distance, supply, and delivery points or the loading on different transmission facilities caused by the transaction under study. The method is based on the assumption that the entire transmission system is used, regardless of the actual facilities that carry the transmission service. The method allocates charges to a transmission user based on an average embedded cost and the magnitude of the user's transacted power.

10.2.2 Contract Path Method

The contract path method is also traditionally used by electric utilities to allocate the fixed transmission cost [Hap94, Ili97]. It is likewise an embedded cost method that does not require power flow calculations. This method is based on the assumption that transmission services can be represented by transmission flows along specified and artificial electrical path throughout the transmission network. The contract path is a physical transmission path between two transmission users that disregards the fact that electrons follow physical paths that may differ dramatically from contract paths. The method ignores power flows in facilities that are not along the identified path. After specifying contract paths, transmission charges will then be assigned using a postage-stamp rate, which is determined either individually for each of the transmission systems or on the average for the entire grid. As a consequence, the recovery of embedded capital costs would be limited to artificial contract paths.

10.2.3 MW-Mile Method

The MW-mile method is an embedded cost method that is also known as a line-by-line method because it considers, in its calculations, changes in MW transmission flows and transmission line lengths in miles [Lim96, Pan00, Shi89, Shi91]. The method calculates charges associated with each wheeling transaction based on the transmission capacity use as a function of the magnitude of transacted power, the path followed by transacted power, and the distance traveled by transacted power. The MW-mile method is also used in identifying transmission paths for a power transaction. As such, this method requires dc power flow calculations. The MW-mile method is the first pricing strategy proposed for the recovery of fixed transmission costs based on the actual use of transmission network.

The method guarantees the full recovery of fixed transmission costs and reasonably reflects the actual usage of transmission systems.

The following algorithm is used in the MW-mile method to estimate the usage of firm transmission services by wheeling transactions:

1. For each transaction t:
    - Use nodal power injections involved in transaction t, calculate transaction-related flows on all network lines using an approximate (dc) power flow model.
    - The magnitude of MW flow on every line is multiplied by its length (in miles) and the cost per MW per unit length of the line (in $/MW-mile), and summed over all the lines.

2. Repeat the process for other transactions.

3. The contribution of transaction t to the total transmission capacity cost is calculated as follows: transmission facility costs are allocated in proportion to the ratio of flow magnitude (absolute value) contributed by transaction t and the sum of absolute flows caused by all transactions, as given by the equation

$$TC_t = TC \times \frac{\sum_{k \in K} c_k L_k MW_{t,k}}{\sum_{t \in T} \sum_{k \in K} c_k L_k MW_{t,k}} \quad (10.1)$$

where

TC_t	=	cost allocated to transaction t
TC	=	total cost of all lines in $
L_k	=	length of line k in mile
c_k	=	cost per MW per unit length of line k
$MW_{t,k}$	=	flow in line k, due to transaction t
T	=	set of transactions
K	=	set of lines

10.2.4 Unused Transmission Capacity Method

The difference in a facility capacity and the actual flow on that facility is called the unused (unscheduled) transmission capacity [Lim96, Kov94, Pan00, Sil98]. To guarantee the full recovery of all embedded costs, it is assumed that all transmission users are responsible to pay for both the actual capacity use and the unused transmission capacity. Accordingly, the following general expression of MW-mile pricing rule is used:

$$TC_t = \sum_{k \in K} C_k \frac{|F_{t,k}|}{\sum_{t \in T}|F_{t,k}|} \qquad (10.2)$$

where

TC_t = cost allocated to transaction t
C_k = embedded cost of facility k
$|F_{t,k}|$ = flow on facility k caused by transaction t
T = set of transactions
K = set of transmission facilities

The pricing rule given by (10.2) ensures the total cost recovery, whether or not the line capacity is fully used. It is also inequitable to some users when they are forced to share the cost of an expensive transmission facility for which only a small portion of the facility capacity is being utilized. In addition, some margin of a line capacity is left unused for maintaining reliability. As such this rule's shortcoming is that it does not motivate an efficient use of the transmission system. To overcome this drawback, it has been suggested that transmission users be charged based on the percentage utilization of the facility capacity, and not based on the sum of flows contributed by all users (i.e., users are charged based on the actual capacity use, and not for the unscheduled capacity). However, this suggestion has a drawback in the sense that it ignores the reliability of the transmission margin and does not ensure the full recovery of the fixed transmission costs. This suggestion uses the following revised MW-mile rule:

$$TC_t = \sum_{k \in K} C_k \frac{|F_{t,k}|}{\overline{F}_k} \qquad (10.3)$$

where $\overline{F}_k$ is the capacity of facility k.

In addition, multi-part pricing rules have been proposed that consider both the utilized facility capacity, and the difference between the total embedded costs and costs recovered by the utilized transmission capacity.

10.2.5 MVA-Mile Method

The MVA-mile method is an extended version of the MW-mile method [Bia98, Pan00]. The extension is proposed to include charges for reactive power flow in addition to charges for real power flow. It has been shown that monitoring both real and reactive power, given the line MVA loading limits and the allocation of reactive power support from generators and transmission facilities, is a better approach to measuring the use of transmission resources. The tracing methods that will be discussed later in this chapter can be used for this purpose. In addition, the sensitivity approach with ac power flow studies can be used to determine the network usage of reactive power flow. Other approaches have been also proposed to decompose network flows into real and reactive components associated with individual transactions.

10.2.6 Counter-flow Method

The counter-flow method argues that transmission users should be charged or credited based on whether their transactions cause flows or counter-flows with regard to the direction of net flows [Lim96, Pan00]. The method suggests that if a particular transaction flows in the opposite direction of the net flow, then the transaction should be credited (i.e., the transaction would pay a negative charge). This suggestion differs from the traditional MW-mile approach and other usage-based allocation pricing rules, where each transaction pays for its usage regardless of the flow's directions. An example of the counter-flow method is zero counter-flow pricing, which proposes that only those that use the transmission facility in the direction of net flow should be charged in proportion to their contributions to the total positive flow. One of the difficulties in using this method is that it would be hard for transmission service providers to arrange payments to users with counter-flows.

10.2.7 Distribution Factors Method

Distribution factors are calculated based on linear load flows [Ng81, Pan00, Rud95, Shi89, Sin98]. In general, generation distribution factors have been used mainly in security and contingency analyses. They have been used to approximately determine the impact of generation and load on transmission flows. In recent years, these factors are suggested as a mechanism to allocate transmission payments in restructured power

TRANSMISSION CONGESTION MANAGEMENT AND PRICING 377

systems, as these factors can efficiently evaluate transmission usage. To recover the total fixed transmission costs, distribution factors can be used to allocate transmission payments to different users. By using these factors, allocation can be attributed to transaction-related net power injections, to generators, or to loads. The distribution factors are given as follows:

(1) **Generation Shift Distribution Factors (GSDFs or A factors):** GSDFs or A factors provide line flow changes due to a change in generation. These factors can be used in determining maximum transaction flows for bounded generation and load injections. GSDFs or A factors are defined as

$$\Delta F_{l-k} = A_{l-k,i} \Delta G_i \qquad (10.4a)$$

$$\Delta G_r = -\Delta G_i \qquad (10.4b)$$

where

ΔF_{l-k} = change in active power flow between buses l and k
$A_{l-k,i}$ = A factor (GSDF) of a line joining buses l and k corresponding to change in generator at bus i
ΔG_i = change in generation at bus i, with the reference bus excluded
ΔG_r = change in generation at the reference bus (generator) r

$A_{l-k,i}$ is calculated using the definition of a reactance matrix and the dc load flow approximation. The A factor measures the incremental use of transmission network by generators and loads (consumers). We also notice that GSDFs are dependent on the selection of reference (marginal) bus and independent of operational conditions of the system.

(2) **Generalized Generation Distribution Factors (GGDFs or D factors):** They determine the impact of each generator on active power flows; thus they can be negative as well. Since GGDFs are based on the dc model, they can only be used for active power flows. GGDFs or D factors are defined as

$$F_{l-k} = \sum_{i=1}^{N} D_{l-k,i} \, G_i \qquad (10.5a)$$

where

$$D_{l-k,i} = D_{l-k,r} + A_{l-k,i} \tag{10.5b}$$

$$D_{l-k,r} = \{F^0_{l-k} - \sum_{\substack{i=1\\i\neq r}}^{N} A_{l-k,i} G_i\} / \sum_{i=1}^{N} G_i \tag{10.5c}$$

and

F_{l-k} = total active power flow between buses l and k

F^0_{l-k} = power flow between buses l and k from the previous iteration

$D_{l-k,i}$ = D factor (GGDF) of a line between buses l and k corresponding to generator at bus i

$D_{l-k,r}$ = GGDF of a line between buses l and k due to the generation at reference bus r

G_i = total generation at bus i

GGDFs measure the total use (not incremental) of transmission network facilities produced by generator injections. GGDFs depend on line parameters, system conditions, and not on the choice of reference bus.

(3) **Generalized Load Distribution Factors (GLDFs or C factors):** are very similar to GGDFs. GGDFs determine the contribution of each load to line flows. GLDFs also allocate charges of the sub-transmission network to loads within a distribution company service area. GLDFs or C factors are defined as:

$$F_{l-k} = \sum_{j=1}^{N} C_{l-k,j} \, L_j \tag{10.6a}$$

where

$$C_{l-k,j} = C_{l-k,r} - A_{l-k,j} \tag{10.6b}$$

$$C_{l-k,r} = \{F^0_{l-k} + \sum_{\substack{j=1\\j\neq r}}^{N} A_{i-k,j} L_j\} / \sum_{j=1}^{N} L_j \tag{10.6c}$$

and,

F_{l-k} = total active power flow between buses l and k

F_{l-k}^0 = power flow on a line between buses l and k from the previous iteration

$C_{l-k,j}$ = C factor (GLDF) of a line between buses l and k corresponding to demand at bus j

$C_{l-k,r}$ = GLDF for a line between buses l and k due to the load at reference bus r

L_j = total demand at bus j

GGDFs are also based on dc power flows. We notice that C factors (GLDFs) measure the total use of transmission network facilities by loads in which loads are seen as negative injections. As in the case of GGDFs, GLDFs depend on line parameters, system conditions, and not on the reference bus location.

10.2.8 AC Power Flow Methods

Many ac-based approaches have been proposed to allocate transmission cost. Among them there are flow sensitivity indices, full ac power flow solutions, and power flow decomposition [Bar99, Hao97, Ili97, Pan00, Par98, Shi91, Zob97]. The ac flow sensitivity indices method uses the same logic as the dc flow distribution factors, but the sensitivity of transmission flows to bus power injections are derived from ac power flow models. The full ac power flow solutions method uses full ac power flow calculations or utilizes optimal power flow studies. In these methods, more detailed cost information is usually required to study the impact of wheeling transactions. The power flow decomposition method would decompose network flows into components associated with individual transactions plus one component to account for the nonlinear nature of power flow model. For each transaction, the algorithm determines real and reactive flow components of the transmission network usage, the net power imbalance, and the contribution of participating generators to real-power-loss compensation.

10.2.9 Tracing Methods

Tracing methods determine the contribution of transmission users to transmission usage [Bia96, Bia97a, Bia97b, Bia98, Kir97, Pan00, Str98]. Tracing methods may be used for transmission pricing and recovering fixed transmission costs. In this section, we discuss two tracing methods, which are recognized as the Bialek's tracing method and the Kirschen's tracing method. Tracing methods are generally based on the so-called proportional sharing principle.

(1) Bialek's Tracing Method

In Bialek's tracing method, it is assumed that nodal inflows are shared proportionally among nodal outflows [Bia96, Bia97a, Bia97b, Bia98, Pan00]. This method uses a topological approach to determine the contribution of individual generators or loads to every line flow based on the calculation of topological distribution factors. This method can deal with both dc power flow and ac power flows; that is, it can be used to find contributions of both active and reactive power flows. Bialek's tracing method considers:

- Two flows in each line, one entering the line and the other exiting the line (to consider losses in line).
- Generation and load at each bus.

The main principle used to trace the power flow will be that of proportional sharing, see Figure 10.1 for an illustration. The figure shows four lines connected to a node. The *outflows* (f_1 and f_2) can be represented in terms of the inflows (f_a and f_b); in other words, we can determine how much of f_1 comes from f_a and how much of f_1 comes from f_b. The same applies to f_2.

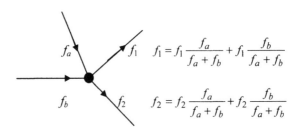

Figure 10.1 Illustration of Proportional Sharing

This method uses either the *upstream-looking* algorithm or the *downstream-looking* algorithm. In the upstream-looking algorithm, the transmission usage/supplement charge is allocated to individual generators and losses are apportioned to loads. In the downstream-looking algorithm, the transmission usage/supplement charge is allocated to individual loads

TRANSMISSION CONGESTION MANAGEMENT AND PRICING

and losses are apportioned to generators. Bialek's tracing method is used to determine how much of a particular generator's output supplies a particular load or how much of a particular load is supplied by a particular generator. Topological distribution factors calculated in this method are always positive; therefore this method would eliminate the counter-flow problem.

To show how this algorithm works, we define the *gross demand* as the sum of a particular load and its allocated part of the total transmission loss. The total gross demand in a system is equal to the total actual generation. Topological distribution factors are given by the following equation in which $D_{ij,k}^g$ refers to the kth generator's contribution to line i–j flow.

$$p_{ij}^g = \frac{p_{ij}^g}{p_i^g} \sum_{k=1}^{n} [A_u^{-1}]_{ik} P_{Gk} = \sum_{k=1}^{n} D_{ij,k}^g P_{Gk}; \quad j \in \alpha_i^d \tag{10.7a}$$

where

$$p_i^g = \sum_{j \in \alpha_i^u} |P_{ij}^g| + P_{Gi}; \quad i = 1, 2, \ldots, n \tag{10.7b}$$

$$[A_u]_{ij} = \begin{cases} 1 & i = j \\ -\dfrac{|P_{ji}|}{P_j} & j \in \alpha_i^u \\ 0 & \text{otherwise} \end{cases} \tag{10.7c}$$

and

p_{ij}^g	=	an unknown gross line flow in line i–j
p_i^g	=	an unknown gross nodal power flow through node i
A_u	=	upstream distribution matrix
P_{Gk}	=	generation in node k
α_i^d	=	set of nodes supplied directly from node i
α_i^u	=	set of buses supplying directly bus i
$D_{ij,k}^g$	=	topological distribution factors

The gross power at any node is equal to the generated power at the node plus the imported power flows from neighboring nodes. The total usage of the network by the kth generator (U_{Gk}) is calculated by summing up the individual contributions (multiplied by line weights) of that generator to line flows. This is given by:

$$U_{Gk} = \sum_{i=1}^{n} \sum_{j \in \alpha_i^d} w_{ij}^g D_{ij,k}^g P_{Gk} = P_{Gk} \sum_{i=1}^{n} \left\{ \frac{[A_u^{-1}]_{ik}}{p_i^g} \sum_{j \in \alpha_i^d} C_{ij} \right\} \quad (10.8)$$

where

C_{ij} = total supplement charge for the use of line i–j

w_{ij}^g = charge per MW of each line i–j

The method can be summarized as follows:

1. Solve power flow (either ac or dc) and define line flows (inflows and outflows).
2. If losses exist, allocate each line's loss as additional loads to both ends of the line.
3. Find matrix A_u.
4. Define generation vector P_G
5. Invert matrix A_u (i.e., A_u^{-1})
6. Find gross power P_g using $P_g = A_u^{-1} P_G$. The gross power at node i is given as $p_i^g = \sum_{k=1}^{n} [A_u^{-1}]_{ik} P_{Gk}$
7. The gross outflow of line i–j, using the proportional sharing principle, is given as

$$p_{ij}^g = \frac{p_{ij}^g}{p_i^g} p_i^g = \frac{p_{ij}^g}{p_i^g} \sum_{k=1}^{n} [A_u^{-1}]_{ik} P_{Gk} = \sum_{k=1}^{n} D_{ij,k}^g P_{Gk}$$

where $D_{ij,k}^g = \frac{p_{ij}^g [A_u^{-1}]_{ik}}{p_i^g} \cong \frac{P_{ij} [A_u^{-1}]_{ik}}{P_i}$

and j is the set of nodes supplied directly from node i.

The *downstream-looking* method that allocates usage charges to individual loads would use the same methodology.

(2) Kirschen's Tracing Method

Kirschen's tracing method is based on a set of definitions for domains, commons, and links [Kir97, Pan00, Str98].

A domain is a set of buses that obtain power from a particular generator. A common is a set of contiguous buses supplied by the same set of generators. Links are branches that interconnect commons. Based on these definitions, the state of system (an acyclic state graph) is represented by a directed graph that consists of commons and links, with directed flows between commons and the corresponding data for generations/loads in commons and flows on links. The method uses a recursive procedure for calculating the contributions of generators (or loads) to commons, links, and loads (or generators), and line flows within each common. For a given common, the method assumes that the proportion of inflow traced to a particular generator is equal to the proportion of outflow traced to the same generator. As in Bialek's tracing method, Kirschen's tracing method can determine contributions from individual generators to line flows, and determine contributions of individual loads to line flows.

Starting from a root common, the method finds recursively the contribution of each common's generation (load) to line flows and consumed loads. The method uses a proportionality assumption to allocate the outflow of a common to contributors of the inflow of a common. By determining the flow in each branch, the method apportions each branch usage among system users that contribute to the branch flow. Usage of a transmission system should be allocated to generators (loads) on the basis of their contribution to each branch flow, that is the usage of a branch must be apportioned among all parties. In this section, generation (load) refers to net generation (load) at a bus. The contributions can be calculated based on traceable contributions of each generator (load) to branch flows.

Kirschen's tracing method is a topological trace method that would answer the following question: *What proportion of the active (reactive) power flow in a branch is contributed by each generator?* The method is applicable to both ac and dc load flow solutions. This traceable allocation method does not rely on a linearized model of the network and is therefore not limited to incremental changes in injections. The method starts by calculating line flows, which in turn provides the flow direction in each branch. Starting from each generator's bus, and based on the flow direction in each line, the method determines the domain of each generator. In this chapter, we simulate a situation with more than one generator at some buses. In this case, the generation at each bus is the sum of generator outputs. The net generation at each bus is used to trace the contribution to line flows. Then, we determine how much each generator contributes to line flows. The domain is a set of buses that are reached by the active (reactive) power produced by a generator. For a system of N_g generator buses, there are N_g domains. After determining domains, the method determines the commons. A common is defined as a group of buses that are reached by the same generators. If G_i refers to the ith generator, the first common (*rank* = 1, *root node*) is the set of buses that are only reached by G_1, the second common (*rank* = 2) is the set of buses that are reached by G_1 and G_2, the third common (*rank* = 3) is the set of buses that are reached by G_1, G_2 and G_3, and so on. By knowing line flows and commons, links between commons are formed. A link is defined as a group of lines (branches) that connect two commons directly. After determining commons and links, the method uses a state graph to calculate different contributions. A state graph is a transformation of the meshed network into an acyclic graph. All that is needed in the state graph are commons, links between commons (with flows), and generation and load in each common. Using the state graph, the method determines how much a generator (load) contributes to loads (generators) and flows in commons and links of a graph.

To calculate the contribution of each generation to commons and line flows, the method calculates the inflow to each common. The inflow to common k is the sum of generation at common k and the flow to common k from other commons with a lower rank j. Mathematically,

$$I_k = g_k + \sum_j F_{jk} \qquad (10.9)$$

where,

$\quad I_k \quad =$ inflow of common k
$\quad g_k \quad =$ net generation in common k
$\quad F_{jk} \quad =$ flow (from j to k) in a link connecting commons j and k

The next step is to recursively calculate relative contributions by each generator to the load and outflow of each common, starting from the root common (that has rank 1). Relative contributions are calculated based on absolute contributions to a common. Let

$\quad R_{ij} \quad =$ relative contribution of common i to the load and the outflow of common j
$\quad A_{ij} \quad =$ absolute inflow contribution of common j to common i

N_c = number of commons
F_{ki} = flow between commons k and i

The elements of absolute contribution matrix (A) and relative contribution matrix (R) are calculated using the following algorithm:

do $j = 1, N_c$
    $A_{jj} = g_j$
    $R_{jj} = A_{jj}/I_j$
enddo

do $i = 1, N_c$
    do $j = i+1, N_c$
        $A_{ij} = 0$
        do $k = 1, j$
            $A_{ji} = 0$
            $R_{ji} = 0$
            $A_{ij} = A_{ij} + R_{ik} F_{kj}$
            $R_{ij} = A_{ij}/I_j$
        enddo
    enddo
enddo

Each element R_{ij} signifies that the generation in common i (g_i) produces ($R_{ij} \times 100$)% of the load consumed in common j (d_j), ($R_{ij} \times 100$)% of the flow in each link originating from common j and ending with other commons[3], and ($R_{ij} \times 100$)% of the flow in each branch between any two buses in common j. Note that the under-diagonal elements of both A and R are zero, which means that a common does not contribute to commons of lower rank. The elements of R will be the basis for the calculation of transmission charges, as we will show for the test system and examples.

---

[3] What is applied to a link is also applied to lines forming this link.

10.2.10 Comparison of Cost Allocation Methods

Table 10.1 summarizes some of the aforementioned methods used in transmission cost allocation.

Table 10.1 Summary of Transmission Cost Allocation Methods

Method	Application	Load Flow Analysis	Payments Based On	Comments
Postage-stamp	Real power generation or load	—	• Magnitude of transacted power • An average embedded cost	Depends on assumption that the entire transmission system is used
Contract path	Real power generation or load	—	• Magnitude of transacted power • An embedded cost	Depends on assumption that the transmission service is restricted to flow along a specified and artificial path
MW-mile	Real power generation or load	dc, ac (usually dc)	• Magnitude of the transacted power • Path followed by the transacted power • Distance traveled by the transacted power	Depends on operational conditions (system configuration)
A factors (GSDFS)	Real power generation or load	dc	Incremental flow	Depends on • System configuration • Selection of reference bus • Power flow directions
D factors (GGDFS)	Real power generation	dc	Total flow	Depends on operational conditions
C factors (GLDFS)	Real power load	dc	Total flow	Depends on operational conditions
Bialek	Real and reactive generation or load	dc, ac	Total flow	Depends on operational conditions
Kirschen	Real and reactive generation or load	dc, ac	Total flow	Depends on operational conditions

10.3 EXAMPLES FOR TRANSMISSION COST ALLOCATION METHODS

Figure 10.2 represents a 5-bus test system with two generators and three loads. Line data are given in Table 10.2. Voltage magnitudes are assumed to be 1.060 and 1.050 at buses 1, and 4, respectively. Bus 1 is assumed the reference bus. The ac load flow solution is shown in Tables 10.3 and 10.4. The dc load flow solution is given in Table 10.5. Table 10.4 also shows the average line flows. The average flow disregards the impact of MW losses in calculating line contributions.

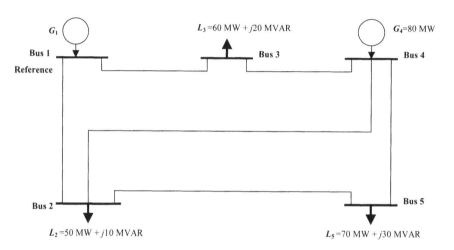

Figure 10.2 5-Bus Test System

Table 10.2 System Data of the 5-Bus Example

Line No.	From	To	R	X	B/2	$C_k L_k$
1	1	2	0.02	0.06	0.030	60
2	1	3	0.08	0.24	0.025	240
3	2	4	0.06	0.18	0.020	280
4	2	5	0.04	0.12	0.015	120
5	3	4	0.01	0.03	0.010	30
6	4	5	0.08	0.24	0.025	240

Table 10.3 Voltage, Angle, Generation and Load at Each bus

Bus	Voltage	Angle (Deg)	P_L	Q_L	P_G	Q_G
1	1.060	0.000	0.000	0.000	103.361	14.523
2	1.035	-2.451	50.000	10.000	0.000	0.000
3	1.042	-2.281	60.000	20.000	0.000	0.000
4	1.050	-1.711	0.000	0.000	80.000	28.570
5	0.999	-4.909	70.000	30.000	0.000	0.000
		Total	180.000	60.000	183.361	43.093

Table 10.4 Line Flow Results of the ac Load Flow

Line	i	j	P_{ij}	P_{ji}	P_{ij}^{ave}	Q_{ij}	Q_{ji}	MVA_{ij}	MVA_{ji}	P_{Loss}	Q_{Loss}
1	1	2	84.343	-83.016	83.680	15.221	-17.819	85.706	84.906	1.328	-2.598
2	1	3	19.018	-18.757	18.888	-0.703	-4.037	19.031	19.186	0.261	-4.740
3	2	4	-9.664	9.734	9.699	-7.772	3.637	12.402	10.391	0.070	-4.135
4	2	5	42.680	-41.888	42.284	15.591	-16.319	45.438	44.955	0.791	-0.728
5	3	4	-41.243	41.420	41.332	-15.963	14.307	44.225	43.821	0.177	-1.656
6	4	5	28.846	-28.112	28.479	10.631	-13.681	30.742	31.264	0.734	-3.050
									Total Loss	3.361	-16.907

Table 10.5 Line Flows for the dc Load Flow Solution

Line No.	From	To	P_{ij}
1	1	2	81.78
2	1	3	18.22
3	2	4	-9.93
4	2	5	41.70
5	3	4	-41.78
6	4	5	28.30

10.3.1 Cost Allocation Using Distribution Factors Method

Table 10.6 shows GSDFs and GGDFs of the test system. Table 10.7 shows transmission usage contributions of each generator based on GGDFs. We used the average real power flows to find generator contributions. These contributions will be used later for comparison with other methods.

TRANSMISSION CONGESTION MANAGEMENT AND PRICING

Table 10.6 A Factors (GSDFs) and D Factors (GGDFs) of the 5-Bus

Line i–j	A Factors					D Factors	
	$A_{ij,1}$	$A_{ij,2}$	$A_{ij,3}$	$A_{ij,4}$	$A_{ij,5}$	$D_{ij,1}$	$D_{ij,4}$
1–2	0.00	-0.8667	-0.5333	-0.60	-0.7778	0.71816	0.11816
1–3	0.00	-0.1333	-0.4667	-0.40	-0.2222	0.26740	0.22740
2–4	0.00	0.0889	-0.3556	-0.40	-0.0741	0.22720	-0.17280
2–5	0.00	0.0444	-0.1778	-0.20	-0.7037	0.31780	0.11780
3–4	0.00	-0.1333	0.5333	-0.40	-0.2222	-0.05089	-0.45089
4–5	0.00	-0.0444	0.1778	0.20	-0.2963	0.06805	0.26805

Table 10.7 Transmission Usage Allocation using GGDFs

Line	i	j	P_{ij}^{ave}	$P_{ij}^{G_1}$	$P_{ij}^{G_4}$
1	1	2	83.6795	74.2261	9.4528
2	1	3	18.8875	28.6813	- 9.8000
3	2	4	9.6990	23.5031	-13.8240
4	2	5	42.2840	32.8465	9.4240
5	3	4	41.3315	5.2597	36.0712
6	4	5	28.4790	7.0334	21.4440

10.3.2 Cost Allocation Using Bialek's Tracing Method

Figure 10.3 shows the lossless power system corresponding to the test system. The average line flows are shown. Line losses are is divided equally as two additional loads at two ends of each line.

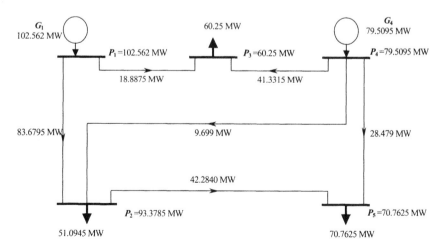

Figure 10.3 Lossless Power Flow

Upstream distribution matrix and its inverse (A_u and A_u^{-1}), nodal generation vector (P_G) and gross power vector (P_g) are calculated and given as follows:

$$A_u = \begin{bmatrix} 1.0000 & 0.0000 & 0.0000 & 0.0000 & 0.0000 \\ -0.8159 & 1.0000 & 0.0000 & -0.1220 & 0.0000 \\ -0.1842 & 0.0000 & 1.0000 & -0.5198 & 0.0000 \\ 0.0000 & 0.0000 & 0.0000 & 1.0000 & 0.0000 \\ 0.0000 & -0.4524 & 0.0000 & -0.3582 & 1.0000 \end{bmatrix}$$

$$A_u^{-1} = \begin{bmatrix} 1.0000 & 0.0000 & 0.0000 & 0.0000 & 0.0000 \\ 0.8159 & 1.0000 & 0.0000 & 0.1220 & 0.0000 \\ 0.1842 & 0.0000 & 1.0000 & 0.5198 & 0.0000 \\ 0.0000 & 0.0000 & 0.0000 & 1.0000 & 0.0000 \\ 0.3691 & 0.4524 & 0.0000 & 0.4134 & 1.0000 \end{bmatrix}$$

$$P_G = \begin{bmatrix} 102.5620 \\ 0.0000 \\ 0.0000 \\ 79.5095 \\ 0.0000 \end{bmatrix}, \quad P_g = \begin{bmatrix} 102.5620 \\ 93.3805 \\ 60.1494 \\ 79.5095 \\ 70.7256 \end{bmatrix}$$

Note that we calculate the unknown gross power vector P_g using $P_g = A_u^{-1} P_G$. Table 10.8 gives the transmission usage allocation based on the Bialek's upstream algorithm.

Table 10.8 Transmission Usage Allocation of Bialek's Upstream Algorithm

Line	i	j	P_{ij}^{ave}	$P_{ij}^{G_1}$	$P_{ij}^{G_4}$
1	1	2	83.6795	83.6795	0
2	1	3	18.8875	18.8875	0
3	2	4	9.6990	0	9.6990
4	2	5	42.2840	37.8924	4.3924
5	3	4	41.3315	0	41.3315
6	4	5	28.4790	0	28.4790

TRANSMISSION CONGESTION MANAGEMENT AND PRICING

10.3.3 Cost Allocation Using Kirschen's Tracing Method

The lossless system is redrawn in Figure 10.4 to include three commons based on Kirschen's tracing method. Common 1 includes bus 1, common 2 includes bus 4, and common 3 includes buses 3, 4, and 5. The system has two links. The first link connects commons 1 and 3, which includes branches 1–2 and 1–3. The second link connects commons 2 and 3, which includes branches 2–4, 2–3 and 2–5. The acyclic diagram of the test system is shown in Figure 10.5. Transmission usage allocations obtained by Kirschen's tracing method are shown in Table 10.9.

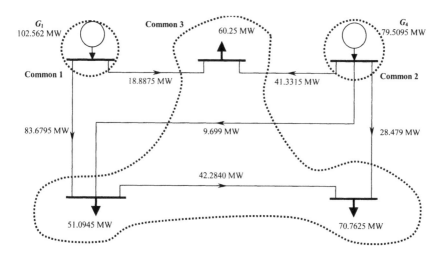

Figure 10.4 Commons of the Test System

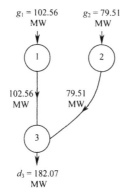

Figure 10.5 Acyclic Diagram of Generator Contributions

Table 10.9 Transmission Usage Allocation of Kirschen's Tracing Method

Line	i	j	P_{ij}^{ave}	$P_{ij}^{G_1}$	$P_{ij}^{G_4}$
1	1	2	83.6795	83.6795	0
2	1	3	18.8875	18.8875	0
3	2	4	9.6990	0	9.6990
4	2	5	42.2840	23.81857	18.46543
5	3	4	41.3315	0	41.3315
6	4	5	28.4790	0	28.4790

10.3.4 Comparing the Three Cost Allocation Methods

Here, we show a comparison of the three allocation methods. As we observe in Tables 10.7, 10.8, and 10.9, the contributions of G_1 and G_2 to line flows differ. The differences are expected to be greater in a larger power system. However, occasionally the results in these tables are the same for certain transmission lines. For example, Bialek's and Kirschen's tracing methods result in different contributions for line 2–5 but the same contributions for lines 1–2, 1–3, 2–4, 3–4 and 3–5. We also notice zero contributions of generators using these two methods. Using GGDFs, none of the two generators gives zero contribution, and the results are different from those using the tracing methods.

Based on Tables 10.7, 10.8, and 10.9, allocations of transmission charges are shown in Table 10.10. The allocations are based on the MW-mile method (Equation 10.1). As Table 10.10 shows, the total transmission charge of G_1 is $607.4309 using GGDFs, $538.1039 using Bialek's tracing method, and $473.657 using Kirschen's method. For every 1 MW, it costs G_1 an average of $5.87707 using GGDFs, $5.20632 using Bialek's tracing method, and $4.58277 using Kirschen's tracing method. For G_4, the total transmission charge is $362.5691 using GGDFs, $431.8961 using Bialek's tracing method, and $496.343 using Kirschen's tracing method. For every 1 MW, it costs G_4 and average of $4.53211 using GGDFs, $5.3987 using Bialek's tracing method, and $6.20429 using Kirschen's tracing method. As in the case of contributions to line loading, we see that different methods give different costs for transmitting power. These differences are expected to be high for more complicated power systems with more transactions.

Table 10.10 Allocation of Transmission Charges of the Three Methods

Line k	Line Cost $	GGDFs Method		Bialek's Method		Kirschen's Method	
		$c_k L_k MW_{1,k}$	$c_k L_k MW_{4,k}$	$c_k L_k MW_{1,k}$	$c_k L_k MW_{4,k}$	$c_k L_k MW_{1,k}$	$c_k L_k MW_{4,k}$
1	60	4453.566	567.168	5020.770	0.000	5020.770	0.000
2	240	6883.512	2352.000	4533.000	0.000	4533.000	0.000
3	280	6580.868	3870.720	0.000	2715.720	0.000	2715.720
4	120	3941.580	1130.880	4547.088	527.088	2858.230	2215.852
5	30	157.791	1082.136	0.000	1239.945	0.000	1239.945
5	240	1688.016	5146.560	0.000	6834.960	0.000	6834.962
Total	970	23705.333	14149.464	14100.858	11317.713	12412.000	13006.479
$\sum_{t \in T}\sum_{k \in K} c_k L_k MW_{t,k}$		37854.797		25418.571		25418.479	
TC_t		607.4309	362.5691	538.1039	431.8961	473.6570	496.3430
Cost ($/MW)		5.87707	4.53211	5.20632	5.39870	4.58277	6.20429

10.4 LMP, FTR, AND CONGESTION MANAGEMENT

A transmission congestion charge is incurred when the system is constrained by physical limits. So a reasonable transmission pricing method should provide some economic signal to reflect the charge due to the physical constraints. One option is to base the change on locational marginal prices. That is, the congestion charge for a specified path is the product of the flow along the path and the price differences between the two terminals of the path.

The transmission congestion charge may skyrocket in some cases, and create a big loss for a market participant. To hedge the risk, the participant can purchase a right to transfer power over a constrained transmission path for a fixed price, which is called a firm transmission right. The holder of such a right receives a credit that counteracts the congestion charge.

In this section, we will introduce the concepts of locational marginal pricing and firm transmission right and give some examples. A transmission congestion management scheme that incorporates an application of these two concepts is presented.

10.4.1 Locational Marginal Price (LMP)

LMP is the marginal cost of supplying the next increment of electric energy at a specific bus considering the generation marginal cost and the physical aspects of the transmission system. LMP is given as

LMP = generation marginal cost + congestion cost
+ cost of marginal losses

Mathematically, LMP at any node in the system is the dual variable (sometimes called a shadow price) for the equality constraint at that node (sum of injections and withdrawals is equal to zero). Or, LMP is the additional cost for providing one additional MW at a certain node.

Using LMP, buyers and sellers experience the actual price of delivering energy to locations on the transmission systems. The difference in LMPs appears when lines are constrained. If the line flow constraints are not included in the optimization problem or if the line flow limits are assumed to be very large, LMPs will be the same for all buses, and this is the marginal cost of the most expensive dispatched generation unit (marginal unit). In this case, no congestion charges apply. However, if any line is constrained, LMPs will vary from bus to bus or from zone to zone, which may cause possible congestion charges.

Example 10.1: Illustration of LMP

Figure 10.6a shows the system under study, where line data are given in Table 10.11, and losses are neglected.

Figures 10.6b and 10.6c show the solutions for two cases: when line 1–3 is unconstrained and when the line is constrained. For the unconstrained case, we note that only generator G_1 is dispatched and it sets the price at 10 \$/MWh. In this case, $LMP_1 = LMP_2 = LMP_3 = 10$ \$/MWh, since the system is unconstrained. The price set by G_1 acts as a market-clearing price.

Table 10.11 Line Data

Line	From	To	Reactance (p.u.)	Limit (p.u.)
1	1	2	0.25	2.0
2	1	3	0.25	2.0
3	2	3	0.25	2.0

TRANSMISSION CONGESTION MANAGEMENT AND PRICING

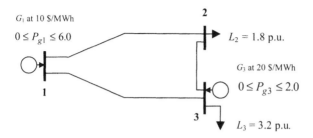

a. System Configuration

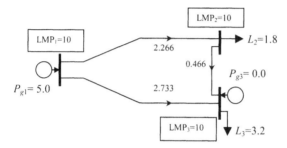

b. Solution When Line Limits are Ignored

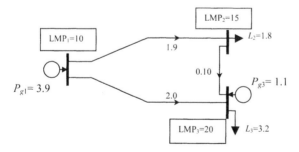

c. Solution When Line Limits are Considered

Figure 10.6 System of Example 10.1

For the constrained case (when line limits are considered), G_3 is also dispatched and accordingly G_1 is ramped down. As a result, LMP values become different at each bus. LMP_2 (15 $/MWh) is obtained as follows: if the demand at bus 2 were to increase by 1 MW (0.01 p.u.), MW would not be completely supplied by G_1 because the constraint in line 1–3 prevents additional power from being generated at bus 1 and delivered to bus 2. Sending additional power over line 1–2 could increase the flow on line 1–3 as well. Therefore, bus 2 must receive the remaining energy from G_3, and thus these generators (G_1 and G_3) become the marginal units for the system as they control the flow on line 1–3.

Power flow from G_1 to bus 2 would split between lines 1–2 and 1–3 (then 3–2). Given line reactances as shown in Table 10.11 (where all lines have the same reactance), 2/3 of any additional power would flow on line 1–2 and 1/3 on line 1–3 (then 3–2). See Figure 10.7, where ΔP_{g1} is the additional power from G_1 required to supply the increase in demand at bus 2. By the same logic, flow from G_3 to bus 2 would split between lines 3–1 (then 1–2) and 3–2. Also, 2/3 of the power would flow on line 3–2 and 1/3 would flow on line 3–1 (then 1–2). See Figure 10.8, where ΔP_{g3} is the additional power from G_3 required to supply the increase in demand at bus 2.

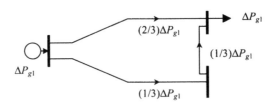

Figure 10.7 Flows due to an Additional Generation of ΔP_{g1}

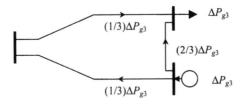

Figure 10.8 Flows due to an Additional Generation of ΔP_{g3}

TRANSMISSION CONGESTION MANAGEMENT AND PRICING

In order to keep the power flow of line 1–3 at thermal limit, G_3 must be turned on to counteract the flow from 1 to 3. Equal amounts of generation on each side of the constraint will cancel out and prevent overflow on line 1–3, that is, as $(1/3)\,\Delta P_{g1} = (1/3)\,\Delta P_{g3}$, or $\Delta P_{g1} = \Delta P_{g3}$. For an increase of 1 MW at bus 2, we would need $\Delta P_{g1} + \Delta P_{g3} = 1$ MW.

This means that in order to keep the flow on line 1–3 within reliability limits and maintain the most economic dispatch, 50% of the additional load at bus 2 is served by G_1 and 50% is served by G_3. The resulting locational price at load bus 2 is:

$$\text{LMP}_2 = (0.5 \times 10\ \$/\text{MWh}) + (0.5 \times 20\ \$/\text{MWh}) = 15\ \$/\text{MWh}$$

Example 10.2: Delivery Factor, Generator Shift Factor, Constraint's Cost, and Decomposing Components of LMPs

(1) **Delivery Factor:** A delivery factor of bus i with respect to bus j as a reference bus (or $DF_{i,j}$) is a measure of the portion of the next MW generation at bus i that is delivered to bus j. For example, $DF_{1,2} = 0.8$ means that of the next 1 MW generation sent from bus 1, 0.8 MW is delivered to bus 2, when bus 2 is the reference bus. Note that $DF_{2,1} = 1/DF_{1,2}$. For $DF_{1,2} = 0.8$, $DF_{2,1} = 1/0.8 = 1.25$. A DF that is larger than 1.0 represents a reduction in losses. Values of DF depend on the choice of the reference bus.

(2) **Generator Shift Factor:** The generator shift factor is defined as the ratio of the change in line flow to the change in generation of the designated bus. A factor GSF_{ik} refers to generation shift factor for bus i on line k. All generation shift factors at the reference bus are equal to zero.

(3) **Constraint's Cost:** In the system shown in Figure 10.9, the line limit connecting the two buses is 100 MW. For this system, the optimal dispatch is shown in Figure 10.10. For the optimal dispatch, the total cost is $10 \times 220 + 15 \times 70 = \$3250$. Now, if we permit the constrained line to increase its limit by 1 MW, the constraint's cost (β) is equal to

$$\beta = \frac{\text{Reduction in total cost}}{\text{Change in contraint's flow}} \qquad (10.10)$$

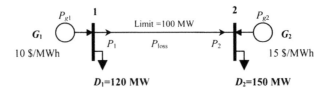

Figure 10.9 System of Example 10.2

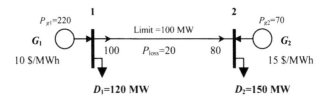

Figure 10.10 Optimal Dispatch

Case 1: Bus 2 is the reference bus. If we let the line limit increase to 101 MW (increase of 1 MW in line limit), then G_1 (the cheapest generator) will pick up 1 MW and G_2 (the most expensive generator) will decrease its output by 0.8 MW. From (10.10), the constraint's cost is equal:

$$\beta = \frac{(0.8)(15) - (1.0)(10)}{1.0} = 2 \text{ \$/MWh}$$

When the line limit is increased by 1 MW, G_1 generates 221 MW and G_2 generates 69.2 MW. The total cost is $10 \times 221 + 15 \times 69.2 = \$3248$ and the difference between this cost and the initial cost is 2 (or 3250 - 3248).

Case 2: Bus 1 is the reference bus. If G_2 decreases its output by 1 MW, the line flow will increase by 1 MW, and G_1 will increase its output by 1/0.8 MW (or 1.25 MW) to compensate. In this case, the constraint cost, from (10.10), is equal to

$$\beta = \frac{(1.0)(15) - (1.25)(10)}{1.0} = 2.5 \text{ \$/MWh}$$

Note that in this case G_1 generates 221.25 MW and G_2 generates 69 MW. Total cost is $10 \times 221.25 + 15 \times 69 = \$3247.5$. The difference between this cost and the initial cost is 2.5 (or 3250 - 3247.5).

TRANSMISSION CONGESTION MANAGEMENT AND PRICING

(4) Decomposing Components of LMPs: At any bus i, LMP_i is composed of three components. The first component is marginal generation price at the reference bus (LMP_i^{ref}), the second component is a loss component (LMP_i^{loss}), and the third component is a congestion component (LMP_i^{cong}). Thus, LMP_i could be expressed as:

$$LMP_i = LMP_i^{ref} + LMP_i^{loss} + LMP_i^{cong} \qquad (10.11a)$$

To calculate the last two components, *DF* and *GSF* are required. The values of the three components are based on the selection of reference bus. The last two components are given as:

$$LMP_i^{loss} = (DF_i - 1)\, LMP^{ref} \qquad (10.11b)$$

$$LMP_i^{cong} = -\sum_{k \in K} GSF_{ik}\, \beta_k \qquad (10.11c)$$

where

DF_i = delivery factor of bus i relative to the reference bus
GSF_{ik} = generation shift factor for bus i on line k
β_k = constraint cost of line k
K = set of congested transmission lines

Now, let's apply these equations to our example to show how they work.

For Case 1 (bus 2 is the reference bus)

Bus 1:

$$LMP^{ref} = 15\, \$/MWh$$
$$LMP_1^{loss} = (DF_{1,2} - 1)\, LMP^{ref} = (0.8 - 1)(15) = -3\, \$/MWh$$
$$LMP_1^{cong} = -GSF_{11}\, \beta_1 = -(1)(2) = -2\, \$/MWh$$
$$LMP_1 = LMP^{ref} + LMP_1^{loss} + LMP_1^{cong} = 15 - 3 - 2 = 10\, \$/MWh$$

Bus 2:

$$\text{LMP}^{ref} = 15\,\$/\text{MWh}$$

$$\text{LMP}_2^{loss} = (DF_{2,2} - 1)\ \text{LMP}^{ref} = (1-1)(15) = 0\,\$/\text{MWh}$$

$$\text{LMP}_2^{cong} = -GSF_{21}\ \beta_1 = -(0)(2) = 0\,\$/\text{MWh}$$

$$\text{LMP}_2 = \text{LMP}^{ref} + \text{LMP}_2^{loss} + \text{LMP}_2^{cong} = 15 + 0 + 0 = 15\,\$/\text{MWh}$$

For Case 2 (bus 1 is the reference bus)

Bus 1:

$$\text{LMP}^{ref} = 10\,\$/\text{MWh}$$

$$\text{LMP}_1^{loss} = (DF_{1,1} - 1)\ \text{LMP}^{ref} = (1-1)(15) = 0\,\$/\text{MWh}$$

$$\text{LMP}_1^{cong} = -GSF_{11}\ \beta_1 = -(0)(2.5) = 0\,\$/\text{MWh}$$

$$\text{LMP}_1 = \text{LMP}^{ref} + \text{LMP}_1^{loss} + \text{LMP}_1^{cong} = 10 + 0 + 0 = 10\,\$/\text{MWh}$$

Bus 2:

$$\text{LMP}^{ref} = 10\,\$/\text{MWh}$$

$$\text{LMP}_2^{loss} = (DF_{2,1} - 1)\ \text{LMP}^{ref} = (1.25-1)(10) = 2.5\,\$/\text{MWh}$$

$$\text{LMP}_2^{cong} = -GSF_{21}\ \beta_1 = -(-1)(2.5) = 2.5\,\$/\text{MWh}$$

$$\text{LMP}_2 = \text{LMP}^{ref} + \text{LMP}_2^{loss} + \text{LMP}_2^{cong} = 10 + 2.5 + 2.5 = 15\,\$/\text{MWh}$$

Example 10.3: AC Analysis of a System with Losses and Congestion

Figure 10.11 shows a system of two buses, where bus 1 is the reference bus. The line admittance is given by $y_{12} = 4.0 - j20.0$ p.u. and equality constraints at two buses are given by

$$P_{G1} - P_{D1} = v_1^2 g_{11} + v_1 v_2 [g_{12}\cos(\delta_1 - \delta_2) + b_{12}\sin(\delta_1 - \delta_2)]$$

$$P_{G2} - P_{D2} = v_2^2 g_{22} + v_2 v_1 [g_{21}\cos(\delta_2 - \delta_1) + b_{21}\sin(\delta_2 - \delta_1)]$$

TRANSMISSION CONGESTION MANAGEMENT AND PRICING

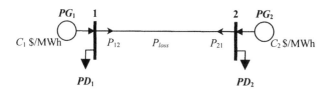

Figure 10.11 System of Example 10.3

The flow from 1 to 2, and 2 to 1 are given by

$$f_{1-2} = v_1^2 g_{11} + v_1 v_2 [g_{12} \cos(\delta_1 - \delta_2) + b_{12} \sin(\delta_1 - \delta_2)]$$

$$f_{2-1} = v_2^2 g_{22} + v_1 v_2 [g_{21} \cos(\delta_2 - \delta_1) + b_{21} \sin(\delta_2 - \delta_1)]$$

The loss in the line connecting the two buses is given by

$$\begin{aligned} P_{loss} &= f_{1-2} + f_{2-1} \\ &= v_1^2 g_{11} + v_2^2 g_{22} + v_1 v_2 [g_{12} \cos(\delta_1 - \delta_2) + b_{12} \sin(\delta_1 - \delta_2) + g_{21} \cos(\delta_2 - \delta_1) + b_{21} \sin(\delta_2 - \delta_1)] \\ &= v_1^2 g_{11} + v_2^2 g_{22} + v_1 v_2 [g_{12} \cos(\delta_1 - \delta_2) + b_{12} \sin(\delta_1 - \delta_2) + g_{21} \cos(\delta_1 - \delta_2) - b_{21} \sin(\delta_1 - \delta_2)] \\ &= v_1^2 g_{11} + v_2^2 g_{22} + 2 v_1 v_2 g_{12} \cos(\delta_1 - \delta_2) \end{aligned}$$

Inequality constraints reflect real power flows between buses, and stability and thermal limits define line limits. Inequality constraints for each flow is given by

$$P_{12}^{min} \leq f_{1-2} = v_1^2 g_{11} + v_1 v_2 [g_{12} \cos(\delta_1 - \delta_2) + b_{12} \sin(\delta_1 - \delta_2)] \leq P_{12}^{max}$$

$$P_{12}^{min} \leq f_{2-1} = v_2^2 g_{22} + v_1 v_2 [g_{21} \cos(\delta_2 - \delta_1) + b_{21} \sin(\delta_2 - \delta_1)] \leq P_{12}^{max}$$

where g_{11}, b_{11}, g_{12}, b_{12}, g_{21}, b_{21}, and b_{22} are contributions to the network admittance matrix for a branch between buses 1 and 2, $g_{12} = g_{21}$ and $b_{12} = b_{21}$. If we consider a constant bus voltage of 1.0 p.u., the former equations will be

$$P_{G1} - P_{D1} = g_{11} + g_{12} \cos(\delta_1 - \delta_2) + b_{12} \sin(\delta_1 - \delta_2)$$

$$P_{G2} - P_{D2} = g_{22} + g_{21} \cos(\delta_2 - \delta_1) + b_{21} \sin(\delta_2 - \delta_1)$$

The flow from 1 to 2, and 2 to 1 are given by:

$$f_{1-2} = g_{11} + g_{12} \cos(\delta_1 - \delta_2) + b_{12} \sin(\delta_1 - \delta_2)$$

$$f_{2-1} = g_{22} + g_{21} \cos(\delta_2 - \delta_1) + b_{21} \sin(\delta_2 - \delta_1)$$

The loss in the line connecting the two buses is given by

$$P_{loss} = g_{11} + g_{22} + 2\, g_{12} \cos(\delta_1 - \delta_2)$$

Inequality constraints is given by

$$P_{12}^{min} \leq g_{11} + g_{12} \cos(\delta_1 - \delta_2) + b_{12} \sin(\delta_1 - \delta_2) \leq P_{12}^{max}$$

$$P_{21}^{min} \leq g_{22} + g_{21} \cos(\delta_2 - \delta_1) + b_{21} \sin(\delta_2 - \delta_1) \leq P_{21}^{max}$$

For this example, the problem is formulated as

$$\min \quad C_1 P_{G1} + C_2 P_{G2}$$

s.t.

$$P_{G1} - P_{D1} - g_{11} - g_{12} \cos(\delta_1 - \delta_2) - b_{12} \sin(\delta_1 - \delta_2) = 0$$

$$P_{G2} - P_{D2} - g_{22} - g_{21} \cos(\delta_2 - \delta_1) - b_{21} \sin(\delta_2 - \delta_1) = 0$$

$$-P_{12}^{max} \leq f_{1-2} = g_{11} + g_{12} \cos(\delta_1 - \delta_2) + b_{12} \sin(\delta_1 - \delta_2) \leq P_{12}^{max}$$

$$-P_{12}^{max} \leq f_{2-1} = g_{22} + g_{21} \cos(\delta_2 - \delta_1) + b_{21} \sin(\delta_2 - \delta_1) \leq P_{12}^{max}$$

$$0 \leq P_{G1} \leq P_{G1}^{max}$$

$$0 \leq P_{G2} \leq P_{G2}^{max}$$

Equality constraints represent power flow balance equation at each bus. Inequality constraints reflect real power flows between buses, and stability and thermal limits define line limits. The Lagrangian is given by:

$$\begin{aligned}
L =\ & C_1 P_{G1} + C_2 P_{G2} \\
& + \lambda_1 (P_{G1} - P_{D1} - g_{11} - g_{12} \cos(\delta_1 - \delta_2) - b_{12} \sin(\delta_1 - \delta_2)) \\
& + \lambda_2 (P_{G2} - P_{D2} - g_{22} - g_{21} \cos(\delta_2 - \delta_1) - b_{21} \sin(\delta_2 - \delta_1)) \\
& + \gamma_1 (g_{11} + g_{12} \cos(\delta_1 - \delta_2) + b_{12} \sin(\delta_1 - \delta_2) - P_{12}^{max}) \\
& + \gamma_2 (g_{22} + g_{21} \cos(\delta_2 - \delta_1) + b_{21} \sin(\delta_2 - \delta_1) - P_{12}^{max}) \\
& + \omega_1 (-g_{11} - g_{12} \cos(\delta_1 - \delta_2) - b_{12} \sin(\delta_1 - \delta_2) + P_{12}^{max}) \\
& + \omega_2 (-g_{22} - g_{21} \cos(\delta_2 - \delta_1) - b_{21} \sin(\delta_2 - \delta_1) + P_{12}^{max}) \\
& + \pi_1 (P_{G1} - P_{G1}^{max}) + \pi_2 (P_{G2} - P_{G2}^{max}) + \psi_1 (-P_{G1} + P_{G1}^{min}) + \psi_2 (-P_{G2} + P_{G2}^{min})
\end{aligned}$$

For the case where bus 1 is the reference bus ($\delta_1 = 0$), the Lagrangian is

TRANSMISSION CONGESTION MANAGEMENT AND PRICING

$$L = C_1 P_{G1} + C_2 P_{G2}$$
$$+ \lambda_1 (P_{G1} - P_{D1} - g_{11} - g_{12} \cos(\delta_2) + b_{12} \sin(\delta_2))$$
$$+ \lambda_2 (P_{G2} - P_{D2} - g_{22} - g_{21} \cos(\delta_2) - b_{21} \sin(\delta_2))$$
$$+ \gamma_1 (g_{11} + g_{12} \cos(\delta_2) - b_{12} \sin(\delta_2) - P_{12}^{max})$$
$$+ \gamma_2 (g_{22} + g_{21} \cos(\delta_2) + b_{21} \sin(\delta_2) - P_{12}^{max})$$
$$+ \omega_1 (-g_{11} - g_{12} \cos(\delta_2) + b_{12} \sin(\delta_2) + P_{12}^{max})$$
$$+ \omega_2 (-g_{22} - g_{21} \cos(\delta_2) - b_{21} \sin(\delta_2) + P_{12}^{max})$$
$$+ \pi_1 (P_{G1} - P_{G1}^{max}) + \pi_2 (P_{G2} - P_{G2}^{max}) + \psi_1 (-P_{G1} + P_{G1}^{min}) + \psi_2 (-P_{G2} + P_{G2}^{min})$$

At the optimal solution, we have

$$\partial L / \partial P_{G1} = 0 = C_1 + \lambda_1 + \pi_1 \quad (10.12a)$$

$$\partial L / \partial P_{G2} = 0 = C_2 + \lambda_2 + \pi_2 \quad (10.12b)$$

$$\partial L / \partial \delta_2 = 0 = \lambda_1 (g_{12} \sin(\delta_2) + b_{12} \cos(\delta_2)) + \lambda_2 (g_{21} \sin(\delta_2) - b_{21} \cos(\delta_2))$$
$$+ \gamma_1 (-g_{12} \sin(\delta_2) - b_{12} \cos(\delta_2)) + \gamma_2 (-g_{21} \sin(\delta_2) + b_{21} \cos(\delta_2))$$
$$+ \omega_1 (g_{12} \sin(\delta_2) + b_{12} \cos(\delta_2)) + \omega_2 (g_{21} \sin(\delta_2) - b_{21} \cos(\delta_2))$$

The last equation can be simplified as

$$\partial L / \partial \delta_2 = 0 = g_{12} \sin(\delta_2) [\lambda_1 + \lambda_2] + b_{12} \cos(\delta_2) [\lambda_1 - \lambda_2]$$
$$+ g_{12} \sin(\delta_2) [-\gamma_1 - \gamma_2] + b_{12} \cos(\delta_2) [-\gamma_1 + \gamma_2] \quad (10.12c)$$
$$+ g_{12} \sin(\delta_2) [\omega_1 + \omega_2] + b_{12} \cos(\delta_2) [\omega_1 - \omega_2]$$

For $g_{12} = g_{21} = -4.0$, $g_{11} = g_{22} = 4.0$, $b_{12} = b_{21} = 20.0$, and $b_{11} = b_{22} = -20.0$, the solution of the system for six cases is shown in Table 10.12. Five of the cases are where bus 1 is the reference bus, and the last case is where bus 2 is the reference bus. To see the impact of the line limit on LMPs, we take different line limits, as shown in Table 10.12.

Table 10.12 Results for Different Cases

Case	Ref. Bus	Line Limit	P_{G1}^{max}	δ_1	δ_2	P_{G1}	P_{G2}	f_{12}	f_{21}	P_{loss}	LMP_1	LMP_2
1	1	20.0	20.0	0.0	-24.73	14.7337	0.0000	8.7337	-8.0000	0.7337	10.000	12.029
2	1	5.0	20.0	0.0	-14.12	11.0000	3.2417	5.0000	-4.7583	0.2417	10.000	20.000
3	1	4.0	20.0	0.0	-11.31	10.0000	4.1554	4.0000	-3.8446	0.1554	10.000	20.000
4	1	2.0	20.0	0.0	-5.68	8.0000	6.0393	2.0000	-1.9607	0.0393	10.000	20.000
5	1	2.0	20.0	0.0	-2.85	7.0000	7.0099	1.0000	-0.9901	0.0099	10.000	20.000
6	2	5.0	5.0	-2.88	0.00	5.0000	9.0101	-1.0000	1.0101	0.0101	20.407	20.000

In the first case, the line limit is high (the line is not constrained), so LMPs will only reflect the effect of line loss. In this case, only P_{G1} is producing energy, and LMP_2 is the same as LMP_1 plus the loss price. The loss price in this case is 2.029 $/MWh.

To see how LMP_2 is obtained, we take the first case. When line limit is high, line constraint is not active, i.e., $\gamma_1 = \gamma_2 = \omega_1 = \omega_2 = 0.0$. From (10.12c) we have

$$g_{12}\sin(\delta_2)[\lambda_1 + \lambda_2] + b_{12}\cos(\delta_2)[\lambda_1 - \lambda_2] = 0.0$$

or

$$-4.0\sin(\delta_2)[\lambda_1 + \lambda_2] + 20.0\cos(\delta_2)[\lambda_1 - \lambda_2] = 0.0$$

or

$$\lambda_2 = \lambda_1 \frac{4.0\sin(\delta_2) - 20.0\cos(\delta_2)}{-4.0\sin(\delta_2) - 20.0\cos(\delta_2)} \tag{10.12d}$$

When $P_{G1} = 14.7337$ (not on limit), we have $\pi_1 = 0.0$.

From (10.12a) $C_1 + \lambda_1 + \pi_1 = 0.0 \Rightarrow \lambda_1 = 10$

For $\delta_2 = -24.73°$, we find, from (10.12d) that

$$\lambda_2 = \lambda_1 \frac{-19.8392}{-16.4924} = (10)(1.20293) = 12.0293$$

For the second case, both generator outputs are within limits (none of them hits the maximum or minimum limit), so $\pi_1 = \pi_2 = 0.0$. From (10.12a), we have

$$0 = C_1 + \lambda_1 + \pi_1 \Rightarrow \lambda_1 = C_1 = 10$$

$$0 = C_2 + \lambda_2 + \pi_2 \Rightarrow \lambda_2 = C_2 = 20$$

For this case line flow is constrained from bus 1 to bus 2, so $\gamma_2 = \omega_1 = \omega_2 = 0.0$, and $\gamma_1 \neq 0.0$. From (10.12c) we have

$$0 = g_{12}\sin(\delta_2)[\lambda_1 + \lambda_2] + b_{12}\cos(\delta_2)[\lambda_1 - \lambda_2] + g_{12}\sin(\delta_2)[-\gamma_1] + b_{12}\cos(\delta_2)[-\gamma_1]$$

$$\gamma_1 = \frac{g_{12}\sin(\delta_2)[\lambda_1 + \lambda_2] + b_{12}\cos(\delta_2)[\lambda_1 - \lambda_2]}{g_{12}\sin(\delta_2) + b_{12}\cos(\delta_2)} = \frac{29.2744 - 193.957}{0.975814 + 19.3975} = -8.08395$$

$$\lambda_2 = \lambda_1 + \lambda_2^{cong} + \lambda_2^{loss} \Rightarrow \lambda_2^{loss} = 20 - 10 - 8.08395 = 1.91605 \text{ $/MWh}$$

Other cases can be analyzed similarly.

10.4.2 LMP Application in Determining Zonal Boundaries

Transmission zones can be defined based on LMPs. In each zone, the LMP of energy is close to the marginal price at all nodes inside the zone. Differences in the marginal price in constrained situations define zonal boundaries, which indicates that the zonal definitions provide both economical and physical perspectives.

Initially, zonal boundaries may be defined based on historical LMPs at different nodes of the system or, equivalently, experience may be applied to identify lines that are expected to be congested. Later, these initially defined zones may be modified (merged or subdivided) based on new LMPs.

Because all nodes in a certain zone have similar marginal prices, all nodes in a zone can be aggregated and thought of as one node. We can represent the price in a zone using average locational marginal price (ALMP). The ALMP in any zone is the average LMP of the nodes in that zone.

Example 10.4: Zone Definition and Boundaries

Figure 10.12 shows a five-bus system with its line parameters given in Table 10.13. Assume that each L_2, L_3, and L_4 is fixed at 300 MW. Table 10.14 shows the generation data at a certain hour, which include initial (preferred) schedules and generation bids at this hour. We assume in this example that the system is lossless.

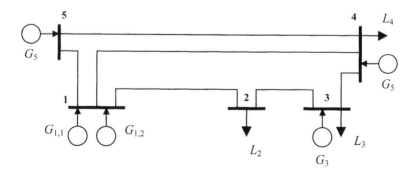

Figure 10.12 Five-bus System

Table 10.13 System Data

Line	From	To	Reactance (Ohm)	Line Limit (MW)
1	1	2	0.0280	350.0
2	1	4	0.0301	160.0
2	1	5	0.0060	380.0
3	2	3	0.0110	120.0
4	3	4	0.0300	230.0
5	4	5	0.0300	240.0

Table 10.14 Generation Data

Bus	Generator	Preferred Generation Schedule (MW)	Adjustment Range Min (MW)	Adjustment Range Max (MW)	Inc/Dec Price ($/MWh)
1	$G_{1,1}$	110	0.0	110	14
1	$G_{1,2}$	100	0.0	100	15
2	G_2	-	-	-	-
3	G_3	90	0.0	520	30
4	G_4	0	0.0	200	30
5	G_5	600	0.0	600	10

The inter-zonal lines are usually defined based on most frequently congested lines. The zone boundaries are confirmed or modified based on calculated LMPs. Initially, we consider lines 1–2, 1–4 and 4–5 are the most frequently congested (inter-zonal lines), which divide the system into two zones. These lines are demanded highly to convey power from injection to extraction points. For this system, let's study two cases:

Case 1: Unconstrained case (no line limit is taken into consideration) with preferred schedules.

Case 2: Constrained case (lines 1–2, 1–4 and 1–5 are constrained)

The results of inter-zonal congestion management for the data in Table 10.14 are shown in Tables 10.15–10.18. Table 10.15 shows the line flows for both cases, where Case 1 (case of preferred schedules) shows that some inter-zonal line flows violate limits (in boldface), while Case 2 shows that flows of inter-zonal lines are within limits. In Table 10.16, Case 2 shows that the preferred generations are re-dispatched to remove inter-zonal line flow violations. The re-dispatched values represent actual generations and deliveries during this hour. The LMPs of the system are calculated and shown in Table 10.17. In this case the system looks like a two-zone system.

Table 10.15 Line Flows

Line	From	To	Limit	Flow Case 1	Flow Case 2
1	1	2	350.00	352.13	350.00
2	1	4	160.00	191.68	160.00
3	1	5	380.00	-333.81	-360.22
4	2	3	120.00	52.13	50.00
5	3	4	230.00	-157.86	-187.17
6	4	5	240.00	-266.19	-239.78

Table 10.16 Re-dispatched Generation

Variable	Case 1 (MW)	Case 2 (MW)
$G_{1,1}$	110.0000	110.0000
$G_{1,2}$	100.0000	39.7781
G_3	90.0006	62.8285
G_4	0.0000	87.3941
G_5	600.0000	600.0000

Table 10.17 LMPs for both Cases

Bus	LMP ($/MWh) Case 1	LMP ($/MWh) Case 2
1	30.0000	15.0000
2	30.0000	30.0000
3	30.0000	30.0000
4	30.0000	30.0000
5	30.0000	17.6593

From Table 10.17, we learn that prices at buses 1 and 5 are very close (15 or 17.6593), and prices at buses 2, 3 and 4 are the same (30.0). This indicates that all buses could be categorized into two zones, with exactly or nearly the same LMP for buses in each zone. Table 10.18 classifies buses based on the zone they are located. Values of LMPs confirm the assumption that the most frequently congested lines divide the system into two zones. Figure 10.13 shows the equivalent zonal system based on the LMPs. Each zone contains buses with nearly the same LMPs, and the lines connected to these zones are inter-zonal lines. After this step, a decision can be made to confirm the initial assumption of zonal boundaries or suggest combining or splitting zones to form new zones.

Table 10.18 Specifying zones based on LMPs

Bus	LMP ($/MWh)	Zone
1	15.0000	1
2	30.0000	2
3	30.0000	2
4	30.0000	2
5	17.6593	1

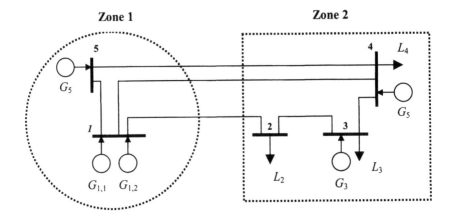

Figure 10.13 Equivalent Zonal System

10.4.3 Firm Transmission Right

A firm transmission right[4] (FTR) is a purchased right that can hedge congestion charges on constrained transmission paths. In other words, it provides FTR owners with the right to transfer an amount of power over a constrained transmission path for a fixed price.

Market participants pay congestion charges under a constrained situation based on LMP differences. These charges arise when the energy demand across a transmission path is more than the capability of transmission lines on that path. Under constrained situations, each participant is charged for congestion based on the MWh value of generation ordered to serve its load. The charge will be based on MWh and the difference in LMPs of injection and extraction points. If it happens that

---

[4] FTR is also called fixed transmission right or financial transmission right.

TRANSMISSION CONGESTION MANAGEMENT AND PRICING 409

a market participant's generation is not exactly equal to its load, it will either purchase or sell energy to the spot market.

Each FTR holder receives a congestion credit in each constrained hour that is proportional to the FTR value. This credit allocation is based on preferred schedules, while congestion charges are based on actual deliveries. From the preferred schedule FTRs, the total congestion credits are calculated and compared with the total congestion charges, which are based on the cost of re-dispatched schedules at each hour. If the total congestion credits are less than or equal to the total congestion charges, the congestion credit for each FTR holder is equal to the one calculated. If there are any extra congestion charges, the extra charges are distributed among market participants at the end of the month. Otherwise, the congestion credit for each FTR will be equal to a share of total congestion charges in proportion to its credit allocation. The insufficiency in hourly congestion charges may be offset by excessive charges in other hours at the end of the accounting month.

If a market participant does not hold a FTR and its contract is not curtailable, this participant will incur a congestion charge and have no mechanism to offset the congestion charge. In comparison, FTR holders will receive a credit that counteracts the congestion charge for the specified path. The credit is computed as follows, where LMP_1 and LMP_2 represent LMPs at starting and ending points of the FTR respectively.

$$\text{FTR credit} = \text{amount of FTR} \times (LMP_1 - LMP_2) \quad (10.13)$$

Although FTRs act as a financial instrument to hedge risk associated with transmission congestion charges, they are advantageous only when the designated path is in the same direction as the congested flow (which also indicates that the LMP of the extraction point is greater than the LMP of the injection point). It may happen that the FTR holder has to pay for having the FTR when the LMP of the extraction point is less than the LMP of the injection point. In this case, the monetary charge would be equal to the MWh value of the FTR multiplied by the difference in LMPs from the point of receipt to the point of delivery.

The following examples show that FTRs can work both ways: as a benefit or as a liability. Because of this, it is important for the holder to estimate the transmission path's economic value before obtaining an FTR.

Example 10.5: FTR in an Unconstrained Case

Load serving entity 2 (LSE_2) has an annual peak load of 300 MW. LSE_2 has contracted 225 MW from $GENCO_1$ and 135 MW from $GENCO_3$ to meet its generation capacity requirement of 360 MW. Because the flow of electricity on the shown system is mainly from bus 1 to bus 2, LSE_2 has obtained an FTR for 225 MW from 1 to 2 to protect itself from congestion on line 1–2. Figure 10.14 shows the system for a given hour (when the load of $LSE_2 = 150$ MW) on a given day without constraints. All loads are met with flow on line 1–2 and kept under the thermal limit. The highest cost generator ($GENCO_1$) running sets the price at 15 $/MWh. No congestion charges apply.

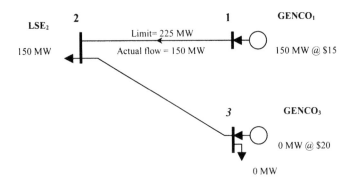

Figure 10.14 Unconstrained System

Example 10.6: FTR as a Benefit in a Constrained Case

Figure 10.15 shows the system of the last example during a given hour on a summer day, when the load of LSE_2 is increased to its peak at 300 MW. The line from 1 to 2 becomes constrained. The extra load cannot be supplied by $GENCO_1$, so generation from $GENCO_3$ is needed to meet the peak load, which makes the system out of economic merit. LMPs at buses 2 and 3 become higher than that of bus 1, creating congestion charges. LSE_2 will be charged for congestion. The amount of the charge is based on the amount of generation used to serve the load. In this case, 225 MWh of generation from $GENCO_1$ and 75 MWh of generation from $GENCO_3$ are used to serve the 300 MWh load for LSE_2. The congestion charge is calculated by multiplying the generation netted with the load (MWh) and the difference between the LMPs of sink and source, or:

Line 1–2: Congestion Charge = 225 MWh × (20 - 15) $/MWh   = $1125

Line 3–2: Congestion Charge = 75 MWh × (20 - 20) $/MWh = $ 0

Total = $1125

TRANSMISSION CONGESTION MANAGEMENT AND PRICING 411

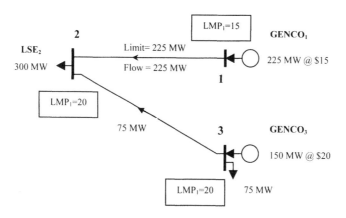

Figure 10.15 Constrained System

Because LSE_2 has obtained an FTR for 225 MW on line 1-2 and energy delivery is consistent with the firm reservations made, the LSE_2 will be reimbursed by congestion credit (during the settlement process). The congestion credit is calculated by multiplying amount of FTR in MW and the LMPs difference between sink bus and source bus, or

Line 1-2: Congestion credit = 225 MWh × (20 - 15) $/MWh = $1125

Net payment of LSE_2 after receiving the congestion credits is

Net Payment = Congestion charges - Congestion credits
= $1125 - $1125 = $ 0

Example 10.7: FTR as a Liability in a Constrained Case

In the system shown in Figure 10.16, we assume that LSE_1 contracted 10 MW from $GENCO_2$ for 20 $/MWh and owns 10 MW of firm point-to-point transmission with an FTR from 2 to 1. The day before the scheduled transaction, LSE_1 decides to supply its load from $GENCO_1$ instead of $GENCO_2$. Figure 10.16 shows a constrained system, where load LSE_1 is supplied by $GENCO_1$. LSE_1 does not deliver the energy on the designated path, and consequently does not receive a congestion charge (the charge in this case would actually have been a credit because the flow of energy would have been opposite the congested flow). Regardless of how the energy was delivered, LSE_1 still holds the FTR for 10 MW from 2 to 1. According to (10.13), this FTR will provide a credit as follows:

FTR Credit = 10 MWh × (15 - 20) $/MWh = - $50

As a result, L_1 must pay $50 for this hour because the price at the sink (extraction) bus is less than the price at source (injection) bus. We notice that in this case the FTR credit actually becomes a charge.

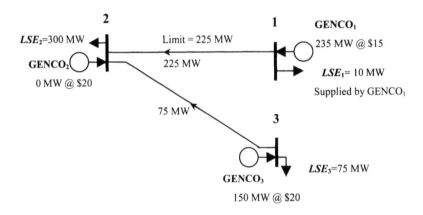

Figure 10.16 Illustration of the of FTR's Liability

10.4.4 FTR Auction

To guarantee the availability of FTRs to all parties on a non-discriminatory basis, there should be a mechanism to permit system users to buy, sell, and trade FTRs. This mechanism is guaranteed by conducting an auction for these rights. Trading these rights in secondary markets can self-arrange access across different paths, create long-term transmission rights, and provide more commercial certainty. This section presents a mathematical model of the FTR auction that covers purchase, sale, and base case FTRs.

The FTRs auction operates as a centralized auction in which market participants submit their bids for the purchase and sale of FTRs. The auction is conducted by the ISO or an auctioneer appointed by the ISO. The main aim of FTRs auction is to address the question: *How should FTRs be reconfigured or awarded to maximize revenues obtained from the auction while keeping the system within limits?*

Based on bids and actual dispatch, the ISO determines LMPs and either buys and sells energy at these prices or charges locational differences on these prices for transmission of power from one location to another. Differences in LMPs are reflections of congestion costs and transmission customer charges based on these locational differences. Total congestion payments collected by the ISO for actual use of the system must always be at least as large as the congestion payments to FTR holders in order to make adequate system revenues and hedge ISO charges. FTRs that are sold via bilateral contracts and not offered for sale in the auction are represented as fixed generations or loads at their points of injection or extraction. Any FTR seller may offer any portion of its FTR for sale, and the rest will be considered as the base case. In the auction, the reference bus should be identified by the ISO. The market-clearing price (MCP) for each FTR in the auction is based on the lowest winning bid made in the auction for that FTR. The auction provides market participants with opportunities to purchase FTRs that would not be available through bilateral transactions in a secondary market. In the auction, LSEs trade FTRs as a result of load variations.

The objective of the ISO is to maximize revenues from FTRs while keeping the system within limits when all FTRs exist simultaneously in the system. Each FTR may be originating from a single bus or many buses and ending with a single bus or multiple buses. The FTR holders may sell their FTRs partially or totally in a secondary market (as bilateral contracts) and different market participants may buy and sell FTRs through an auction. On the other hand, the auction can be seen as a short-term reshaping of FTRs where a participant may acquire an FTR different from those offered in the auction. Although FTRs acquired at an auction are effective for a shorter time period than those acquired with a firm transmission service, they still hedge against congestion similar to FTRs of firm transmission service.

There are four ways to acquire FTRs:

1. Network integration service customers acquire FTRs up to the value of their peak loads from capacity resources to their aggregate loads.

2. Firm point-to-point service customers acquire FTRs from source to destination.

3. FTRs may be traded monthly through an auction conducted by the ISO or an auctioneer replacing the ISO.

4. FTR holders may trade with other market participants in secondary markets (bilateral transactions) without participating in the FTRs auction.

To purchase a certain FTR in the auction, bidders provide the following information: maximum amount of FTR the bidder is willing to pay for, bid price, and points of injection and extraction. To sell a certain FTR in the auction, bidders provide the following information: maximum amount of FTR the bidder is willing to be paid for, bid price, and points of injection and extraction.

The formulation of auction identifies possible contracts and decides which contracts would maximize revenues for transmission network use. For single-injection single-withdrawal of an FTR, suppose that the MW power associated to the FTR of bidder i to be transmitted between buses m and n is α. This represents an injection (generation) of α MW at bus m and an extraction (load) of α MW at bus n. The FTRs auction ought to maximize revenues while satisfying system limits. The constraints set in the auction would represent line flow limits in pre- and post-contingency conditions. Mathematically, this FTR is represented by a vector with two elements: 1 at the bus of injection and -1 at the bus of withdrawal. The solution to this problem will yield optimal FTR awards and MCPs for bids.

For a general case with multiple injection and withdrawal points, if some injections or withdrawals are not included in the auction or only a portion of a certain right is entered in an auction, a generalized model is necessary. In this model, we need to define a mapping matrix (instead of a vector as in the first case) to map the awarded FTRs to points of injection and extraction. The sum of all elements of this matrix in each column is zero which indicates that the total injection is equal to the total withdrawals, or the total generation balances the total load.

To ensure that the system is simultaneously feasible, the base case should also be taken into consideration. The base case is the situation where some of the FTRs enter the auction as fixed values. The FTRs not entering the auction are considered as base case loads and generations. When the auction takes place, these base case values should be represented in system equations. In the FTR auction, any new FTR is modeled as injections and withdrawals at corresponding buses. For an outage, the equations representing FTR injections and withdrawals should be modified.

With the foregoing introduction, we may summarize the FTRs auction problem as follows: to determine the highest valued combination of FTRs to be awarded in the auction as judged by bids that are simultaneously feasible while taking into consideration pre-contingency and post-contingency transmission limits, previously awarded FTRs (base case values), power flow balance equations, and FTRs limits. This problem can be detailed as:

- Objective function of the auction
  The objective is to maximize revenues from FTRs (FTRs awarded to purchasers and taken from sellers). The objective may otherwise be stated as to determine the highest valued combination of FTRs to be awarded in the auction as judged by bids that are simultaneously feasible.

- Control Variables
  MW values of FTRs to be purchased or sold

- Fixed Values
  Base case net injections at each bus representing previously awarded FTRs and those FTRs that sellers like to keep for themselves.

- Constraints
  - *Equality constraints:* Simultaneously feasible security-constrained power flow equations. The equations take into account the base case values, FTRs to be purchased and FTRs to be sold.
  - *Inequality constraints*: (1) FTR limits and (2) line limits.

- Auction Output
  - *Feasible FTRs (purchased and sold)* as a set of winning bids: The ISO or the auctioneer announces winning bidders for FTRs and then the quantity of each FTR awarded in the auction is determined.
  - *Prices of feasible FTRs:* A purchaser of FTR will pay MCP and a seller of FTR will be paid MCP. The ISO or the auctioneer announces market price of each FTR awarded in the auction and the market price of an FTR between each bus and the reference bus. The MCP of each FTR is determined by the opportunity cost of that FTR in the auction; that is, the difference in the market prices for power on the two sides of a congested line determines the MCP of an FTR between the two sides. All buyers and sellers of FTRs between the same injection and withdrawal points will pay the same price.

Market prices of FTRs are regarded as interdependent because the price of every FTR is determined by the marginal dispatched FTR (bid). The marginal displaced FTR is the most valuable FTR that cannot be awarded because it would not be feasible simultaneously. Net auction revenues will be allocated by the ISO to transmission owners (TOs) based on a given criterion. The equality constraints used in the auction are represented by a system of equation with three components: base case, FTRs to be purchased, and FTRs offered for sale.

The process of purchase and sale of different FTRs in the auction is shown in Figure 10.17, where purchasers and sellers can make bilateral contracts through secondary markets before entering the auction. Also, they can do that after the auction is finished.

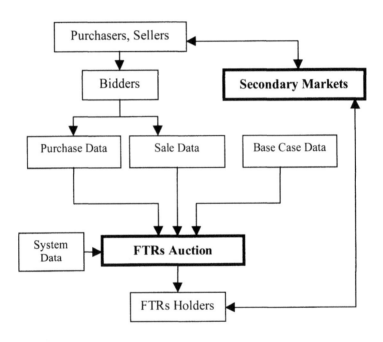

Figure 10.17 Procedure in the FTRs Auction

As seen in Figure 10.17, the first step in the auction is that bidders pass their data to the ISO or an auctioneer that replaces the ISO. The data include purchase and sale bids, or minimum and maximum MW values of FTRs for sale or purchase and their prices, and point(s) of injections and extractions. Also, base case values should be passed on to the ISO or the auctioneer. Along with these data, system parameters such as line parameters and line limits are passed on. The reference bus should be identified in the auction. Once FTR holders are identified by the auction, they can sell, buy, or trade FTRs through secondary markets.

TRANSMISSION CONGESTION MANAGEMENT AND PRICING

Example 10.8: FTRs Auction

A 4-bidder 3-bus system is shown in Figure 10.18. Injections, withdrawals, and base case values are also shown. For this test system, Tables 10.19 to 10.21 show system data, bidders' injections and withdrawals, base case values, and bid prices, respectively.

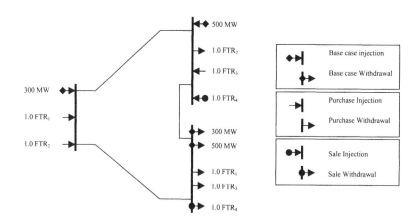

Figure 10.18 Base Case Purchase and Sale FTR Values

Table 10.19 System Data

Line	From	To	Reactance (Ohm)	Line Limit (MW)
1	1	2	0.02	400
2	1	3	0.04	325
3	2	3	0.01	500

Table 10.20 Bidders Injections and Withdrawals, and Base Case Values

Bus	Base Case Values (MW)	Bidder 1 FTR_1	Bidder 2 FTR_2	Bidder 3 FTR_3	Bidder 4 FTR_4
1	300.00	1.0	1.0	0.0	0.0
2	500.00	0.0	-1.0	+ 1.0	+ 1.0
3	-800.00	-1.0	0.0	- 1.0	- 1.0
Sum	0.00	0.0	0.0	0.0	0.0

Table 10.21 Bidders' Data

Bidder	FTR_{min}	FTR_{max}	Bid Price
1	0	200	12.0
2	0	300	10.0
3	0	100	15.0
4	0	175	11.0

The dc power flow equations of this system can be represented as:

$$\sum_{j=1}^{Npur} M_j \, FTR_j - \sum_{m=Npur+1}^{Npur+Nsale} M_m \, FTR_m - \sum_{n=1}^{Ninj} M_{Bn} \, P_n = B \, \delta \qquad (10.14)$$

where the first, second, and third terms in the left side of the equation refer to purchase, sale, and base case FTRs, respectively. These terms can be expressed as follows:

Base Case:

$$\sum_{n=1}^{Ninj} M_{Bn} P_n = \sum_{n=1}^{2} M_{Bn} P_n = \begin{bmatrix} 1 \\ 0 \\ -1 \end{bmatrix}(300) + \begin{bmatrix} 0 \\ 1 \\ -1 \end{bmatrix}(500) = \begin{bmatrix} 1 & 0 \\ 0 & 1 \\ -1 & -1 \end{bmatrix}\begin{bmatrix} 300 \\ 500 \end{bmatrix} = \begin{bmatrix} 300 \\ 500 \\ -800 \end{bmatrix}$$

Purchase of FTRs:

$$\sum_{j=1}^{Npur} M_j FTR_j = \sum_{j=1}^{3} M_j FTR_j = \begin{bmatrix} 1 \\ 0 \\ -1 \end{bmatrix} FTR_1 + \begin{bmatrix} 1 \\ -1 \\ 0 \end{bmatrix} FTR_2 + \begin{bmatrix} 0 \\ 1 \\ -1 \end{bmatrix} FTR_3$$

$$= \begin{bmatrix} 1 & 1 & 0 \\ 0 & -1 & 1 \\ -1 & 0 & -1 \end{bmatrix}\begin{bmatrix} FTR_1 \\ FTR_2 \\ FTR_3 \end{bmatrix} = \begin{bmatrix} FTR_1 + FTR_2 \\ -FTR_2 + FTR_3 \\ -FTR_1 - FTR_3 \end{bmatrix}$$

Sale of FTRs:

$$\sum_{m=Npur+1}^{Npur+Nsale} M_m \, FTR_m = \sum_{m=3+1}^{4} M_m \, FTR_m = \begin{bmatrix} 0 \\ 1 \\ -1 \end{bmatrix} FTR_4 = \begin{bmatrix} 0 \\ FTR_4 \\ -FTR_4 \end{bmatrix}$$

TRANSMISSION CONGESTION MANAGEMENT AND PRICING 419

The equations that represent power flow equations are

$$\begin{bmatrix} B_{11} & B_{12} & B_{13} \\ B_{21} & B_{22} & B_{23} \\ B_{31} & B_{32} & B_{33} \end{bmatrix} \begin{bmatrix} \delta_1 \\ \delta_2 \\ \delta_3 \end{bmatrix} = \begin{bmatrix} FTR_1 + FTR_2 \\ -FTR_2 + FTR_3 \\ -FTR_1 - FTR_3 \end{bmatrix} - \begin{bmatrix} 0 \\ FTR_4 \\ -FTR_4 \end{bmatrix} + \begin{bmatrix} 300 \\ 500 \\ -800 \end{bmatrix}$$

which can be written as

$$-75\delta_1 + 50\delta_2 + 25\delta_3 + FTR_1 + FTR_2 + 300 = 0.0$$
$$50\delta_1 - 150\delta_2 + 100\delta_3 - FTR_2 + FTR_3 - FTR_4 + 500 = 0.0$$
$$25\delta_1 + 100\delta_2 - 125\delta_3 - FTR_1 - FTR_3 + FTR_4 - 800 = 0.0$$

If we assume that bus 3 is the reference bus ($\delta_3 = 0$), the FTR auction can be written as

max $\quad 12FTR^1 + 10FTR^2 + 15FTR^3 + 11FTR^4$

s.t.

$$-75\delta_1 + 50\delta_2 + 25\delta_3 + FTR_1 + FTR_2 + 300 = 0.0$$
$$50\delta_1 - 150\delta_2 + 100\delta_3 - FTR_2 + FTR_3 - FTR_4 + 500 = 0.0$$
$$25\delta_1 + 100\delta_2 - 125\delta_3 - FTR_1 - FTR_3 + FTR_4 - 800 = 0.0$$
$$-400 \leq 50\delta_1 - 50\delta_2 \leq 400$$
$$-350 \leq 25\delta_1 - 25\delta_2 \leq 350$$
$$-500 \leq 100\delta_1 - 100\delta_2 \leq 500$$
$$\delta_3 = 0$$
$$0 \leq FTR_1 \leq 200$$
$$0 \leq FTR_2 \leq 300$$
$$0 \leq FTR_3 \leq 100$$
$$0 \leq FTR_4 \leq 175$$

We will study this example using two cases:

- Case 1: Line limits are ignored (or assume that line limits are very large).
- Case 2: Line limits are considered.

For both cases, FTR awards, line flows, and MCPs are shown in Tables 10.22 to 10.24. Table 10.24 shows MCPs of different FTRs where each price is given with respect to the reference bus (bus 3). For example, the MCP of a one-unit FTR from bus 1 to bus 3 is $12 (or 0.0 - (- 12.0)). Note that in the first case, all bidders are awarded up to their desired maximum values and this solution gives the maximum revenue of FTRs. Also in this case, there is no cost for any FTRs, since the system is unconstrained (MCPs are zero). The line flows in Table 10.22 show that line limits are violated for Case 1, and maintained for Case 2. For the constrained case, some bidders get less than the maximum they bid in the auction, and the constrained lines impose MCPs on different bidders. Each bidder's gain depends on the bid prices and line constraints.

Table 10.22 Line Flows

Line	From	To	Line Limit (MW)	Flow (MW) Case 1	Flow (MW) Case 2
1	1	2	400	439.29	400
2	1	3	325	360.71	325
3	2	3	500	564.29	500

Table 10.23 FTR Award

Bidder	FTR (MW) Case 1	FTR (MW) Case 2
1	200	125
2	300	300
3	100	75
4	175	175

Table 10.24 MCPs

Bus	Price Case 1	Price Case 2
1	0.0	-12.0
2	0.0	-15.0
3	0.0	0.0

10.4.5 Zonal Congestion Management

Here, we present a scheme for congestion management that incorporates locational marginal prices and fixed transmission rights. The scheme is summarized in the following steps:

Step 1: Define firm transmission reservations and the associated FTRs for each market participant.

The firm transmission service is reserved and scheduled between specified points of receipt and delivery where reservations are based on available transfer capability (ATC) calculations. In this reservation process, the generations are designated to serve designated loads. For each transmission service request, path-name, points of delivery and receipt, and capacity should be identified. Firm transmission reservations can be long term (for one year or more) or short term (less than one year). Network customers acquire transmission services on annual basis and are allowed to utilize their network resources to serve their local loads located in the control area. Network customers are allocated FTR up to their annual peak load. The path for each network service customer FTR is defined from designated network resource(s) to designated customer load. Firm point-to-point service customers obtain FTRs for the associated reservations where the path of the FTR is equivalent to the path of the firm reservation from source to sink. The MW value of each firm point-to-point is equal to the MW of firm transmission service being provided and the duration of each firm point-to-point FTR is the same as the associated firm transmission service.

Step 2: Perform SFT for FTRs.

Different FTRs are modeled as generations at points of injection and loads at points of extraction. When the results of simultaneous feasibility test (SFT) show that FTRs are not feasible, the FTRs will be reduced in proportion to their MW values until SFT gives a feasible solution. When a zonal scheme is used, all other injections and withdrawals associated with inter-zonal FTRs inside each zone will be modeled in the SFT but will not be taken into consideration during the congestion credit calculation. The FTRs will pass the feasibility test if inter-zonal line flows are less than the thermal limits of lines.

Step 3: Make inter-zonal contracts available for congestion management.

After applying SFT, loads and generators bids associated with those FTRs that passed SFT, in addition to loads and generators bids associated with intra-zonal contracts, would be available for inter-zonal congestion management. The inter-zonal congestion management output includes re-dispatched generations and loads as well as calculated energy locational marginal prices at all nodes in the system.

Step 4: Define zonal boundaries.

After determining the LMP at each node (from the last step), we determine the zonal boundaries by considering LMPs that are similar within a zone. We then calculate the ALMP in each zone

Step 5: Calculate congestion charges and congestion credits for market participants.

This step is based on zonal ALMPs which also includes the calculation of energy sales to and purchases from spot market. We use the ALMP in each zone to allocate congestion credits to users of inter-zonal lines. Also, use ALMP to find congestion charges based on actual deliveries to different loads according to the re-dispatch solution.

Step 6: Make intra-zonal congestion management.

In each congested zone, we apply a modified AC-OPF to adjust preferred schedules. The main goal is to minimize the absolute MW of re-dispatch, taking into account the cost of re-dispatch as determined by the submitted incremental and decremental price bids. For each zone in the network, the congestion management is performed separately. If any generator or load at any bus in any zone is not involved in congestion management and does not submit incremental/decremental bids, we set its minimum and maximum limits to the preferred scheduled values. Figure 10.19 and Tables 10.25 and 10.29 illustrate the proposed scheme.

Figure 10.19 shows an 8-bus system with its line parameters given in Table 10.25 and contracted generators shown in Table 10.26. The contracted generators identify the amount of generation contracted by each load based on the annual peak load. Ideally, the contracted generation should be in balance with the load, but sometimes, contracted generation units are not selected during re-dispatch process, or the load is less than the contracted generation, which means that the generation and load are not in balance.

TRANSMISSION CONGESTION MANAGEMENT AND PRICING

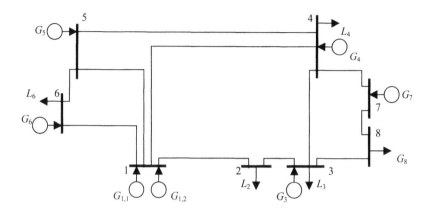

Figure 10.19 8-Bus System

Table 10.25 Line Data of the 8-bus System

Line	From	To	Reactance (Ohm)	Line limit (MW)
1	1	2	0.0300	280
2	1	4	0.0300	140
3	1	5	0.0065	380
4	2	3	0.0100	120
5	3	4	0.0300	230
6	4	5	0.0300	200
7	5	6	0.0200	300
8	6	1	0.0250	250
9	7	4	0.0150	250
10	7	8	0.0220	340
11	8	3	0.0180	240

Table 10.26 Designated Generation to Serve Loads

Load (MW)	Peak Load (MW)	Firm Transmission Reservation (MW)	Contracted Generation to Serve Load (MW)
L_2	320	320	G_5 (320)
L_3	350	350	G_3 (70), G_5 (280)
L_4	420	420	$G_{1,1}$ (90), G_3 (80), G_4 (250)
L_6	550	550	G_7 (550)
L_8	250	250	G_6 (250)
Total	1890	1890	1890

If there is a shortage or surplus in generation from the contracted units, the energy will be interchanged with the energy spot market. In other words, when a load MWh at a certain hour is less than the contracted energy, the extra energy is sold to the spot market at the generation LMP. When a generator is not dispatched or dispatched at less than the contracted energy, a load contracted for this generator will purchase energy from the spot market at the load's LMP.

Load or generation LMP is assumed to be the same as corresponding ALMP. We assumed in this example that the system is lossless and that the firm transmission reservation for each load is equal to the peak load. Table 10.27 shows the generation and load data at a certain hour. These data represent preferred schedules as well as generation and load bids at this hour. If congestion exists during real-time operation, schedules will be adjusted by the ISO. If any generator is generating more than the preferred value, the surplus is sold to the spot market, and if any generator is generating less than the preferred schedule, the shortage will be purchased from the spot market. Generators are either contracted or owned by network service customers who also may have loads.

To ensure that the FTRs associated with fixed transmission reservations are revenue adequate, we apply the SFT to Table 10.26 where the generation at each bus is equal to the sum of contacted generations at that bus to serve different loads. For example, G_5 is serving two loads; $L_2 = 320$ MW and $L_3 = 280$ MW for a total of 600 MW, so, in SFT, G_5 is represented by an injection of 600 MW at bus 5. Also, $G_{1,1}$ is only contracted for 90 MW to serve L_4 and $G_{1,2}$ is not contracted to serve any load, so 90 MW represents generation at bus 1 in the SFT.

Table 10.27 Generation and Load Data

Bus	Preferred Generation Schedule (MW)	Adjustment Range (MW) Min	Adjustment Range (MW) Max	Inc/Dec Price ($/MWh)	Preferred Load Schedule (MW)	Adjustment Range (MW) Min	Adjustment Range (MW) Max	Dec Price ($/MWh)
1	50	0.0	110	14	-	-	-	-
1	0	0.0	100	15	-	-	-	-
2	-	-	-	-	300	300	300	41
3	200	0.0	520	30	300	300	300	41
4	200	0.0	250	30	300	300	300	41
5	600	0.0	600	10	-	-	-	-
6	200	0.0	400	20	300	300	300	42
7	200	0.0	200	20	-	-	-	-
8	-	-	-	-	250	250	250	45
Total	1450				1450			

TRANSMISSION CONGESTION MANAGEMENT AND PRICING

Likewise, all FTRs are modeled as generations at the points of injection and loads at the points of extraction. After representing these FTRs, we run a dc load flow. The feasibility of these FTRs is guaranteed by ensuring that a flow in each inter-zonal line is below its thermal limit. The data used for SFT are shown in Table 10.28. Note that in SFT, total generation should be equal to total load which is the same as the total contracted generation.

Table 10.28 Data of Generation and Loads for SFT

Bus	Generation (MW)	Load (MW)
1	90	-
2	-	320
3	150	350
4	250	420
5	600	-
6	250	550
7	550	-
8	-	250
Total	1890	1890

The SFT results are given in Table 10.29. The results show that FTRs are feasible and no inter-zonal line is carrying power above its thermal limit. The inter-zonal lines are defined initially based on a similar situation and are confirmed or modified later based on calculated LMPs in the inter-zonal congestion management. Initially, we consider lines 1–2, 1–4 and 4–5 to be most frequently congested (inter-zonal lines). These lines are demanded highly to convey power from injection to extraction points.

The results of inter-zonal congestion management for the data in Table 10.27 are shown in Tables 10.30 and 10.31. Table 10.30 shows the line flows indicating that all inter-zonal line flows are within limits. This means that generation units with the adjustment ranges given in Table 10.27 are capable of alleviating congestion on intra-zonal lines. The results also show that no intra-zonal line is overloaded. So no further intra-zonal management will be performed. Table 10.31 shows that the preferred generations and loads are re-dispatched to remove inter-zonal line flow violations. The results obtained in Table 10.31 will be used for the calculation of congestion charges because the re-dispatched values represent actual generations and deliveries during this hour.

Table 10.29 Line Flows for SFT

Line	From	To	Line Limit (MW)	Flow (MW)
1	1	2	280	249.36
2	1	4	140	38.93
3	1	5	380	-289.77
4	2	3	120	-70.64
5	3	4	230	-186.88
6	4	5	200	-101.71
7	5	6	300	208.52
8	6	1	250	-91.48
9	7	4	250	216.24
10	7	8	340	333.76
11	8	3	240	83.76

Table 10.30 Line Flows after Inter-zonal Congestion Management

Line	From	To	Line Limit (MW)	Flow (MW)
1	1	2	280	280.00
2	1	4	140	140.00
3	1	5	380	-258.81
4	2	3	120	-20.00
5	3	4	230	-132.67
6	4	5	200	-196.07
7	5	6	300	145.12
8	6	1	250	-48.81
9	7	4	250	-8.73
10	7	8	340	208.73
11	8	3	240	-41.27

Table 10.31 Re-dispatched Generation and Loads for Inter-zonal Congestion Management

Bus	Adjusted Generation Schedules (MW)	Adjustment Load Schedule (MW)
1	110	-
1	100	-
2	-	300
3	228.604	300
4	105.320	300
5	600	-
6	106.076	300
7	200	-
8	-	250
Total	1450	1450

TRANSMISSION CONGESTION MANAGEMENT AND PRICING

The LMPs of the system are calculated from the inter-zonal congestion management optimization problem and shown in Table 10.32. In this table, LMP values are either close to 20 or close to 30, which indicates that all buses are categorized into two zones with the same LMP for buses in each zone. This observation confirms the assumption that the most frequently congested lines divide the system into two zones. Figure 10.20 shows the equivalent zonal system based on the LMPs. Each zone contains buses with nearly the same LMPs, and the lines connected to these zones are inter-zonal lines. After this step, a decision can be made to confirm the initial assumption of zonal boundaries to combine or split the zones to form new zones.

Table 10.32 Specifying Zones Based on LMPs

Bus	LMP ($/MWh)	Zone
1	19.03	1
2	30.00	2
3	30.00	2
4	30.00	2
5	20.78	1
6	20.00	1
7	30.00	2
8	30.00	2

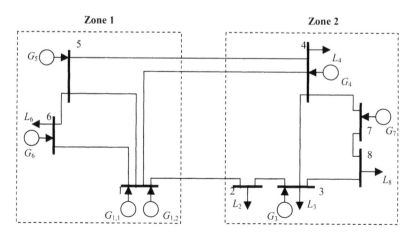

Figure 10.20 The Equivalent 2-zone System Based on LMPs

The FTRs associated with firm transmission reservations are shown in Table 10.33. Figure 10.21 shows the designated paths for inter-zonal FTRs. The FTRs that will be used for congestion credit allocations in the zonal scheme are inter-zonal FTRs, where $ALMP_1 = 19.94$ \$/MWh and $ALMP_2 = 30.00$ \$/MWh.

The transmission congestion credit allocations will be based on inter-zonal FTRs. The credits are shown in Table 10.34. For example, we see in the first case that an FTR of 90 MW from zone 1 to zone 2 that represents the contracted generation from $G_{1,1}$ to serve load L_4 has a congestion credit of 90 MWh x (30.00 − 19.94) \$/MWh = \$905.40. The 30.00 \$/MWh and 19.94 \$/MWh represent the ALMPs of delivery and receipt zones, respectively. We notice that for the last FTR in Table 10.34, the congestion credit is negative, which means that it is a liability (charge) because it has a direction opposite to the direction of flows in the inter-zonal lines. Later, we will see that the holder of this FTR will be credited partially for this liability from the negative congestion charge (credit) for this holder.

Table 10.33 FTR Data

FTR (MW)	From Bus	To Bus	From Zone	To Zone
90	1	4	1	2
70	3	3	2	2
80	3	4	2	2
250	4	4	2	2
320	5	2	1	2
280	5	3	1	2
250	6	8	1	2
550	7	6	2	1
Total 890				

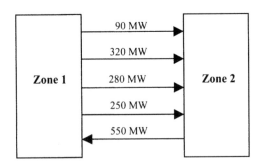

Figure 10.21 Inter-zonal FTRs

TRANSMISSION CONGESTION MANAGEMENT AND PRICING

Table 10.34 Transmission Congestion Credits Based on Inter-zonal FTRs

FTR Value (MWh)	From Zone	To Zone	ALMP$_i$ ($/MWh)	ALMP$_j$ ($/MWh)	Congestion Credit ($)
90	1	2	19.94	30.00	905.40
320	1	2	19.94	30.00	3219.20
280	1	2	19.94	30.00	2816.80
250	1	2	19.94	30.00	2515.00
550	2	1	30.00	19.94	-5533.00
				Total	3923.40

The transmission congestion charges will be calculated based on actual deliveries from designated generators to the associated loads. We use inter-zonal lines to calculate these charges as shown in Table 10.35. Each congestion charge in the last column of Table 10.35 is calculated based on the actual generation supplied from the contracted generator. For example, L_2 contracted for 320 MW from G_5 but L_2 at this hour is 300 MW which means that 300 MW is the actual delivery from G_5 to L_2 with the extra 20 MW sold to the spot market at the generator ALMP, which is 19.94 $/MWh. The congestion charge is calculated as 300 multiplied by the difference in ALMPs of the generation and the load.

Table 10.35 Transmission Congestion Charges

Market Participant	Generation Supplied*	ALMP$_i$ ($/MWh)	ALMP$_j$ ($/MWh)	Congestion Charge ($)
L_2	G_5 (300.00)	30.00	19.94	3018.00
L_3	G_3 (20.00)	30.00	30.00	0.00
	G_5 (280.00)	30.00	19.94	2816.80
L_4	$G_{1,1}$ (90.00)	30.00	19.94	905.40
	G_3 (80.00)	30.00	19.94	804.80
	G_4 (105.32)	30.00	30.00	0.00
L_6	G_7 (200.00)	19.94	30.00	-2012.00
L_8	G_6 (106.08)	30.00	19.94	1067.12
			Total	6600.12

* Actual deliveries from designated generators to associated loads.

Another example is the case of L_3 where it contracted for 280 MW from G_5 and 70 MW from G_3, but L_3 at this hour is 300 MW which means that L_3 will take 280 MW from G_5 and 20 MW from G_3 and the extra 50 MW will be sold to the spot market. The congestion charges for this load are separated in the last column for this load. We notice that one of the congestion charges is negative, which means that it is a credit (this is a credit for the FTR holder who has negative credit in Table 10.34). Congestion charge for any entity can be a credit if the flow in the inter-zonal lines caused by the contract held by this entity opposes the direction of flows in the congested inter-zonal lines.

Energy interchanges with the energy spot market represents sales and purchases for each load and each generator. For example, L_2 contracted with G_5 for 320 MW and L_3 contracted for 280 MW from G_5 and 70 MW from G_3. After re-dispatched values are obtained from the inter-zonal management, G_5 is generating 600 MW that covers both contracted generations (320 for L_2 and 280 for L_3). But L_2 in this hour is 300 MW which means that 20 MW will be available for sale to the spot market at the LMP of G_5. The net energy purchases and sales for loads should be equal to the net sales and purchases for generators. In Tables 10.36 and 10.37 we separated sales from purchases.

We notice that the net sale is equal to the net purchase. The sale revenue and purchase payments of all parties are calculated in these tables. Any load will buy energy from the spot market at the load ALMP, and any generator will sell energy to the spot market at the generator ALMP. These purchases and sales from and to the spot market are based on the ALMPs, which means that congestion charges are included in these sales and purchases. In other words, these revenues and payments should enter the calculations of the total congestion charges of the control area.

Table 10.36 Energy Sales of Loads and Generators

Market Participant	Energy Sale (MWh)	ALMP ($/MWh)	Sale Revenue ($)
L_2	20.00	30.00	600.00
L_3	50.00	30.00	1500.00
$G_{1,1}$	20.00	19.94	398.80
$G_{2,1}$	100.00	19.94	1994.00
G_3	78.60	30.00	2358.00
Total	268.60		6850.80

Table 10.37 Energy Purchases of Loads and Generators

Market Participant	Energy Purchase (MWh)	ALMP ($/MWh)	Purchase Payment ($)
L_4	24.68	30.00	740.40
L_6	100.00	19.94	1994.00
L_8	143.92	30.00	4317.60
Total	268.60		7052.00

All energy sales to the energy spot market are sold at the ALMP of the generator zone, and all energy purchased from the energy spot market at the load ALMP of the load zone, that is, energy purchase from and sale to the spot market imply congestion charges. The total of congestion charges implicitly paid in the energy spot market equals the total purchase payments minus total sales revenues. The total congestion charges of the system are ($6,600.12 + $7,052.00 - $6,850.80) = $6,801.32. Since $6,801.32 is greater than $3,923.40 (total congestion credits to FTRs holders), all FTR holders are credited the monetarily values of the associated FTRs, and the extra money obtained from congestion charges is used later for any deficiency in congestion charges to cover congestion credits for another hour. The results of the inter-zonal congestion management (see Table 10.30) show that no intra-zonal line limit is violated, and this is expected because the intra-zonal congestion is of low possibility. For our example, there is no need to apply intra-zonal congestion management to any zone.

10.5 A COMPREHENSIVE TRANSMISSION PRICING SCHEME

In this section, we propose a comprehensive transmission pricing scheme that incorporates curtailment prioritization, contributions of generators and loads to line flows, and FTRs and LMPs to manage congestion and settle transmission charges and credits.

In the proposed scheme, transmission cost is composed of two parts: transmission service cost and transmission congestion cost. Transmission service cost is to cover the transmission revenue requirement of transmission owners (TOs), and transmission congestion cost is to cover the cost of congestion (when the transmission network is constrained) and marginal losses. A detailed account of transmission revenue requirements is beyond the scope of this book. What will be described is how to allocate the transmission service cost to meet the revenue requirement. In the proposed scheme, the allocation of the transmission cost is based on Kirschen's tracing method. The cost of congestion and marginal losses is calculated based on LMP differences. In this scheme, we also propose allocating the transmission congestion cost based on Kirschen's tracing method. Corresponding to the two parts of transmission cost, for a market participant, the transmission charge includes the transmission service (usage) charge and the transmission congestion charge.

In the proposed scheme, as congestion occurs, we use prioritization and then curtailment processes to eliminate the congestion. To lock in the advance transmission prices, market participants may hold FTRs as financial hedges against the congestion costs. In the settlement, FTR holders can get transmission credits based on the FTR shares and LMP differences of the FTR path.

In the proposed scheme, market participants have three options to use in the transmission network: buyers and sellers can buy and sell energy through the pool (bidding transactions), they can schedule use of the transmission system for bilateral transactions, or they can schedule the use of transmission system for multilateral (as group) transactions. A bilateral contract is a two-party (GENCO-DISCO or pairwise-contract) transaction, and a multilateral contract is purchase and sale agreement among several GENCOs and DISCOs.

In the following, we first outline the proposed transmission pricing scheme. Then, we discuss congestion management with prioritization, followed by a discussion on how to calculate transmission usage and congestion charge and FTR credits. Numerical examples will be presented to illustrate the proposed scheme.

10.5.1 Outline of the Proposed Transmission Pricing Scheme

To incorporate the curtailment prioritization, contributions of generators, and loads to line flows, FTRs, and LMPs, we propose the following algorithm. This algorithm is designated to solve the congestion and allocate transmission usage charges and credits (see Figure 10.22):

Step 1: Transaction data (initial schedules, generation limits of pool, bilateral and multilateral transactions, and bid prices for pool participants) are passed to the ISO along with curtailment data (willingness to pay and multilateral curtailment weights).

Step 2: The ISO determines whether or not the desired schedules would result in transmission violations:

> *If yes*, the ISO uses the curtailment data to determine the final schedules that result in no transmission violations for scheduled generators and loads. This step also determines LMPs, which reflect constrains and transmission losses.

TRANSMISSION CONGESTION MANAGEMENT AND PRICING

If no, the desired schedules are accepted without any changes. In this case all LMPs will be the same. If transmission losses are considered, LMPs will differ slightly to reflect losses (see the dotted line in Figure 10.22).

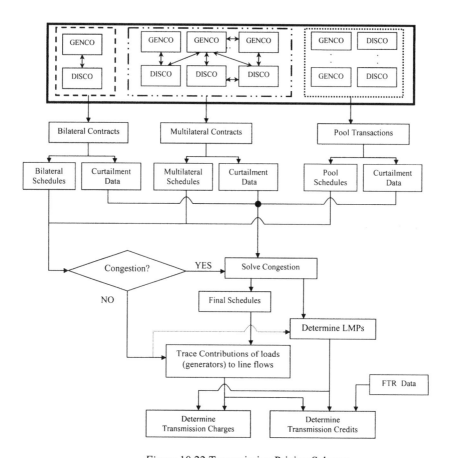

Figure 10.22 Transmission Pricing Scheme

Step 3: The ISO traces contributions of generators (loads) to line flows (using Kirschen's tracing algorithm): Given the final schedules for generators and loads in step 2, the ISO determines line flows, net generation and net load at each bus, domains of net generations (loads), commons of net generations (loads), and links (with their flows). Using the state graph concept, the ISO then apportions each branch usage among net generations (loads), and the result is used to find contributions of each generator (load). The ISO will then determine users that contribute to line flows.

Step 4: The ISO uses contributions to line flows and LMPs to allocate transmission usage charges, congestion charges, and loss charges. If generators are to pay for system usage, the ISO should calculate contributions of generators to branch flows (how much each generator contributes to line flows). On the other hand, if loads are responsible for system usage charges, the ISO should calculate contributions of loads to branch flows (how much of a line flow is consumed by each load).

Step 5: The ISO uses FTR data[5], LMPs, and line flow contributions to allocate transmission usage credits. When an entity holds an FTR along a path, this entity should be paid by other entities that use this path. The payment should be based on the MW value of the FTR, difference in LMPs between the two ends of the FTR path, and the contribution of each entity to the flow on that path.

10.5.2 Prioritization of Transmission Dispatch

In previous work [Fan99a, Fan99b] prioritization has been considered in restructured power systems where pool, bilateral and multilateral transactions exist simultaneously. This method ensures a non-discriminatory curtailment priority of transactions based on the participants' willingness to pay. Here, we extend the prioritization and curtailment processes [Fan99a, Fan99b] to include active and reactive power flows.

Prioritization indicates how transaction curtailment strategies ought to be used by the ISO. In other words, to solve congestion, the ISO should make curtailment decisions based on competition among market participants. The idea is that consumers may be willing to make an extra payment to avoid curtailment, the more the payment is, the smaller the curtailment will be. Hence, the ISO would add terms to the pool's objective function to represent extra payments to avoid curtailment.

The method so far proposed in [Fan99a, Fan99b] consists of three priority procedures for curtailments: free mode, pool protection mode, and contract protection mode. In the free mode, all participants compete for the

---

[5] MW values, start and end points (FTR paths, from m to n), and holder of each FTR (a GENCO, a DISCO or a transmission provider).

TRANSMISSION CONGESTION MANAGEMENT AND PRICING   435

usage of congested lines by paying extra to avoid curtailment. In the pool protection mode, curtailment will be on bilateral/multilateral contracts first, and then on pool demands, as necessary. In the contract protection mode, curtailment will be on the pool demand first, and then on the bilateral/multilateral contracts. In this discussion, we use the free mode because we believe it is the mode that can guarantee non-discriminatory and competitive access to the transmission system. The free mode is used for calculating the final schedule of each transaction after resolving the congestion.

There are two proposals for loss compensation [Fan99a, Fan99b]: (1) the ISO provides this service from the pool, and (2) based on transmission loss factors (published by the ISO), each transaction provides additional power to cover losses. Here, we opt to use the first alternative.

In the following, we present the mathematical formulation for a transmission dispatch with prioritization. In the formulation, a superscript refers to the type of transaction (*PL* refers to pool and *T* refers to bilateral/multilateral transaction), variables G, D, and L refer to active power generation, demand, and losses, respectively, and variables Q and R refer to reactive power generation and demand, respectively. Equality constraints include the total injected power at each node, bilateral/multilateral power balance equations, and bilateral/multilateral curtailment equations.

The total injected active (reactive) power at each bus i is equal to the power injected at bus i due to pool transactions, power injected at bus i due to bilateral/multilateral transactions to cover load, plus power injected at bus i due to bilateral/multilateral transactions to compensate transmission loss. Likewise, the total extracted active (reactive) power at each bus j is equal to the power extracted at bus j due to pool transactions, plus power extracted at bus i due to bilateral/multilateral transactions. In general, these relations are represented as

$$P_i = G_i^{PL} + \sum_{k \in K} G_{k,i}^T + \sum_{k \in K} L_{k,i}^T \tag{10.15a}$$

$$Q_i = Q_i^{PL} + \sum_{k \in K} Q_{k,i}^T + \sum_{k \in K} Q_{k,i}^L \tag{10.15b}$$

$$D_j = D_j^{PL} + \sum_{k \in K} D_{k,j}^T \tag{10.15c}$$

$$R_j = R_j^{PL} + \sum_{k \in K} R_{k,j}^T \tag{10.15d}$$

where $i \in I_G$ (set of generator buses), and $j \in J_D$ (set of load buses), k is bilateral/multilateral index, and K is the set of bilateral and multilateral transactions. The last term in (10.15a) and (10.15b) is skipped when a pool provides the loss compensation.

The bilateral/multilateral power balance and curtailment are represented as:

$$\sum_{i \in I_G} G_{k,i}^T = \sum_{j \in J_D} D_{k,j}^T \tag{10.16a}$$

$$D_{k,j}^T = \text{function of } G_{k,1}^T, G_{k,2}^T, \ldots, G_{k,i}^T, \ldots \tag{10.16b}$$

Equation (10.16a) indicates that each transaction must balance out its demand (considering that the pool provides the loss compensation) and that the sum of generator outputs in a multilateral contract balances out the sum of loads in that transaction. For each multilateral transaction, (10.16b) represents each generator's increment when this transaction is curtailed. We assume a linear relation between a transaction's demand and the sum of transaction's generator output. Equations (10.16) will be included as equality constraints in the problem formulation. Note that for a bilateral transaction, (10.16a) will be the same as (10.16b) when there are one generator and one load in the system.

As in the conventional OPF formulation, inequality constraints include limits on pool power, voltage levels, and line overloads. When all transactions provide additional power for loss compensation, the objective function should include an additional term to cover the cost of transmission losses compensation for bilateral/multilateral contracts. Equality constraints will include equations for additional power to be provided by each bilateral/multilateral participant and the pool to compensate for loss. In this formulation, we assume that the ISO will compensate for transmission losses from the pool's generation.

The problem is formulated as follows. In this formulation, equality constraints (10.18–10.20) represent: active power at each generator bus i (10.18a), reactive power at each generator bus i (10.18b), active power at each load bus j (Equation 10.18c), reactive power at each load bus j (10.18d), group power balance (10.19a), group curtailment (Equation 10.19b), net injection of real power at each bus (10.20a), and net injection of reactive power at each bus (10.20b). The inequality constraints

TRANSMISSION CONGESTION MANAGEMENT AND PRICING

(10.21–10.22) represent the following: real power flows between buses (10.21), limits on active pool power generation (10.22a), limits on reactive pool power generation (10.22b), limits on active bilateral/multilateral power generation (10.22c), limits on reactive bilateral/multilateral power generation (10.22d), and voltage limits (Equation 10.22e).

$$\max \sum_{i \in S_D^{PL}} D_i^{PL} C_{Di}^{PL} - \sum_{i \in S_G^{PL}} G_i^{PL} C_{Gi}^{PL} - \sum_{i \in S_D^{PL}} \gamma_i^{PL}(D_i^{PL}) - \sum_{k \in K} \sum_{i \in S_G^T} \gamma_{k,i}^T(G_{k,i}^T) \quad (10.17)$$

s.t.
$$P_i = G_i^{PL} + \sum_{k \in K} G_{k,i}^T + \sum_{k \in K} L_{k,i}^T \quad (10.18a)$$

$$Q_i = Q_i^{PL} + \sum_{k \in K} Q_{k,i}^T + \sum_{k \in K} Q_{k,i}^L \quad (10.18b)$$

$$D_j = D_j^{PL} + \sum_{k \in K} D_{k,j}^T \quad (10.18c)$$

$$R_j = R_j^{PL} + \sum_{k \in K} R_{k,j}^T \quad (10.18d)$$

$$\sum_{i \in I_G} P_{k,i}^T = \sum_{j \in J_D} D_{k,j}^T \quad (10.19a)$$

$$D_{k,j}^T = \omega_{k,j}^M \sum_{i \in S_{k,G}^M} G_{k,i}^T \quad (10.19b)$$

$$v_i \sum_m [v_m[g_{im} \cos(\delta_i - \delta_m) + b_{im} \sin(\delta_i - \delta_m)]] = P_i - D_i \quad (10.20a)$$

$$v_i \sum_m [v_m[g_{im} \sin(\delta_i - \delta_m) - b_{im} \cos(\delta_i - \delta_m)]] = Q_i - R_i \quad (10.20b)$$

$$\underline{f_l} \le f_l \le \overline{f_l} \quad (10.21a)$$

$$f_l = v_j^2 g_{jj} + v_j v_m [g_{jm} \cos(\delta_j - \delta_m) + b_{jm} \sin(\delta_j - \delta_m)] \quad (10.21b)$$

$$\underline{G}_i^{PL} \le G_i^{PL} \le \overline{G}_i^{PL} \quad (10.22a)$$

$$\underline{Q}_i^{PL} \le Q_i^{PL} \le \overline{Q}_i^{PL} \quad (10.22b)$$

$$\underline{G}_{k,i}^T \le G_{k,i}^T \le \overline{G}_{k,i}^T \quad (10.22c)$$

$$\underline{Q}_{k,i}^T \le Q_{k,i}^T \le \overline{Q}_{k,i}^T \quad (10.22d)$$

$$\underline{v}_i \le v_i \le \overline{v}_i \quad (10.22e)$$

where

$$C_{Gi}^{PL} = a_{Gi} G_i^{PL} + b_{Gi} (G_i^{PL})^2$$

$$C_{Di}^{PL} = a_{Di} D_i^{PL} + b_{Di} (D_i^{PL})^2$$

$$\gamma_i^{PL}(D_i^{PL}) = W_{k,i}^{PL} (D_i^{PL,0} - D_i^{PL})^2$$

$$\gamma_{k,i}^{T}(G_{k,i}^{T}) = W_{k,i}^{T} (G_{k,i}^{T,0} - G_{k,i}^{T})^2$$

$(D_i^{PL,0} - D_i^{PL})$ = pool demand shortfall

$(G_{k,i}^{T,0} - G_{k,i}^{T})$ = bilateral/multilateral supply shortfall

k = bilateral/multilateral transaction index

K = set of bilateral/multilateral transactions

i, j, m = bus index

l = line index for the line connecting buses j–m

$D_i^{PL,0}$ = initial (preferred) schedule of pool demand at bus i

$G_{k,i}^{T,0}$ = initial (preferred) schedule of bilateral/multilateral transaction k at bus i

γ = curtailment function reflecting willingness to pay for avoiding curtailment

W_i^{PL} = willingness to pay by pool transaction to avoid curtailment

$W_{k,i}^{T}$ = willingness to pay by bilateral/multilateral transaction k to avoid curtailment

C_{Gi}^{PL} = bid price of a pool generator at bus i

C_{Di}^{PL} = bid price of a pool load at bus i

a_{Gi}, b_{Gi} = linear and nonlinear coefficients of a pool generator bid price at bus i

a_{Di}, b_{Di} = linear and nonlinear coefficients of a pool load bid price at bus i

$\omega_{k,j}^M$	=	curtailment weight of the kth multilateral contract's load at bus j
$S_{k,G}^M$	=	set of buses where the kth multilateral contract's generators exist
S_D^{PL}	=	set of load buses of the pool transactions
S_G^{PL}	=	set of generation buses of the pool transactions
S_D^T	=	set of load buses of the bilateral/multilateral transactions
S_G^T	=	set of generation buses of the bilateral/multilateral transactions

10.5.3 Calculation of Transmission Usage and Congestion Charges and FTR Credits

We use Kirschen's tracing method to calculate the contribution of each generation and load on line flow, based on which MW-mile method is used to calculate transmission charge.

10.5.3.1 Calculation of Transmission Usage Charge

Let f_{m-n,g_i} (f_{m-n,d_i}) refer to the contribution of each net generation (load) of common i to each line flow f_{m-n}, then the contribution of each single generator (load) that exists in common i at bus j (G_j or D_j) is given by:

$$f_{m-n,G_j} = \frac{G_j}{g_i} \times f_{m-n,g_i} \tag{10.23a}$$

$$f_{m-n,D_j} = \frac{D_j}{d_i} \times f_{m-n,d_i} \tag{10.23b}$$

where, g_i and d_i are the net generation and net load in common i, respectively, and G_j, D_j refer to any single generation or load at any bus j in common i. See Figure 10.23 for an illustration.

If D_{m-n} is the length of line m–n in miles, and R_{m-n} is the required revenue per unit length of line m–n ($/mile), the MW-mile method uses the following equation to find the MW-mile charge for line m–n (C_{m-n,G_j}) corresponding to G_j:

$$C_{m-n,G_j} = \frac{f_{m-n,G_j} D_{m-n} R_{m-n}}{f_{m-n}} \quad (10.24)$$

Let z_{m-n} be the required revenue of line m–n in $, such that $z_{m-n} = D_{m-n} R_{m-n}$. For usage of all lines, G_j would pay

$$\begin{aligned} C_{G_j} &= \sum_{\text{all lines}} C_{m-n,G_j} \\ &= \sum_{\text{all lines}} \frac{f_{m-n,G_j} D_{m-n} R_{m-n}}{f_{m-n}} \\ &= \sum_{\text{all lines}} \frac{f_{m-n,G_j} z_{m-n}}{f_{m-n}} \end{aligned} \quad (10.25)$$

Total payments for transmission usage by all participating generators is given by

$$C_{Gt} = \sum_{j \in S_G} \sum_{\text{all lines}} \frac{f_{m-n,G_j} z_{m-n}}{f_{m-n}} \quad (10.26)$$

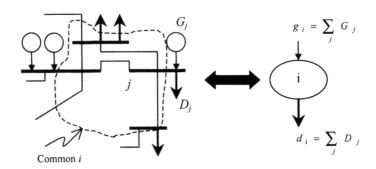

Figure 10.23 Splitting the Contribution of Net Generation (load) in a Common

10.5.3.2 Calculation of Transmission Congestion Charge

Let D_{D_j,G_i} be the contribution of a generator at bus i (G_i) to a load at bus j (D_j). Using LMPs, we find the charges that each load D_j is paying for congestion and transmission losses using the following relation:

$$Ch_{D_j} = \sum_{j \in S_G} D_{D_j,G_i} \times (\text{LMP}_j - \text{LMP}_i) \qquad (10.27)$$

We find the congestion charges of D_j by knowing the contribution of each generator to this load, and then adding the product of these contributions and the difference in LMPs between load and generator buses. D_{D_j,G_i} is found either by the contribution of each generator to each load, or by the contribution of each load to each generator. Both approaches give the same results as shown in Appendix F.

Total charges collected by the ISO for a line connecting buses m and n is calculated using any of the following relations (see the example in Appendix F):

$$\begin{aligned}
Ch_{m-n} &= \sum_{j \in S_G} Ch_{m-n,Gj} \\
&= \sum_{j \in S_G} f_{m-n,G_j} (\text{LMP}_n - \text{LMP}_m) \qquad (10.28a) \\
&= f_{m-n} (\text{LMP}_n - \text{LMP}_m)
\end{aligned}$$

$$\begin{aligned}
Ch_{m-n} &= \sum_{j \in S_D} Ch_{m-n,Dj} \\
&= \sum_{j \in S_D} f_{m-n,D_j} (\text{LMP}_n - \text{LMP}_m) \qquad (10.28b) \\
&= f_{m-n} (\text{LMP}_n - \text{LMP}_m)
\end{aligned}$$

where S_G and S_D are sets of generators and sets of loads in the system under study, and G_j and D_j are the generation and load of a certain transaction at any bus j. We use either (10.28a) or (10.28b) to calculate charges for congestion and losses for a certain line m–n. The first equation tells us that total congestion and loss charges of a certain line can be calculated by adding the charges for individual contributions of generators to that line. The charge for a contribution to line m–n equals the product of

f_{m-n,G_j} and the difference in LMPs at the two ends of that line. This is because a contribution to a line flow can be seen as injecting the contribution at one end of the line (generation) and extracting it at the other end (load). If we opt to use 10.28a, we would first find the contributions of each load to line flows using a procedure similar to that of Section 10.2.9 (Kirschen's tracing algorithm). Appendix F gives a comprehensive example for different alternatives.

10.5.3.3 Calculation of FTR Credit

As shown in Figure 10.24, when a participant X holds an FTR between points $m-n$ ($FTR_{m-n,x}$), the participant is entitled to a credit ($Cr_{m-n,x}$) as

$$Cr_{m-n,x} = FTR_{m-n,x} \times (LMP_n - LMP_m) \tag{10.29}$$

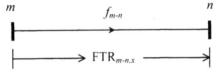

Figure 10.24 Total Flow and FTR between m and n

If there are more than one FTR holder on line $m-n$, then the total FTR credits for line $m-n$ is

$$Cr_{m-n} = \sum_x FTR_{m-n,x} (LMP_n - LMP_m) \tag{10.30}$$

The holder X's credit comes from Ch_{m-n} charges (in Step 4) for this line. If FTR_{m-n} is less than f_{m-n} (or $Cr_{m-n} < Ch_{m-n}$), that is, the FTR on $m-n$ is less than the total flow on line $m-n$, then collected charges for this line are adequate to cover FTR credits on this line, and each holder gets $Cr_{m-n,x}$ calculated from (10.29). In this case, the extra credit will be paid to the transmission line owner. Otherwise, FTR_{m-n} will be larger than f_{m-n} and the owner of line $m-n$ should pay the difference. The ISO will manage these transactions in either case. For instance, in Figure 10.25, $Ch_{m-n} = \$1000$. If $FTR_{m-n,1} = 40MW$ and $FTR_{m-n,2} = 50MW$, then $Cr_{m-n} = \$900$. In this case, the ISO will get \$1000 from charges, pay \$900 to FTR holders (\$400 to holder 1, \$500 to holder 2), and the extra \$100 (\$1000 − \$900) will be paid to the owner of line $m-n$. Now let's consider another situation where $FTR_{m-n,1} = 60MW$, $FTR_{m-n,2} = 50MW$, and $FTR_{m-n} = 110MW$ which is larger than f_{m-n}. In this case, the ISO will get $1,000 from charges, pay $1100 to FTR holders ($600 to holder 1, $500 to holder 2), and the $100 deficit ($1000 − $1100) will be paid by the owner of line $m–n$.

10.5.4 Numerical Example

The proposed framework is applied to a 6-bus test system shown in Figure 10.26. The system shows three types of transactions: pool, bilateral, and multilateral (group).

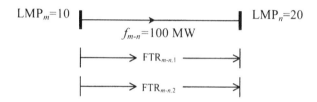

Figure 10.25 Example of Calculating Charges and Credits

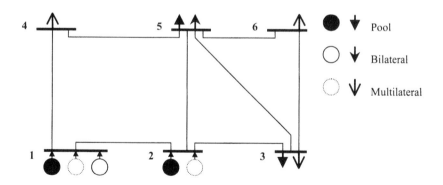

Figure 10.26 6-Bus Test System with Three Types of Transactions

Tables 10.38 to 10.41 provide system data, pool transactions, bilateral transactions, and multilateral transactions, respectively. In these tables, G, D, Type, Min, Max, a, b, and Pref refer to generation, load, type of transaction, minimum value, maximum value, linear and nonlinear coefficients of the bid price, and preferred values, respectively. Table 10.42 shows the curtailment weights of multilateral contract loads at each bus. Table 10.43 shows voltage limits. All values are given in per unit and bus number 1 is the reference bus.

Table 10.38 System Data

Line	m	n	Resistance	Reactance	Limit	$z_{m\text{-}n}$
1	1	2	0.03	0.10	1.0	$100.0
2	1	4	0.025	0.06	1.2	$ 60.0
3	2	3	0.025	0.08	1.4	$ 80.0
4	2	5	0.02	0.05	1.3	$ 50.0
5	3	5	0.02	0.10	1.0	$100.0
6	3	6	0.02	0.10	1.0	$100.0
7	4	5	0.02	0.08	1.0	$ 80.0
8	5	6	0.01	0.05	1.0	$ 50.0

Table 10.39 Pool Data

Bus	Type	Min	Max	a	b	Pref	W_i^{PL}
1	G	0.0	2.0	6.00	0.06	-	-
2	G	0.0	2.0	3.00	0.03	-	-
3	D	0.0	1.0	9.00	0.00	1.0	20.0
5	D	0.0	0.8	10.00	0.00	0.8	20.0

Table 10.40 Bilateral Contract Data

Bus	Type	Min	Max	Pref	$W_{1,i}^T$
1	G	0.0	1.0	1.0	-
3	D	0.0	1.0	1.0	15.0

Table 10.41 Multilateral Contract Data

Bus	Type	Min	Max	Pref	$W_{2,i}^T$
1	G	0.0	1.0	1.0	-
2	G	0.0	1.0	1.0	-
3	D	0.0	0.5	0.5	15.0
4	D	0.0	1.0	1.0	15.0
6	D	0.0	0.5	0.5	20.0

TRANSMISSION CONGESTION MANAGEMENT AND PRICING 445

Table 10.42 Multilateral Curtailment Weights

Bus	Type	Transaction	Weight
3	D	Multilateral	0.25
4	D	Multilateral	0.50
6	D	Multilateral	0.25
		Total	1.00

Table 10.43 Bus Voltages

Bus i	$\underline{v}_i$	$\overline{v}_i$
1	1.02	1.02
2	1.04	1.04
3	0.95	1.05
4	1.05	1.05
5	0.95	1.05
6	0.95	1.05

If initial schedules submitted by the three types of transactions are honored by the ISO, they would cause congestion (see the bold letters in Table 10.44). Based on the optimization problem 10.17–10.22, the ISO solves the congestion problem. The solutions are shown in Tables 10.45–10.47. Table 10.45 gives generator outputs and loads based on prioritization of transactions, which reflect the willingness to pay to avoid curtailments. Line flows are given in Table 10.46. LMPs are given in Tables 10.47. Some notes on the results follow.

Table 10.44 Line Flows of Initial Schedules

Line	i	j	Limit	P_{ij}	P_{ji}
1	1	2	1.0	0.648	0.636
2	1	4	1.2	1.569	1.540
3	2	3	1.4	1.773	1.740
4	2	5	1.3	1.650	1.626
5	3	5	1.0	-0.576	-0.579
6	3	6	1.0	-0.218	-0.218
7	4	5	1.0	0.512	0.508
8	5	6	1.0	0.723	0.721

Table 10.45a Optimization Results (Generation Values)

Bus	Type	Transaction	Min	Max	Pref	Final Value
1	G	Pool	0.0	2.0	1.0	0.3359
2	G	Pool	0.0	2.0	0.8	1.2958
1	G	Bilateral	0.0	1.0	1.0	0.7755
1	G	Multilateral	0.0	1.0	1.0	0.5006
2	G	Multilateral	0.0	1.0	1.0	1.0000
		Total Generation			4.8	3.9078

Table 10.45b Optimization Results (Load Values)

Bus	Type	Transaction	Min	Max	Pref	Final Value
3	D	Pool	0.0	1.0	1.0	0.8058
5	D	Pool	0.0	0.8	0.8	0.6670
3	D	Bilateral	0.0	1.0	1.0	0.7755
3	D	Multilateral	0.0	0.5	0.5	0.3751
4	D	Multilateral	0.0	1.0	1.0	0.7503
6	D	Multilateral	0.0	0.5	0.5	0.3751
		Total Load			4.8	3.7488

Table 10.46 Line Flows

Line	i	j	Limit	P_{ij}	P_{ji}	Loss
1	1	2	1.0	0.4120	-0.4042	0.0078
2	1	4	1.2	1.2000	-1.1442	0.0558
3	2	3	1.4	1.4000	-1.3537	0.0463
4	2	5	1.3	1.3000	-1.2611	0.0389
5	3	5	1.0	-0.4355	0.4393	0.0038
6	3	6	1.0	-0.1672	0.1679	0.0007
7	4	5	1.0	0.3939	-0.3910	0.0029
8	5	6	1.0	0.5458	-0.5431	0.0027
					Total Loss	0.1590

Table 10.47 LMPs

Bus i	LMP_i
1	10.03143
2	10.77453
3	16.76740
4	15.60574
5	15.31915
6	15.86894

TRANSMISSION CONGESTION MANAGEMENT AND PRICING

In Table 10.45, the total generation (3.9078) is larger than the total load (3.7488) by 0.159 p.u. which is the transmission loss compensation generated by the reference bus (pool generation) as we assume that the ISO compensates for transmission losses.

In each of the bilateral and multilateral transactions, the generation balances the load. In other words, if we let $k = 1$ refer to the bilateral contracts, and $k = 2$ refer to multilateral contracts, then, from Table 10.45, we have $G_{1,1}^T = D_{1,3}^T = 0.7755$ for bilateral contracts and $G_{2,1}^T + G_{2,2}^T = D_{2,3}^T + D_{2,4}^T + D_{2,6}^T = 1.5$ for multilateral contracts.

Loads in multilateral transactions fulfill the curtailment given in (10.19b) and weights given in Table 10.42, that is,

$$D_{2,3}^T = 0.25(G_{2,1}^T + G_{2,2}^T)$$

$$D_{2,4}^T = 0.50(G_{2,1}^T + G_{2,2}^T)$$

$$D_{2,6}^T = 0.25(G_{2,1}^T + G_{2,2}^T)$$

Table 10.46 shows line flows corresponding to final schedules. In this table, some lines are forced to carry the maximum limit (bold letters). Note that the total transmission loss is equal to the difference between generation and load in Table 10.45.

If transmission loss is the only component taken into consideration, LMPs will differ slightly to reflect marginal losses. The difference in LMPs in Table 10.47 represents both congestion and transmission losses.

With the optimization results, we trace the contribution of each generator or load to line flows (see the example in the Appendix F for details). To trace the contribution of each party to the line usage, we use line flows (Table 10.46) and net bus generation and load in Table 10.48. The power generated by G_1 reaches all buses (domain of G_1 includes all buses), and the power generated by G_2 reaches buses 2, 3, 5, and 6 (domain of G_2 includes buses 2, 3, 5 and 6). So the system under study has two commons, as seen in Table 10.49 and Figure 10.27. Figure 10.27 also shows the net generation and net load at each bus.

Table 10.48 Net Bus Generation and Load

Bus i	G_i	D_i
1	1.6120	0.0000
2	2.2958	0.0000
3	0.0000	1.9564
4	0.0000	0.7503
5	0.0000	0.6670
6	0.0000	0.3751
Total	3.9078	3.7488
$\Sigma G_i - \Sigma D_i =$		0.1590

Table 10.49 Common of Each Bus

Bus	1	2	3	4	5	6
Common	1	2	2	1	2	2

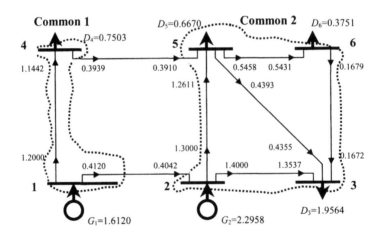

Figure 10.27 System's Two Commons

The system under study in Figure 10.27 can be represented in a stategraph form as in Figure 10.28, where g_1 and g_2 refer to net generation in commons 1 and 2, respectively, and d_1 and d_2 refer to net loads in commons 1 and 2, respectively. Lines 1–2 and 4–5 can be combined to form one link between the two commons, and the other lines are internal to commons, i.e., line 1–4 is internal to common 1, and lines 2–3, 2–5, 3–5, 3–6, and 5–6 are internal to common 2. The inflow (I_1 and I_2), generation (g_1 and g_2) and load (d_1 and d_2) in each common are shown in Table 10.50.

TRANSMISSION CONGESTION MANAGEMENT AND PRICING

The flow between commons 1 and 2 in Figure 10.28 is the net flow on lines 1–2 and 4–5, which form link 1–2.

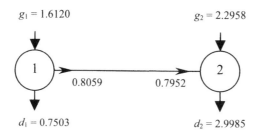

Figure 10.28 System's State Graph

Table 10.50 Inflows, Generation and Load for Each Common

Common k	Inflow I_k	Generation g_i	Load d_i
1	1.6120	1.6120	0.7503
2	3.1017	2.2958	2.9985

Using Table 10.50 and Figure 10.28, the absolute contribution matrix of the common's generation to common inflows (A) and the relative contribution matrix of the common's generation to common's loads and outflows (R) are given by

$$A = \begin{bmatrix} 1.6120 & 0.8059 \\ 0.0000 & 2.2958 \end{bmatrix}, \quad R = \begin{bmatrix} 1.0000 & 0.2598 \\ 0.0000 & 0.7402 \end{bmatrix}$$

The contribution of each common's generation (g_1 and g_2) to each line flow (f_{m-n,g_i}) is shown in Table 10.51. Since g_1 is composed of three generators G_1^{PL}, $G_{1,1}^T$, $G_{2,1}^T$, and g_2 is composed of G_2^{PL} and $G_{2,2}^T$, we find the contribution of each generator to branch flows by obtaining the proportion of each generator to the commons generation and multiplying the result by the contribution of common's generation to the line flow. For example, the generation of common 1 is 1.6120 and G_1^{PL} =0.3359, since g_1 contributes 0.4120 to the flow of line 1–2, then the contribution of G_1^{PL} to this flow equals $(0.3359/1.6120) \times (0.4120) = 0.0859$. The contributions of each generation to the line flows and to each load are detailed in Tables 10.52 and 10.53, respectively.

In Table 10.52, P, B, and M refers to Pool, Bilateral, and Multilateral, respectively. The number in bracket refers to bus number. Also, $(P, 1) \equiv G_1^{PL}$, $(P, 2) \equiv G_2^{PL}$, $(B, 1) \equiv G_{1,1}^T$, $(M, 1) \equiv G_{2,1}^T$, and $(M, 2) \equiv G_{2,2}^T$.

Table 10.51 Contributions of Common Generation to Line Flows

Line No.	From Bus m	To Bus n	Total Flow $f_{m\text{-}n}$	Contribution from	
				g_1	g_2
1	1	2	0.4120	0.4120	0.0000
2	1	4	1.2000	1.2000	0.0000
3	2	3	1.4000	0.3638	1.0362
4	2	5	1.3000	0.3378	0.9622
5	3	5	-0.4355	-0.1132	-0.3223
6	3	6	-0.1672	-0.0434	-0.1238
7	4	5	0.3939	0.3939	0.0000
8	5	6	0.5458	0.1418	0.4040

Table 10.52 Contribution of each Generator to Line Flows

Line No.	From Bus m	To Bus n	Total Flow	Contribution from Generation				
				(P, 1)	(P, 2)	(B, 1)	(M, 1)	(M, 2)
1	1	2	0.4120	0.0859	0.0000	0.1982	0.1279	0.0000
2	1	4	1.2000	0.2500	0.0000	0.5773	0.3727	0.0000
3	2	3	1.4000	0.0758	0.5849	0.1750	0.1130	0.4514
4	2	5	1.3000	0.0704	0.5431	0.1625	0.1049	0.4191
5	3	5	-0.4355	-0.0236	-0.1819	-0.0544	-0.0351	-0.1404
6	3	6	-0.1672	-0.0091	-0.0699	-0.0209	-0.0135	-0.0539
7	4	5	0.3939	0.0821	0.0000	0.1895	0.1223	0.0000
8	5	6	0.5458	0.0296	0.2280	0.0682	0.0440	0.1760

Table 10.53 Contribution of Bus Generation to Loads

Load	Value	Type	Contribution from Generation (Type, Bus)					Total
			(P, 1)	(P, 2)	(B, 1)	(M, 1)	(M, 2)	
1	0.8058	(P, 3)	0.0436	0.3366	0.1007	0.0650	0.2598	0.8058
2	0.6670	(P, 5)	0.0361	0.2787	0.0834	0.0538	0.2150	0.6670
3	0.7755	(B, 3)	0.0420	0.3240	0.0969	0.0626	0.2500	0.7755
4	0.3751	(M, 3)	0.0203	0.1567	0.0469	0.0303	0.1209	0.3751
5	0.7503	(M, 4)	0.1563	0.0000	0.3610	0.2330	0.0000	0.7503
6	0.3751	(M, 6)	0.0203	0.1567	0.0469	0.0303	0.1209	0.3751

TRANSMISSION CONGESTION MANAGEMENT AND PRICING

To show how values in Table 10.53 are calculated, we take the pool's load located at bus 3, which is 0.8058 p.u. (or $D_3^{PL} = 0.8058$) and located in common 2 as shown in Figure 10.24. The pool's generator located at bus 1, or (P, 1), is in common 2 and it generates 0.3359 p.u. (see Table 10.32). In the R matrix, the net generation of common 1 (g_1) contributes 25.98% to the net load of common 2 (d_2). This means that 25.98% of 0.8058 (or 0.2093 p.u.) comes from g_1. But g_1 is composed of three single generators ((P, 1), (B, 1), (M, 1)). Their total is 1.6120 p.u as shown in Figure 10.25. The contribution of (P, 1) is (0.3359/1.6120) × (0.2093) = 0.0436, which is the same under (P, 1) and next to the load number 1 in Table 10.53.

From the results of Table 10.52 and Equations (10.24–10.25), the transmission usage charges that each participant i would pay for using each line and for using all lines are shown in Table 10.54. Congestion and loss charges are shown in Table 10.55. The payments given in Table 10.54 are collected by the ISO and distributed among transmission owners based on the required revenues of their lines. The payments shown in Table 10.55 are also collected by the ISO and distributed among transmission owners and FTR holders.

Table 10.54 Calculation of Transmission Usage Charges

Line	m	n	Payment for Contribution from Generator (Type, Bus):					Total line Payment
			(P, 1)	(P, 2)	(B, 1)	(M, 1)	(M, 2)	
1	1	2	20.84	0.00	48.11	31.05	0.00	100.0
2	1	4	12.50	0.00	28.87	18.63	0.00	60.0
3	2	3	4.33	33.42	10.00	6.46	25.79	80.0
4	2	5	2.71	20.89	6.25	4.03	16.12	50.0
5	3	5	5.41	41.78	12.50	8.07	32.24	100.0
6	3	6	5.41	41.78	12.50	8.07	32.24	100.0
7	4	5	16.67	0.00	38.49	24.84	0.00	80.0
8	5	6	2.71	20.89	6.25	4.03	16.12	50.0
			$70.58	$158.76	$162.97	$105.18	$122.51	$ 620.0

Table 10.55 Congestion and Transmission Loss Charges

Load No.	Load Value	(Type, Bus)	Contribution from Generation:					Charge in $
			(P, 1)	(P, 2)	(B, 1)	(M, 1)	(M, 2)	
1	0.8058	(P, 3)	29.39	201.74	67.85	43.80	155.69	498.47
2	0.6670	(P, 5)	19.10	126.64	44.09	28.46	97.73	316.02
3	0.7755	(B, 3)	28.28	194.16	65.30	42.15	149.84	479.73
4	0.3751	(M, 3)	13.68	93.91	31.58	20.39	72.47	232.03
5	0.7503	(M, 4)	87.15	0.00	201.21	129.88	0.00	418.24
6	0.3751	(M, 6)	11.86	79.83	27.37	17.67	61.61	198.34
						Total Congestion Charges		2142.83

To show how values in Table 10.55 are calculated, we take load number 1, which is a pool's load and located at bus 3 (D_3^{PL}) and has a value of 0.8058 p.u. The LMP at bus 3 is equal to 16.76740 $/MWh. The contribution of the pool's generator located at bus 1 (P, 1) or G_1^{PL} to this load is 0.0436 p.u. (or $D_{D_3^{PL}, G_1^{PL}} = 0.0436$), and the LMP at bus 1 is equal to 10.03143 $/MWh. For this contribution, the associated congestion charge is equal to

$$Ch_{D_3^{PL}, G_1^{PL}} = D_{D_3^{PL}, G_1^{PL}} \times (LMP_3 - LMP_1)$$
$$= 0.0436 \times (16.7674 - 10.03143) \times 100 = \$29.39$$

To show how congestion charges are distributed, we assume that x_1 and x_2 are holders with FTRs shown in Table 10.56. From (10.28) and (10.29), congestion and loss charges for each line and credits for FTR holders are calculated as shown in Table 10.57 (see bold lines).

Table 10.56 FTRs

Holder	FTR	From Bus m	To Bus n
x_1	1.000	2	3
	1.000	2	5
x_2	0.800	1	4

Table 10.57 FTR Credits

Branch No.	m	n	Congestion Charges	FTR	Credits in $	Paid to	Difference
1	1	2	30.326	-	0.00		30.326
2	1	4	653.365	0.8	445.945	x_2	207.420
3	2	3	825.128	1.0	599.287	x_1	225.841
4	2	5	581.961	1.0	454.462	x_1	127.449
5	3	5	63.346	-	0.00		63.346
6	3	6	15.054	-	0.00		15.054
7	4	5	-11.247	-	0.00		-11.247
8	5	6	29.933	-	0.00		29.933

The charges are collected by the ISO, and the net payments are distributed in one of three ways: the ISO may pay them to transmission owners (TOs), to FTR holders, or to generators that would generate counter-flows. Whatever is collected from generators for a certain line is originally paid to the TO of that line, but if there is an FTR holder for the

TRANSMISSION CONGESTION MANAGEMENT AND PRICING

line, the holder will be paid from the collected amount. In the case where the line charges are not enough to cover the FTR credits for this line, the TO of this line will pay the difference to the FTR holder. When the congestion charges for a certain line are larger than the FTR credits, the extra credit (the surplus in Table 10.57) is paid to the TO of this line. On the other hand, when the difference is negative, the TO will pay the difference to the ISO.

10.6 CONCLUSIONS

Electricity restructuring poses major changes for the three main components of the vertically integrated monopoly. Restructuring will mean new types of unbundling, coordination, and rules to guarantee competition and non-discriminatory open access to all users. The new requirements of electric utility restructuring will necessitate the design of new entities that can coordinate these components and these new components are intended to foster competition among different parties. One of the main objectives in the electric industry's restructuring is to bring fairness and open access to the transmission network. Implementing fair rules that allocate price and transmission use fulfills this notion of fairness in the industry.

Knowing whether or not, and to what extent, each power system user contributes to the usage of a particular system component is necessary for a restructured power system to operate economically and efficiently and for the guarantee of open access to all system users. Moreover, the pricing of transmission must provide adequate economic signals that allow the system to grow. As such, in a restructured environment, it is necessary to develop and use reasonable and fair pricing rules that allocate the costs of transmission services based on actual use by different power system brokers. The pricing rules must also allow for congestion management, which is a new ingredient in this environment.

This chapter presents the main existing approaches for transmission pricing. It also presents an overview of recent techniques used for designing equitable access fees to recover fixed transmission costs. Among these methods, are some recent methods used to determine a generator's (load's) contribution to a line power flow and a consumer load.

LMPs and FTRs are being widely utilized in many power restructuring models to resolve problems associated with transmission congestion and energy pricing. Therefore, this chapter exclusively considers these key subjects.

Since transmission congestion is still a major research topic in today's restructuring, this chapter introduces a hybrid scheme for congestion management that combines the best features of the inter-zonal/intra-zonal scheme and FTR. The scheme presented in the chapter utilizes LMP for defining zonal boundaries. The proposed unified scheme facilitates the way in which we deal with congestion charges and congestion credits. In addition, the chapter describes a generalized mathematical model of the FTR auction that guarantees its availability to all parties on a non-discriminatory basis, where system users are permitted to buy, sell, and trade FTRs through an auction. Trading of FTRs in secondary markets is also addressed in this chapter.

In this chapter we also include a comprehensive scheme for the ISO that modifies preferred schedules, traces participants' contributions, and allocates transmission usage and congestion charges. In this scheme, we assume a general restructuring model where pool, bilateral, and multilateral contracts exist concurrently. In this scheme, the preferred schedules are adjusted on a non-discriminatory basis to keep the system within its limits. That is, we use "willingness to pay to avoid curtailment" as a criterion to reshape the initial schedules in times of congestion and we find feasible final schedules. With final schedules and LMPs (by-products of the schedule's adjustment), transmission congestion and losses are calculated. A flow-based tracing method is then utilized to allocate the transmission charges. The FTR holders' credits are calculated based on line flows and LMPs. Examples for a testing system are presented to illustrate the proposed scheme.

Appendix A

List of Symbols

The list of symbols listed here are used in Chapter 4 (PBUC) and Chapter 8 (SCUC).

Symbol	Definition
$B(i,t)$	Power purchase of unit i at time t
$C_i(.)$	Cost function of unit i, $C_i(x) = a(i) + b(i)x + c(i)x^2$
$C_{fi}(.)$	Fuel consumption function of unit i, $C_{fi}(x) = a_f(i) + b_f(i)x + c_f(i)x^2$
$C_{ei}(.)$	Emission function of unit i, $C_{ei}(x) = a_e(i) + b_e(i)x + c_e(i)x^2$
$DR(i)$	Ramp down rate limit of unit i
$E\{.\}$	Expected value
E^{max}	Maximum total emission allowance
EMA	Area emission limit
EMS	System emission limit
$F(i,t)$	Profit of unit i at time t
$f_i(P_0(i,t))$	Profit from bilateral contract of unit i at time t
$F^{min}(FT)$	Minimum total fuel consumption for fuel type FT

$F^{max}(FT)$	Maximum total fuel consumption for fuel type FT
$F^{min}(i)$	Minimum total fuel consumption for unit i
$F^{max}(i)$	Maximum total fuel consumption for unit i
$I(i,t)$	Commitment state of unit i at time t
$I(i,t)^{(k)}$	Commitment state of unit i at time t in previous iteration
$m_0(i,t)$	Incremental bid slope of unit i at time t
$m_b(i,t)$	Bid slope for power purchase of unit i at time t
$m_g(i,t)$	Bid slope for generation of unit i at time t
$m_n(i,t)$	Bid slope for non-spinning reserve of unit i at time t
$m_r(i,t)$	Bid slope for spinning reserve of unit i at time t
$MSR(i)$	Maximum sustained ramp rate of unit i (MW/min)
N_b	Number of buses
N_g	Number of units
N_t	Number of time periods
nc	Number of possible line outages in the contingency case
$N(i,t)$	Non-spinning reserve of unit i at time t
$N_{min}(i,t)$	Minimum non-spinning reserve of unit i at time t
$N_{max}(i,t)$	Maximum non-spinning reserve of unit i at time t
$N(i,t)^{(k)}$	Non-spinning reserve of unit i at time t in the previous iteration
$N^{min}(t)$	Minimum total non-spinning reserve at time t
$N^{max}(t)$	Maximum total non-spinning reserve at time t
$P(i,t)$	Generation of unit i at time t
$P_{min}(i,t)$	Minimum generation of unit i at time t
$P_{max}(i,t)$	Maximum generation of unit i at time t
$P(i,t)^{(k)}$	Generation of unit i at time t in previous iteration
$P_0(i,t)$	Bilateral contract of unit i at time t
$P_{gmin}(i)$	Minimum generation of unit i
$P_{gmax}(i)$	Maximum generation of unit i

APPENDIX A

$P^{min}(t)$	Minimum total generation at time t
$P^{max}(t)$	Maximum total generation at time t
P_{km}	Power flow of line k–m
$P_{km}{}^{min}$	Lower limit for power flow of line k–m
$P_{km}{}^{max}$	Upper limit for power flow of line k–m
$P_D(t)$	Total system real power load demand at time t
$Q_D(t)$	Total system reactive power load demand at time t
$Q_{gmax}(i)$	Maximum reactive power unit i can provide
$q(i)$	Quick start capability of unit i
r_{jt}	Real power interruption at bus j in time t
$r_o(i,t)$	Contribution of unit i to operating reserve at time t
$r_s(i,t)$	Contribution of unit i to spinning reserve at time t
$R_o(t)$	System operating reserve requirement at time t
$R_s(t)$	System spinning reserve requirement at time t
$R(i,t)$	Spinning reserve of unit i at time t
$R_{min}(i,t)$	Minimum spinning reserve of unit i at time t
$R_{max}(i,t)$	Maximum spinning reserve of unit i at time t
$R(i,t)^{(k)}$	Spinning reserve of unit i at time t in previous iteration
$R^{min}(t)$	Minimum total spinning reserve at time t
$R^{max}(t)$	Maximum total spinning reserve at time t
$S(i,t)$	Start-up cost of unit i at time t
$S_f(i,t)$	Start-up fuel of unit i at time t
$S_e(i,t)$	Start-up emission of unit i at time t
$T^{off}(i)$	Minimum OFF time of unit i
$T^{on}(i)$	Minimum ON time of unit i
$UR(i)$	Ramp-up rate limit of unit i
$X^{off}(i,t)$	Time duration for which unit i has been OFF at time t
$X^{on}(i,t)$	Time duration for which unit i has been ON at time t

APPENDIX A

A	Sensitivity coefficient matrix of steady state transmission constraints
$E(j)$	Sensitivity coefficient matrix of contingency transmission constraints for line outage j
$F_1(V)$	Reactive power function of V for units
$F_2(V)$	Reactive power function of V for buses
F^S	Penalty vector for steady state flow constraints
f^S	Steady state flow limit vector
$F^C(j)$	Penalty vector for contingency flow constraints in case of line outage j
$f^C(j)$	Flow limit vector for line contingency j
G^S	Penalty vector for steady state voltage constraints
g^S	Steady state reactive limit vector
$G^C(j)$	Penalty vector for voltage constraints in line outage j
$g^C(j)$	Reactive limit vector for line contingency j
$[J'']$	Jacobian matrix
$[J_1'']$	Jacobian matrices for generator buses
$[J_2'']$	Jacobian matrices for load buses
ΔQ_T	Reactive power increment vector corresponding to transformers
ΔQ_G	Reactive power increment vector corresponding to generating units
$Q_G(t)$	Reactive power generation vector at time t
$Q_G^{min}(t)$	Reactive power generation vector lower limit at time t
$Q_G^{max}(t)$	Reactive power generation vector upper limit at time t
$Q_L(t)$	Reactive power load vector at time t
r	Real power interruption vector
T	Transformer tap vector
T^{min}	Transformer tap lower limit vector

APPENDIX A

T^{max}	Transformer tap upper limit vector
V	System voltage vector
V^{min}	System voltage lower limit vector
V^{max}	System voltage upper limit vector
V^0	System voltage vector based on initial commitment state
ΔV^{min}	$V^{min} - V^0$
ΔV^{max}	$V^{max} - V^0$
α_i	Integrated labor starting-up cost and equipment maintenance cost of unit i
β_i	Starting-up cost of unit i from cold conditions
τ_i	Time constant that characterizes unit i cooling speed
$\lambda_g(t)$	Lagrangian multiplier for energy constraint at time t
$\lambda_r(t)$	Lagrangian multiplier for spinning reserve constraint at time t
$\lambda_n(t)$	Lagrangian multiplier for non-spinning reserve constraint at time t
λ_e	Lagrangian multiplier for system emission constraint
$\lambda_f(FT)$	Lagrangian multiplier for fuel constraint of fuel type FT
$\lambda_{fu}(i)$	Lagrangian multiplier for fuel constraint of unit i
$\lambda(t)$	Lagrangian multiplier for load balance constraint
$\mu_o(t)$	Lagrangian multiplier for operating reserve constraint
$\mu_s(t)$	Lagrangian multiplier for spinning reserve constraint
λ_f	Lagrangian multiplier for fuel constraint
λ_a	Lagrangian multiplier for area emission limit
λ_s	Lagrangian multiplier for system emission limit
$\rho_{gm}(t)$	Forecasted market price for energy at time t (for system)

$p_{nm}(t)$	Forecasted market price for non-spinning reserve at time t (for the system)
$p_{rm}(t)$	Forecasted market price for spinning reserve at time t (for the system)
$p_{gm}(i,t)$	Forecasted market price for energy at time t (for unit i)
$p_{nm}(i,t)$	Forecasted market price for non-spinning reserve at time t (for unit i)
$p_{rm}(i,t)$	Forecasted market price for spinning reserve at time t (for unit i)
ε_t	Upper limit of expected unserved energy at time t

Appendix B

MATHEMATICAL DERIVATION

B.1 DERIVATION OF PROBABILITY DISTRIBUTION

Suppose that the probability distribution of a variable P is $f(P)$ and its expected value is k^*. Consider another variable P' that follows a probability distribution similar to P and has an expected value of k. Assume that the probability distribution of P' is $f(P-\delta)$, where δ is selected so that

$$E(P') = \int_{-\infty}^{+\infty} P \times f(P-\delta)dP = k \tag{B.1}$$

The significance of δ is to shift $f(P)$ and ensure that the expected value is k. After some manipulation, we can get a simple formula for δ:

$$\begin{aligned}\int_{-\infty}^{+\infty} P \times f(P-\delta)dP &= \int_{-\infty}^{+\infty} (Q+\delta) \times f(Q)dQ \\ &= \int_{-\infty}^{+\infty} Q \times f(Q)dQ + \int_{-\infty}^{+\infty} \delta \times f(Q)dQ = k^* + \delta\end{aligned} \tag{B.2}$$

where

$Q = P - \delta$, $k^* = \int_{-\infty}^{+\infty} Q \times f(Q)dQ = \int_{-\infty}^{+\infty} P \times f(P)dP$ is the expected value of the P, and $\int_{-\infty}^{+\infty} f(Q)dQ = 1$.

Combining (B.1) and (B.2), we have

$$\delta = k - k^* \tag{B.3}$$

B.2 LAGRANGIAN AUGMENTATION WITH INEQUALITY CONSTRAINTS

For a constrained minimization problem with equality constraints, we write

$$\min f(x)$$

s.t. $h(x) = 0$

The Lagrangian function is

$$L = f(x) + \lambda h(x)$$

And the augmented Lagrangian function is

$$L_a = f(x) + \lambda h(x) + \frac{1}{2} c h^2(x)$$

The last term $\frac{1}{2} c h^2(x)$ can be thought of as an extra penalty function (c is the penalty factor); it can improve the convexity of the Lagrangian function. For an inequality-constrained minimization problem, we have

$$\min f(x)$$

s.t. $g(x) \leq 0$

Here, we first convert the inequality constraint into an equality constraint

$$g(x) + z = 0$$

where z is a non-negative number $z \geq 0$. The original inequality constrained minimization problem is converted into an equality constrained minimization problem:

$$\min f(x)$$

s.t. $g(x) + z = 0$, $z \geq 0$

The augmented Lagrangian function is

$$L_a = f(x) + \mu\bigl(g(x) + z\bigr) + \frac{1}{2} c \bigl(g(x) + z\bigr)^2$$

To minimize the augmented Lagrangian function with respect to the non-negative z, we have

APPENDIX B

$$\frac{\partial L_a}{\partial z} = \mu + c(g(x) + z) = 0$$

$$z \geq 0$$

From the first equation, we have

$$z = -\left(\frac{\mu}{c} + g(x)\right)$$

From the second equation, we have

$$z = \max\left\{0, -\left(\frac{\mu}{c} + g(x)\right)\right\}$$

So, if $g(x) > -\frac{\mu}{c}$, which means $\frac{\mu}{c} + g(x) > 0$,

then $z = \max\left\{0, -\left(\frac{\mu}{c} + g(x)\right)\right\} = 0$,

and the augmented Lagrangian function is

$$L_a = f(x) + \mu g(x) + \frac{1}{2}cg^2(x)$$

If $g(x) \leq -\frac{\mu}{c}$, which means $\frac{\mu}{c} + g(x) \leq 0$

Then $z = \max\left\{0, -\left(\frac{\mu}{c} + g(x)\right)\right\} = -\left(\frac{\mu}{c} + g(x)\right)$

And the augmented Lagrangian function is

$$L_a = f(x) + \mu(g(x) + z) + \frac{1}{2}c(g(x) + z)^2$$

$$= f(x) + \mu\left(g(x) - \left(\frac{\mu}{c} + g(x)\right)\right) + \frac{1}{2}c\left(g(x) - \left(\frac{\mu}{c} + g(x)\right)\right)^2$$

$$= f(x) - \frac{\mu^2}{c} + \frac{1}{2}c\frac{\mu^2}{c^2}$$

$$= f(x) - \frac{\mu^2}{2c}$$

Combining the two cases, we have

$$L_a = \begin{cases} f(x) + \mu g(x) + \frac{1}{2}cg^2(x) & \text{if } g(x) > -\frac{\mu}{c} \\ f(x) - \frac{\mu^2}{2c} & \text{if } g(x) \leq -\frac{\mu}{c} \end{cases}$$

We define the augmentation term $P(x)$ as

$$P(x) = \begin{cases} \mu g(x) + \frac{1}{2}cg^2(x) & \text{if } g(x) > -\frac{\mu}{c} \\ -\frac{\mu^2}{2c} & \text{if } g(x) \leq -\frac{\mu}{c} \end{cases}$$

Figure B.1 shows $P(x)$ as a function of $g(x)$.

APPENDIX B

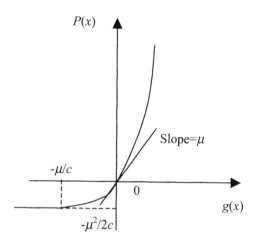

Figure B.1 $P(x)$ as a Function of $g(x)$

For a double-sided inequality constraint

$$g_{min} \leq g(x) \leq g_{max}$$

The augmentation term for $g(x) \leq g_{max}$ is

$$P_1(x) = \begin{cases} \mu_1(g(x) - g_{max}) + \dfrac{1}{2}c(g(x) - g_{max})^2 & \text{if } g(x) - g_{max} > -\dfrac{\mu_1}{c} \\ -\dfrac{\mu_1^2}{2c} & \text{if } g(x) - g_{max} \leq -\dfrac{\mu_1}{c} \end{cases}$$

The augmentation term for $g_{min} \leq g(x)$ is

$$P_2(x) = \begin{cases} \mu_2(g_{min} - g(x)) + \dfrac{1}{2}c(g_{min} - g(x))^2 & \text{if } g_{min} - g(x) > -\dfrac{\mu_2}{c} \\ -\dfrac{\mu_2^2}{2c} & \text{if } g_{min} - g(x) \leq -\dfrac{\mu_2}{c} \end{cases}$$

APPENDIX B

Combining the two, we have

$$P(x) = \begin{cases} \mu(g(x)-g_{max})+\frac{1}{2}c(g(x)-g_{max})^2 & \text{if } g(x)-g_{max} > -\frac{\mu}{c} \\ \frac{-\mu^2}{2c} & \text{if } g(x)-g_{max} \le -\frac{\mu}{c} \text{ and } g_{min}-g(x) \le -\frac{\mu}{c} \\ \mu(g_{min}-g(x))+\frac{1}{2}c(g_{min}-g(x))^2 & \text{if } g_{min}-g(x) > -\frac{\mu}{c} \end{cases}$$

Or

$$P(x) = \begin{cases} \mu(g(x)-g_{max})+\frac{1}{2}c(g(x)-g_{max})^2 & \text{if } g(x) > g_{max} - \frac{\mu}{c} \\ \frac{-\mu^2}{2c} & \text{if } g_{min} + \frac{\mu}{c} \le g(x) \le g_{max} - \frac{\mu}{c} \\ \mu(g_{min}-g(x))+\frac{1}{2}c(g_{min}-g(x))^2 & \text{if } g_{min}-g(x) > -\frac{\mu}{c} \end{cases}$$

Figure B.2 shows $P(x)$ as a function of $g(x)$.

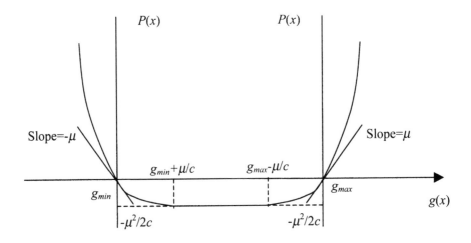

Figure B.2 P(x) as a Function of g(x)

Appendix C
RTS Load Data

The RTS data is excerpted from [Rts96].

Table C.1 Weekly Peak Load in Percent of Annual Peak

Week	Peak Load	Week	Peak Load	Week	Peak Load	Week	Peak Load
1	86.2	14	75.0	27	75.5	40	72.4
2	90.0	15	72.1	28	81.6	41	74.3
3	87.8	16	80.0	29	80.1	42	74.4
4	83.4	17	75.4	30	88.0	43	80.0
5	88.0	18	83.7	31	72.2	44	88.1
6	84.1	19	87.0	32	77.6	45	88.5
7	83.2	20	88.0	33	80.0	46	90.9
8	80.6	21	85.6	34	72.9	47	94.0
9	74.0	22	81.1	35	72.6	48	89.0
10	73.7	23	90.0	36	70.5	49	94.2
11	71.5	24	88.7	37	78.0	50	97.0
12	72.7	25	89.6	38	69.5	51	100.0
13	70.4	26	86.1	39	72.4	52	95.2

Table C.2 Daily Peak Load in Percent of Weekly Peak

Day	Peak Load
Monday	93
Tuesday	100
Wednesday	98
Thursday	96
Friday	94
Saturday	77
Sunday	75

Table C.3 Hourly Peak Load in Percent of Daily Peak

Hour	Winter Weeks 1-8 & 44-52		Summer Weeks 18-30		Spring/Fall Weeks 9-17 & 31-43	
	Weekday	Weekend	Weekday	Weekend	Weekday	Weekend
12-1 am	67	78	64	74	63	75
1-2	63	72	60	70	62	73
2-3	60	68	58	66	60	69
3-4	59	66	56	65	58	66
4-5	59	64	56	64	59	65
5-6	60	65	58	62	65	65
6-7	74	66	64	62	72	68
7-8	86	70	76	66	85	74
8-9	95	80	87	81	95	83
9-10	96	88	95	86	99	89
10-11	96	90	99	91	100	92
11-noon	95	91	100	93	99	94
Noon-1pm	95	90	99	93	93	91
1-2	95	88	100	92	92	90
2-3	93	87	100	91	90	90
3-4	94	87	97	91	88	86
4-5	99	91	96	92	90	85
5-6	100	100	96	94	92	88
6-7	100	99	93	95	96	92
7-8	96	97	92	95	98	100
8-9	91	94	92	100	96	97
9-10	83	92	93	93	90	95
10-11	73	87	87	88	80	90
11-12	63	81	72	80	70	85

Appendix D

Example Systems Data

D.1 5-UNIT SYSTEM

Table D.1 Unit Characteristics

Unit	Pmax	Pmin	af	bf	cf	Min ON	Min OFF	IniT	Ramp Up	Ramp Dn	St MBtu
1	335	125	1169.88	2.6998	0.00753	10	-1	-1	14	14	500
2	232	150	779.64	0.9286	0.01560	10	-1	-1	10	10	300
3	260	50	636.46	3.8260	0.00890	10	-1	-1	15	15	500
4	440	160	669.12	7.9215	1.7E-06	10	-1	-1	25	25	603
5	250	130	590.72	2.4435	0.00906	10	-1	-1	12	12	500

Where
  $Pmax$: Maximum capacity (MW)
  $Pmin$: Minimum capacity (MW)
  af, bf, cf: Fuel consumption coefficient (MBtu, MBtu/MW, MBtu/MW^2)
  Min ON: Minimum ON time (Hour)
  Min OFF: Minimum OFF time (Hour)
  IniT: Initial operational time (Hour, minus means OFF)
  Ramp Up: Ramp-up rate (MW/min)
  Ramp Dn: Ramp-down rate (MW/min)
  St MBtu: Start-up MBtu

Table D.2 Energy Price for Case 1-1

Hour	1-5	6-23	24
Energy Price ($/MWh)	20	48	20

Table D.3 Energy Price for Case 1-2

Hour	1-5	6	7	8	9	10	11
Energy Price ($/MWh)	20	24	26	28	30	32	34
Hour	12	13	14	15	16-23		24
Energy Price ($/MWh)	36	38	40	42	48		20

Table D.4 Dispatch Results for Case 1-1

Unit / Hour	1	2	3	4	5
1-5	0	0	0	0	0
6-23	335	232	260	440	250
24	0	0	0	0	0

Table D.5 Dispatch Results for Case 1-2

Unit / Hour	1	2	3	4	5
1-5	0	0	0	0	0
6	0	210	0	0	250
7	0	232	257	0	250
8	335	232	260	0	250
9-23	335	232	260	440	250
24	0	0	0	0	0

Table D.6 Dispatch Results for Case 2-1

Unit / Hour	1	2	3	4	5
1-2	0	0	0	0	0
3-4	0	0	0	0	130
5	0	0	0	0	180
6	335	232	260	440	240
7-23	335	232	260	440	250
24	0	0	0	0	223

APPENDIX D

Table D.7 Dispatch Results for Case 2-2

Unit / Hour	1	2	3	4	5
1-5	0	0	0	0	0
6-7	335	232	260	440	130
8	335	232	260	440	180
9	335	232	260	440	240
10-23	335	232	260	440	250
24	0	0	0	0	190

Table D.8 Dispatch Results for Case 4-2 and Case 4-3

Unit / Hour	1	2	3	4	5
1-13	0	0	0	0	0
14	0	0	0	0	179
15	0	0	0	0	192
16-23	0	0	0	0	225
24	0	0	0	0	0

Table D.9 Dispatch Results for Case 5-1

Unit / Hour	1	2	3	4	5
1-5	0	0	0	0	0
6	0	210	0	0	250
7	0	232	257	0	250
8	335	232	260	0	250
9-23	335	232	260	440	250
24	0	0	0	0	0

Table D.10 Dispatch Results for Case 5-2

Unit / Hour	1	2	3	4	5
1-5	0	0	0	0	0
6	0	210	0	0	250
7	0	232	257	0	250
8	335	232	260	0	250
9-21	335	232	260	440	250
22-23	335	232	260	0	0
24	0	0	0	0	0

D.2 36-UNIT SYSTEM

Table D.11 Unit Characteristics[1]

Unit	Pmax	Pmin	af	bf	cf	Min ON	Min OFF	IniT	Ramp Up	Ramp Dn	St MBtu
1	12.00	2.40	24.3891	25.5472	0.02533	1	-1	-1	12.00	12.00	0
2	20.00	4.00	118.9083	37.9637	0.01561	1	-1	-1	20.00	20.00	30
3	20.00	4.00	118.4576	37.7770	0.01359	1	-1	-1	20.00	20.00	30
4	20.00	4.00	118.9083	37.9637	0.01161	1	-1	-1	20.00	20.00	30
5	20.00	4.00	119.4576	38.7770	0.01059	1	-1	-1	20.00	20.00	30
6	20.00	4.00	117.7551	37.5510	0.01199	1	-1	-1	20.00	20.00	30
7	20.00	4.00	118.1083	37.6637	0.01261	1	-1	-1	20.00	20.00	30
8	76.00	15.20	81.8259	13.5073	0.00962	3	-2	3	38.00	38.00	80
9	76.00	15.20	81.1364	13.3272	0.00876	3	-2	3	38.00	38.00	80
10	76.00	15.20	81.2980	13.3538	0.00895	3	-2	3	38.00	38.00	80
11	76.00	15.20	81.6259	13.4073	0.00932	3	-2	3	38.00	38.00	80
12	100.00	25.00	217.8952	18.0000	0.00623	4	-2	-3	50.00	50.00	100
13	100.00	25.00	219.7752	18.6000	0.00599	4	-2	3	50.00	50.00	100
14	100.00	25.00	218.3350	18.1000	0.00612	4	-2	3	50.00	50.00	100
15	100.00	25.00	216.7752	18.2800	0.00588	4	-2	-3	50.00	50.00	100
16	100.00	25.00	218.7752	18.2000	0.00598	4	-2	-3	50.00	50.00	100
17	100.00	25.00	216.7752	17.2800	0.00578	4	-2	-3	50.00	50.00	100
18	100.00	25.00	218.7752	19.2000	0.00698	4	-2	-3	50.00	50.00	100
19	155.00	54.25	143.0288	10.7154	0.00473	5	-3	5	77.50	77.50	200
20	155.00	54.25	143.3179	10.7367	0.00481	5	-3	5	77.50	77.50	200
21	155.00	54.25	143.5972	10.7583	0.00487	5	-3	5	77.50	77.50	200
22	197.00	68.95	259.1310	23.0000	0.00259	5	-4	-4	98.50	98.50	300
23	197.00	68.95	259.6490	23.1000	0.00260	5	-4	-4	98.50	98.50	300
24	197.00	68.95	260.1760	23.2000	0.00263	5	-4	-4	98.50	98.50	300
25	197.00	68.95	260.5760	23.4000	0.00264	5	-8	-1	98.50	98.50	300
26	197.00	68.95	261.1760	23.5000	0.00267	5	-8	-1	98.50	98.50	300
27	197.00	68.95	260.0760	23.0400	0.00261	5	-4	-4	98.50	98.50	300
28	350.00	140.00	176.0575	10.8416	0.00150	8	-5	10	175.00	175.00	500
29	350.00	140.00	177.0575	10.8616	0.00153	8	-5	10	175.00	175.00	500
30	350.00	140.00	176.0575	10.6616	0.00143	8	-5	10	175.00	175.00	500
31	350.00	140.00	177.9575	10.9616	0.00163	8	-5	10	175.00	175.00	500
32	400.00	100.00	310.0021	7.4921	0.00194	8	-5	10	200.00	200.00	800
33	400.00	100.00	311.9102	7.5031	0.00195	8	-5	10	200.00	200.00	800
34	400.00	100.00	312.9102	7.5121	0.00196	8	-5	10	200.00	200.00	800
35	400.00	100.00	314.9102	7.5321	0.00197	8	-5	10	200.00	200.00	800
36	400.00	100.00	313.9102	7.6121	0.00199	8	-5	10	200.00	200.00	800

---

[1] See D.1 for explanation of the abbreviations.

APPENDIX D

Table D.12 Market Prices for Case 1

Hour	Energy Price ($/MWh)	Spinning Reserve Price ($/MW.Hour)	Non-Spinning Reserve Price ($/MW.Hour)
1	11.74	11.75	11.75
2	7.70	7.71	7.71
3	1.99	2.08	2.08
4	0.00	0.10	0.10
5	3.00	3.09	3.09
6	14.08	14.09	14.09
7	13.98	15.08	15.08
8	16.29	17.39	17.39
9	18.60	19.70	19.70
10	21.00	22.10	22.10
11	24.25	25.35	25.35
12	24.40	25.50	25.50
13	22.01	27.01	27.01
14	23.32	24.42	24.42
15	24.00	33.12	33.12
16	29.34	35.59	35.59
17	28.12	37.24	37.24
18	24.75	33.87	33.87
19	24.51	25.61	25.61
20	21.45	22.55	22.55
21	16.45	17.55	17.55
22	8.59	9.69	9.69
23	6.00	7.00	7.00
24	0.50	0.51	0.51

Table D.13 Dispatch Results for Case 1

Hour\Unit	0	1	2	3	4	5	6	7	8	9	10	11	12	13	14	15	16	17	18	19	20	21	22	23	24
1	0	0	0	0	0	0	0	0	0	0	0	0	0	0	0	1	1	1	1	0	0	0	0	0	0
2-7	0	0	0	0	0	0	0	0	0	0	0	0	0	0	0	0	0	0	0	0	0	0	0	0	0
8-11	1	0	0	0	0	0	0	1	1	1	1	1	1	1	1	1	1	1	1	1	1	1	0	0	0
12	0	0	0	0	0	0	0	0	0	0	1	1	1	1	1	1	1	1	1	1	1	1	0	0	0
13-14	1	1	0	0	0	0	0	0	0	0	1	1	1	1	1	1	1	1	1	1	1	1	0	0	0
15	0	0	0	0	0	0	0	0	0	0	1	1	1	1	1	1	1	1	1	1	1	1	0	0	0
16	0	0	0	0	0	0	0	0	0	0	1	1	1	1	1	1	1	1	1	1	1	1	0	0	0
17	0	0	0	0	0	0	0	0	0	0	1	1	1	1	1	1	1	1	1	1	1	1	0	0	0
18	0	0	0	0	0	0	0	0	0	0	1	1	1	1	1	1	1	1	1	1	1	1	0	0	0
19-21	1	0	0	0	0	0	1	1	1	1	1	1	1	1	1	1	1	1	1	1	1	1	1	0	0
22	0	0	0	0	0	0	0	0	0	0	0	0	0	0	0	1	1	1	1	1	0	0	0	0	0
23	0	0	0	0	0	0	0	0	0	0	0	0	0	0	0	1	1	1	1	1	0	0	0	0	0
24	0	0	0	0	0	0	0	0	0	0	0	0	0	0	0	1	1	1	1	1	0	0	0	0	0
25	0	0	0	0	0	0	0	0	0	0	0	0	0	0	0	1	1	1	1	1	0	0	0	0	0
26	0	0	0	0	0	0	0	0	0	0	0	0	0	0	0	1	1	1	1	1	0	0	0	0	0
27	0	0	0	0	0	0	0	0	0	0	0	0	0	0	0	1	1	1	1	1	0	0	0	0	0
28-36	1	0	0	0	0	0	1	1	1	1	1	1	1	1	1	1	1	1	1	1	1	1	1	0	0

Table D.14 Dispatch Results for Case 2

Hour\Unit	0	1	2	3	4	5	6	7	8	9	10	11	12	13	14	15	16	17	18	19	20	21	22	23	24
1	0	0	0	0	0	0	0	0	0	0	0	1	1	1	1	1	1	1	1	1	0	0	0	0	0
2-7	0	0	0	0	0	0	0	0	0	0	0	0	0	0	0	0	0	0	0	0	0	0	0	0	0
8-11	1	1	0	0	0	0	1	1	1	1	1	1	1	1	1	1	1	1	1	1	1	1	1	0	0
12	0	0	0	0	0	0	0	0	1	1	1	1	1	1	1	1	1	1	1	1	1	1	0	0	0
13-14	1	1	0	0	0	0	0	0	1	1	1	1	1	1	1	1	1	1	1	1	1	1	0	0	0
15	0	0	0	0	0	0	0	0	1	1	1	1	1	1	1	1	1	1	1	1	1	1	0	0	0
16	0	0	0	0	0	0	0	0	1	1	1	1	1	1	1	1	1	1	1	1	1	1	0	0	0
17	0	0	0	0	0	0	0	1	1	1	1	1	1	1	1	1	1	1	1	1	1	1	0	0	0
18	0	0	0	0	0	0	0	0	1	1	1	1	1	1	1	1	1	1	1	1	1	1	0	0	0
19-21	1	1	1	0	0	0	1	1	1	1	1	1	1	1	1	1	1	1	1	1	1	1	1	1	0
22	0	0	0	0	0	0	0	0	0	1	1	1	1	1	1	1	1	1	1	1	0	0	0	0	0
23	0	0	0	0	0	0	0	0	0	1	1	1	1	1	1	1	1	1	1	1	0	0	0	0	0
24	0	0	0	0	0	0	0	0	0	1	1	1	1	1	1	1	1	1	1	1	0	0	0	0	0
25	0	0	0	0	0	0	0	0	0	1	1	1	1	1	1	1	1	1	1	1	0	0	0	0	0
26	0	0	0	0	0	0	0	0	0	1	1	1	1	1	1	1	1	1	1	1	0	0	0	0	0
27	0	0	0	0	0	0	0	0	0	1	1	1	1	1	1	1	1	1	1	1	0	0	0	0	0
28-36	1	1	1	1	1	1	1	1	1	1	1	1	1	1	1	1	1	1	1	1	1	1	1	1	0

APPENDIX D

Table D.15 Dispatch Results for Case 3

Unit \ Hour	0	1	2	3	4	5	6	7	8	9	10	11	12	13	14	15	16	17	18	19	20	21	22	23	24
1	0	0	0	0	0	0	0	0	0	1	1	1	1	1	1	1	1	1	1	1	1	0	0	0	0
2-7	0	0	0	0	0	0	0	0	0	0	0	0	0	0	0	0	0	0	0	0	0	0	0	0	0
8-11	1	1	1	0	0	1	1	1	1	1	1	1	1	1	1	1	1	1	1	1	1	1	1	1	0
12	0	0	0	0	0	0	1	1	1	1	1	1	1	1	1	1	1	1	1	1	1	1	1	0	0
13-14	1	1	0	0	0	0	1	1	1	1	1	1	1	1	1	1	1	1	1	1	1	1	1	0	0
15	0	0	0	0	0	0	1	1	1	1	1	1	1	1	1	1	1	1	1	1	1	1	1	0	0
16	0	0	0	0	0	0	1	1	1	1	1	1	1	1	1	1	1	1	1	1	1	1	1	0	0
17	0	0	0	0	0	0	1	1	1	1	1	1	1	1	1	1	1	1	1	1	1	1	1	0	0
18	0	0	0	0	0	0	1	1	1	1	1	1	1	1	1	1	1	1	1	1	1	1	1	0	0
19-21	1	1	1	1	1	1	1	1	1	1	1	1	1	1	1	1	1	1	1	1	1	1	1	1	1
22	0	0	0	0	0	0	0	1	1	1	1	1	1	1	1	1	1	1	1	1	1	1	0	0	0
23	0	0	0	0	0	0	0	1	1	1	1	1	1	1	1	1	1	1	1	1	1	1	0	0	0
24	0	0	0	0	0	0	0	1	1	1	1	1	1	1	1	1	1	1	1	1	1	1	0	0	0
25	0	0	0	0	0	0	0	0	1	1	1	1	1	1	1	1	1	1	1	1	1	1	0	0	0
26	0	0	0	0	0	0	0	0	1	1	1	1	1	1	1	1	1	1	1	1	1	1	0	0	0
27	0	0	0	0	0	0	0	1	1	1	1	1	1	1	1	1	1	1	1	1	1	1	0	0	0
28-36	1	1	1	1	1	1	1	1	1	1	1	1	1	1	1	1	1	1	1	1	1	1	1	1	1

Table D.16 Dispatch Results for Case 4

Unit \ Hour	0	1	2	3	4	5	6	7	8	9	10	11	12	13	14	15	16	17	18	19	20	21	22	23	24
1	0	0	0	0	0	0	1	1	1	1	1	1	1	1	1	1	1	1	1	1	1	1	0	0	0
2-7	0	0	0	0	0	0	0	0	0	0	0	0	0	0	0	0	1	1	0	0	0	0	0	0	0
8-11	1	1	1	1	1	1	1	1	1	1	1	1	1	1	1	1	1	1	1	1	1	1	1	1	1
12	0	0	0	0	0	1	1	1	1	1	1	1	1	1	1	1	1	1	1	1	1	1	1	1	0
13-14	1	1	1	0	0	1	1	1	1	1	1	1	1	1	1	1	1	1	1	1	1	1	1	1	0
15	0	0	0	0	0	1	1	1	1	1	1	1	1	1	1	1	1	1	1	1	1	1	1	1	0
16	0	0	0	0	0	1	1	1	1	1	1	1	1	1	1	1	1	1	1	1	1	1	1	1	0
17	0	1	1	1	1	1	1	1	1	1	1	1	1	1	1	1	1	1	1	1	1	1	1	1	0
18	0	0	0	0	0	1	1	1	1	1	1	1	1	1	1	1	1	1	1	1	1	1	1	1	0
19-21	1	1	1	1	1	1	1	1	1	1	1	1	1	1	1	1	1	1	1	1	1	1	1	1	1
22	0	0	0	0	0	0	1	1	1	1	1	1	1	1	1	1	1	1	1	1	1	1	1	0	0
23	0	0	0	0	0	0	1	1	1	1	1	1	1	1	1	1	1	1	1	1	1	1	1	0	0
24	0	0	0	0	0	0	1	1	1	1	1	1	1	1	1	1	1	1	1	1	1	1	1	0	0
25	0	0	0	0	0	0	0	0	1	1	1	1	1	1	1	1	1	1	1	1	1	1	1	0	0
26	0	0	0	0	0	0	0	0	1	1	1	1	1	1	1	1	1	1	1	1	1	1	1	0	0
27	0	0	0	0	0	0	1	1	1	1	1	1	1	1	1	1	1	1	1	1	1	1	1	0	0
28-36	1	1	1	1	1	1	1	1	1	1	1	1	1	1	1	1	1	1	1	1	1	1	1	1	1

D.3 6-UNIT SYSTEM

Table D.17 Unit Characteristics[2]

Unit	Pmax	Pmin	af	bf	cf	Min ON	Min OFF	IniT	Ramp Up	Ramp Dn	St MBtu
1	50	5	118.82	27.896	0.0143	1	-1	-1	50	50	0
2	50	10	118.11	24.664	0.0126	1	-1	-1	50	50	0
3	150	10	218.34	18.100	0.0081	5	-5	-5	75	75	50
4	200	20	142.73	10.694	0.0046	8	-8	8	100	100	50
5	350	140	176.06	10.662	0.0014	8	-5	10	175	175	500
6	400	100	313.91	7.612	0.0020	8	-5	10	200	200	800

Table D.18 Market Prices

Hour	Energy Price ($/MWh)	Spinning Reserve Price ($/MW.Hour)	Non-Spinning Reserve Price ($/MW.Hour)
1	15.74	16.74	17.04
2	11.70	12.70	13.00
3	5.99	6.99	7.29
4	4.00	5.00	5.30
5	7.00	8.00	8.30
6	18.08	19.08	19.38
7	17.98	18.98	19.28
8	20.29	21.29	21.59
9	22.60	23.60	23.90
10	25.00	26.00	26.30
11	28.25	29.25	29.55
12	28.40	29.40	29.70
13	26.01	27.01	27.31
14	27.32	28.32	28.62
15	28.00	29.00	29.30
16	33.34	34.34	34.64
17	32.12	33.12	33.42
18	28.75	29.75	30.05
19	28.51	29.51	29.81
20	25.45	26.45	26.75
21	20.45	21.45	21.75
22	12.59	13.59	13.89
23	10.00	11.00	11.30
24	4.50	5.50	5.80

---

[2] See D.1 for explanation of the abbreviations.

Table D.19 Bilateral Contract

Unit \ Hour	1	2	3	4	5	6	Total
1-10	0	0	0	0	150	250	400
11-22	0	0	80	80	150	250	560
23-24	0	0	0	0	150	250	400

D.4 MODIFIED IEEE 30-BUS SYSTEM

Table D.20 Generators Data (Type 1)

Participant	Bus No	Cost coefficients a b c	Min MW	Max MW	Start up cost	Min ON & OFF Time	Initial State	Ramp up & down rate MW/h
A	30	187.364 49.327 0.0243	10	20	70	3 -2	-2	20
	24	128.820 39.889 0.0163	5	20	30	1 -1	-1	20
	11	118.820 37.889 0.0143	5	20	30	1 -1	-1	20
	2	218.335 18.100 0.0061	10	80	100	4 -2	-2	40
	8	81.298 13.353 0.0089	10	50	80	3 -2	3	25
	5	81.136 13.327 0.0087	10	50	80	3 -2	3	25
	1	142.734 10.694 0.0046	20	100	200	5 -3	5	50
B	13	287.136 19.327 0.0103	10	70	95	4 -2	-2	35
	15	230.000 18.300 0.0071	10	60	90	4 -2	4	30

Table D.21 Load Data at Hour 18 (Daily Peak Loads)

Bus No.	Load (MW)	Bus No.	Load (MW)
1	0.0	16	3.5
2	21.7	17	9.0
3	2.4	18	3.2
4	67.6	19	9.5
5	34.2	20	2.2
6	0.0	21	17.5
7	22.8	22	0.0
8	30.0	23	3.2
9	0.0	24	8.7
10	5.8	25	0.0
11	0.0	26	3.5
12	11.2	27	0.0
13	0.0	28	0.0
14	6.2	29	2.4
15	8.2	30	10.6

Participant A's total load = 227.2 MW
Participant B's total load = 56.2 MW

Table D.22 Network Data

Branch No	From-to	R p.u.	X p.u.	Rating MW
1	1 - 2	0.0192	0.0575	30.0
2	1 - 3	0.0452	0.1852	30.0
3	2 - 4	0.0570	0.1737	30.0
4	3 - 4	0.0132	0.0379	30.0
5	2 - 5	0.0472	0.1983	30.0
6	2 - 6	0.0581	0.1763	30.0
7	4 - 6	0.0119	0.0414	30.0
8	5 - 7	0.0460	0.1160	30.0
9	6 - 7	0.0267	0.0820	30.0
10	6 - 8	0.0120	0.0420	30.0
11	6 - 9	0.0000	0.2080	30.0
12	6 - 10	0.0000	0.5560	30.0
13	9 - 11	0.0000	0.2080	30.0
14	9 - 10	0.0000	0.1100	30.0
15	4 - 12	0.0000	0.2560	65.0
16	12 - 13	0.0000	0.1400	65.0
17	12 - 14	0.1231	0.2559	32.0
18	12 - 15	0.0662	0.1304	32.0
19	12 - 16	0.0945	0.1987	32.0
20	14 - 15	0.2210	0.1997	16.0
21	16 - 17	0.0824	0.1932	16.0
22	15 - 18	0.1070	0.2185	16.0
23	18 - 19	0.0639	0.1292	16.0
24	19 - 20	0.0340	0.0680	32.0
25	10 - 20	0.0936	0.2090	32.0
26	10 - 17	0.0324	0.0845	32.0
27	10 - 21	0.0348	0.0749	30.0
28	10 - 22	0.0727	0.1499	30.0
29	21 - 22	0.0116	0.0236	30.0
30	15 - 23	0.1000	0.2020	16.0
31	22 - 24	0.1150	0.1790	30.0
32	23 - 24	0.1320	0.2700	16.0
33	24 - 25	0.1885	0.3292	30.0
34	25 - 26	0.2544	0.3800	30.0
35	25 - 27	0.1093	0.2087	30.0
36	28 - 27	0.0000	0.3960	30.0
37	27 - 29	0.2198	0.4153	30.0
38	27 - 30	0.3202	0.6027	30.0
39	29 - 30	0.2399	0.4533	30.0
40	8 - 28	0.0636	0.2000	30.0
41	6 - 28	0.0169	0.0599	30.0

APPENDIX D

D.5 118-BUS SYSTEM

Table D.23 Line Data for 118-bus System

Line	From Bus	To Bus	R (p.u.)	X (p.u.)	Line	From Bus	To Bus	R (p.u.)	X (p.u.)
1	1	2	0.0303	0.0999	47	35	37	0.011	0.0497
2	1	3	0.0129	0.0424	48	33	37	0.0415	0.142
3	4	5	0.0018	0.008	49	34	36	0.0087	0.0268
4	3	5	0.0241	0.108	50	34	37	0.0026	0.0094
5	5	6	0.0119	0.054	51	38	37	0	0.0375
6	6	7	0.0046	0.0208	52	37	39	0.0321	0.106
7	8	9	0.0024	0.0305	53	37	40	0.0593	0.168
8	8	5	0	0.0267	54	30	38	0.0046	0.054
9	9	10	0.0026	0.0322	55	39	40	0.0184	0.0605
10	4	11	0.0209	0.0688	56	40	41	0.0145	0.0487
11	5	11	0.0203	0.0682	57	40	42	0.0555	0.183
12	11	12	0.006	0.0196	58	41	42	0.041	0.135
13	2	12	0.0187	0.0616	59	43	44	0.0608	0.2454
14	3	12	0.0484	0.16	60	34	43	0.0413	0.1681
15	7	12	0.0086	0.034	61	44	45	0.0224	0.0901
16	11	13	0.0223	0.0731	62	45	46	0.04	0.1356
17	12	14	0.0215	0.0707	63	46	47	0.038	0.127
18	13	15	0.0744	0.2444	64	46	48	0.0601	0.189
19	14	15	0.0595	0.195	65	47	49	0.0191	0.0625
20	12	16	0.0212	0.0834	66	42	49	0.0715	0.323
21	15	17	0.0132	0.0437	67	42	49	0.0715	0.323
22	16	17	0.0454	0.1801	68	45	49	0.0684	0.186
23	17	18	0.0123	0.0505	69	48	49	0.0179	0.0505
24	18	19	0.0112	0.0493	70	49	50	0.0267	0.0752
25	19	20	0.0252	0.117	71	49	51	0.0486	0.137
26	15	19	0.012	0.0394	72	51	52	0.0203	0.0588
27	20	21	0.0183	0.0849	73	52	53	0.0405	0.1635
28	21	22	0.0209	0.097	74	53	54	0.0263	0.122
29	22	23	0.0342	0.159	75	49	54	0.073	0.289
30	23	24	0.0135	0.0492	76	49	54	0.0869	0.291
31	23	25	0.0156	0.08	77	54	55	0.0169	0.0707
32	26	25	0	0.0382	78	54	56	0.0027	0.0095
33	25	27	0.0318	0.163	79	55	56	0.0049	0.0151
34	27	28	0.0191	0.0855	80	56	57	0.0343	0.0966
35	28	29	0.0237	0.0943	81	50	57	0.0474	0.134
36	30	17	0	0.0388	82	56	58	0.0343	0.0966
37	8	30	0.0043	0.0504	83	51	58	0.0255	0.0719
38	26	30	0.008	0.086	84	54	59	0.0503	0.2293
39	17	31	0.0474	0.1563	85	56	59	0.0825	0.251
40	29	31	0.0108	0.0331	86	56	59	0.0803	0.239
41	23	32	0.0317	0.1153	87	55	59	0.0474	0.2158
42	31	32	0.0298	0.0985	88	59	60	0.0317	0.145
43	27	32	0.0229	0.0755	89	59	61	0.0328	0.15
44	15	33	0.038	0.1244	90	60	61	0.0026	0.0135
45	19	34	0.0752	0.247	91	60	62	0.0123	0.0561
46	35	36	0.0022	0.0102	92	61	62	0.0082	0.0376

Table D.23 Line Data for 118-bus System (Continued)

Line	From Bus	To Bus	R (p.u.)	X (p.u.)	Line	From Bus	To Bus	R (p.u.)	X (p.u.)
93	63	59	0	0.0386	140	90	91	0.0254	0.0836
94	63	64	0.0017	0.02	141	89	92	0.0099	0.0505
95	64	61	0	0.0268	142	89	92	0.0393	0.1581
96	38	65	0.009	0.0986	143	91	92	0.0387	0.1272
97	64	65	0.0027	0.0302	144	92	93	0.0258	0.0848
98	49	66	0.018	0.0919	145	92	94	0.0481	0.158
99	49	66	0.018	0.0919	146	93	94	0.0223	0.0732
100	62	66	0.0482	0.218	147	94	95	0.0132	0.0434
101	62	67	0.0258	0.117	148	80	96	0.0356	0.182
102	65	66	0	0.037	149	82	96	0.0162	0.053
103	66	67	0.0224	0.1015	150	94	96	0.0269	0.0869
104	65	68	0.0014	0.016	151	80	97	0.0183	0.0934
105	47	69	0.0844	0.2778	152	80	98	0.0238	0.108
106	49	69	0.0985	0.324	153	80	99	0.0454	0.206
107	68	69	0	0.037	154	92	100	0.0648	0.295
108	69	70	0.03	0.127	155	94	100	0.0178	0.058
109	24	70	0.0022	0.4115	156	95	96	0.0171	0.0547
110	70	71	0.0088	0.0355	157	96	97	0.0173	0.0885
111	24	72	0.0488	0.196	158	98	100	0.0397	0.179
112	71	72	0.0446	0.18	159	99	100	0.018	0.0813
113	71	73	0.0087	0.0454	160	100	101	0.0277	0.1262
114	70	74	0.0401	0.1323	161	92	102	0.0123	0.0559
115	70	75	0.0428	0.141	162	101	102	0.0246	0.112
116	69	75	0.0405	0.122	163	100	103	0.016	0.0525
117	74	75	0.0123	0.0406	164	100	104	0.0451	0.204
118	76	77	0.0444	0.148	165	103	104	0.0466	0.1584
119	69	77	0.0309	0.101	166	103	105	0.0535	0.1625
120	75	77	0.0601	0.1999	167	100	106	0.0605	0.229
121	77	78	0.0038	0.0124	168	104	105	0.0099	0.0378
122	78	79	0.0055	0.0244	169	105	106	0.014	0.0547
123	77	80	0.017	0.0485	170	105	107	0.053	0.183
124	77	80	0.0294	0.105	171	105	108	0.0261	0.0703
125	79	80	0.0156	0.0704	172	106	107	0.053	0.183
126	68	81	0.0018	0.0202	173	108	109	0.0105	0.0288
127	81	80	0	0.037	174	103	110	0.0391	0.1813
128	77	82	0.0298	0.0853	175	109	110	0.0278	0.0762
129	82	83	0.0112	0.0366	176	110	111	0.022	0.0755
130	83	84	0.0625	0.132	177	110	112	0.0247	0.064
131	83	85	0.043	0.148	178	17	113	0.0091	0.0301
132	84	85	0.0302	0.0641	179	32	113	0.0615	0.203
133	85	86	0.035	0.123	180	32	114	0.0135	0.0612
134	86	87	0.0283	0.2074	181	27	115	0.0164	0.0741
135	85	88	0.02	0.102	182	114	115	0.0023	0.0104
136	85	89	0.0239	0.173	183	68	116	0.0003	0.0041
137	88	89	0.0139	0.0712	184	12	117	0.0329	0.14
138	89	90	0.0518	0.188	185	75	118	0.0145	0.0481
139	89	90	0.0238	0.0997	186	76	118	0.0164	0.0544

APPENDIX D

Table D.24 Load Bus Data and Generator Bus Data

Bus	Load Factor	Gen Capacity (100 MW)	Bid ($/MWh)	Bus Index	Load Factor	Gen Capacity (100 MW)	Bid ($/MWh)
1	0.1	0.8	15	31	0.9	1.2	17
2	0.1	1	10	32	0.1	0	0
3	0.1	0	0	33	0.4	0.8	21
4	0.4	1	20	34	1.5	0	0
5	0.4	0	0	35	0.2	0	0
6	0.1	0	0	36	0.2	0	0
7	0.1	0	0	37	0.1	0	0
8	0.4	0.4	21	38	0.5	0	0
9	0.5	0	0	39	0.1	0	0
10	0.1	1	15	40	0	1	23
11	0.1	0	0	41	0	0	0
12	0.7	1.2	10	42	0.2	1.6	17
13	0.8	0	0	43	0	0	0
14	0.2	0	0	44	0	0	0
15	0.5	0	0	45	0	0	0
16	0.1	0	0	46	0.8	1.4	21
17	0.1	0	0	47	0	0	0
18	0.1	0	0	48	0.1	0	0
19	0.3	0	0	49	0.3	1	10
20	0.5	0.5	12	50	0.1	0	0
21	0.1	0	0	51	0.5	0	0
22	0.4	0	0	52	0	0	0
23	0.6	0	0	53	0	0	0
24	0.1	0	0	54	0	1	23
25	0	1.1	12	55	0	0	0
26	0.4	3.5	11	56	0.5	0	0
27	0.1	0.8	10	57	0	0	0
28	0.1	0	0	58	0.6	0	0
29	0.1	0	0	59	0	1.5	22
30	1.5	0	0	60	0	0	0

Table D.24 Load Bus Data and Generator Bus Data (Continued)

Bus	Load Factor	Gen Capacity (100 MW)	Bid ($/MWh)	Bus Index	Load Factor	Gen Capacity (100 MW)	Bid ($/MWh)
61	0	0.5	16	90	0	1.6	10
62	1	0	0	91	0	0.8	11
63	1	0	0	92	0	0	0
64	0	0	0	93	0.2	0	0
65	0	1	14	94	0.2	0	0
66	1	1	17.5	95	0.4	0	0
67	0	0	0	96	1.1	1	10
68	0	0	0	97	0.5	0	0
69	1.5	1	11	98	0.4	0.5	22
70	0.1	0	0	99	0.5	1.8	21
71	0	0	0	100	0	3	17
72	0	2	16	101	0	0	0
73	1	1	11	102	0	0	0
74	0.1	0	0	103	0.1	1.8	11
75	0.1	0	0	104	0.1	0	0
76	0.25	0	0	105	0.1	0	0
77	0.7	0	0	106	0.1	0	0
78	0	0	0	107	0.1	1.4	11
79	0	0	0	108	0.1	0	0
80	1.2	0	0	109	0.1	0	0
81	0	0	0	110	0.1	0	0
82	1.5	0	0	111	0.1	1.8	11
83	0	0	0	112	0.2	1	10
84	0.85	0	0	113	0	1.2	12
85	0	0	0	114	0.3	0	0
86	0	0	0	115	0.1	0	0
87	0	0.4	11	116	0.1	1.2	11
88	0	0	0	117	0.1	0	0
89	0	1	10	118	0.2	0	0

Appendix E

Game Theory Concepts

In this appendix, we present some theoretical aspects of game theory that would be followed by players in a restructured energy market.

E.1 EQUILIBRIUM IN NON-COOPERATIVE GAMES

The *normal form* of an N-player game [Fer96a] consists of a set of N players, N strategy sets X_i, $i=1,...,N$ and the N-tuple payoff function $G(X_1,...,X_N)$. The value $G_i(x_1,...,x_N)$ is the payoff function of player i when player 1 plays the mixed strategy $x_1 \in X_1$, ... , player N plays the strategy $x_N \in X_N$. We only consider *pure* strategies. Each player can only choose one strategy.

An N-tuple of strategies $x_1, ... , x_N$ is an *equilibrium N-tuple* if

$$G_i(x_1, ... ,y_i, ... , x_N) \leq G_i(x_1,..., x_N)$$

for all i and for all strategies y_i for player i. Hence, the player departing from the mixed strategy in the N-tuple at least does no better.

In a two-player game, the two-dimensional set $G = \{G_1(X_1,X_2), G_2(X_1,X_2)\}$ is called the *non-cooperative payoff region* [Fer96a]. The points in G are called *payoff pairs*. If (u,v) and (u',v') are two payoff pairs, (u,v) *dominates* (u',v') if $u \geq v$ and $v \geq v'$.

Payoff pairs that are not dominated by any other pair are said to be *Pareto optimal*. Clearly, rational cooperating players in a power pool would never play so that their payoffs are not Pareto optimal.

E.2 CHARACTERISTIC FUNCTION

The *max-min value* or *characteristic function* gives each player a pessimistic estimate of how much payoff can be expected. In a two-player game, the max-min value gives player 1 (the row player) the expected payoff by assuming that player 2 (the column player) will act to minimize player 1's payoff.

$$v_1 = \max_{X_1} \min_{X_2} G_1(X_1, X_2)$$

where X_1 and X_2 range over all mixed strategies for players 1 and 2 respectively. The max-min value for player 2 is defined in a similar way.

The characteristic functions of the coalition and counter-coalition are computed as follows: Let m_{ij} be an element of the row player's payoff matrix. The row player's problem is to compute the row value

$$v_r = \min_{j} \sum_{i=1}^{m} p_i m_{ij}$$

and an optimal mixed strategy [Fer96a] $p = (p_1, \dots, p_m)$ with $\Sigma p_i = 1$.

A constant c is added to all entries large enough that $m_{ij} + c > 0$ for all i,j. The problem is formulated now as a function of variable $y_i = p_i / v$ for $1 \le i \le m$; the problem can be formulated as a linear program:

$$\min \ y_1 + \dots + y_m$$

$$\text{s.t.} \ \sum_{i=1}^{m} m_{ij} y_i \ge 1 \quad \text{for} \quad 1 \le j \le n$$

The column's player problem in the non-cooperative game is computed in the same way after transposing the column player's payoff matrix.

In a two-player game, the assumption that player 2 will try to minimize 1's payoff is probably false. Player 2 will try instead to maximize his payoffs; however, v_1 gives player 1 a lower bound (pessimistic estimate) on his payoff.

E.3 N-PLAYERS COOPERATIVE GAMES

As indicated above, a *coalition* is a subset of players that is formed in order to coordinate strategies and to agree on how the total payoff is to be divided among its members.

The set of all N players is denoted by P. Given a coalition $S \subseteq P$, the *counter-coalition* to S is

$$S^c = P - S = \{P_i \in P : P_i \notin S\}$$

In general, in a game with N players, there are 2^N coalitions.

The game between any coalition S and its counter-coalition S^c is a non-cooperative game. The rows in the matrix correspond to the set of strategies available to S and the columns correspond to those available to S^c. The max-min value for the coalition is called the *characteristic function* of S and is denoted $v(S)$.

When a coalition is formed, the payoff is distributed among players in the coalition. The amounts going to each player form an N-tuple X of real numbers. Two conditions are usually required from an N-tuple in order to be likely to actually happen in the game [Fer96a]:

- *Individual rationality*

$$x_i \geq v(\{P_i\})$$

- *Collective rationality*

$$\sum_{i=1}^{N} x_i = v(P)$$

Individual rationality requires that the coalition give any player a higher payoff than what the player can obtain on his own. A N-tuple of payments that satisfies both of these conditions is called an *imputation*.

The *core* of a game consists in all imputations that are not dominated by any other imputation through any coalition. Hence, if an imputation X is at the core, there is no group of players which has a reason to form a coalition and replace X with a different imputation.

Every imputation in the core of the game is an *efficient* allocation. The standard definition of efficient allocation is Pareto optimality. If the allocation is efficient, there is no player who can do better without making some other player worse off. Mathematically, X is in the core if and only if

$$\sum_{P_i \in S} x_i \geq v(S)$$

for every coalition S.

The concept of a core of the game has an important limitation: the core of the game may be of any size; it may even be empty, meaning that there are no stable coalitions. Whatever coalition is formed, there is some incentive for a subgroup to desert the coalition.

In a restructured power system as presented in Chapter 6, the objective from the pool coordinator's perspective should be to make the distribution of savings among players at the core of the game in the case of the grand coalition.

E.4 GAMES WITH INCOMPLETE INFORMATION

One may distinguish between games with complete information (c-games) and games with incomplete information (i-games). In an i-game, some of the players are not certain of the characteristics of some of the other parties; players may lack information about the other player's payoff functions, strategies available to other players, the amount of information that other players have on various aspects of the game and so on. That is to say, players would lack full information on the mathematical structure of the game.

Incomplete information is considered by modeling the player's unknown characteristics as *types* of player. The type of a player embodies any information that is not common to *all* players. An i-game may be used in situations where each player is uncertain about other players' payoff, but also in situations where each player is uncertain about other players' knowledge.

In any given play of an i-game each player knows his type. However, a player searching for his best strategy may need to determine the actions of the other players for other states, since he may be misinformed about the state of the game.

Another distinction is made between games with perfect information and games with imperfect information. This distinction is based on the amount of information that players have about the moves made in earlier stages of the game. In games with perfect information–for instance, chess–all players have full information on all moves made in early stages. In

APPENDIX E

games with imperfect information players only have partial information about the moves made in earlier stages.

Equilibrium of i-games may be derived from Nash equilibrium conditions. In Nash equilibrium, each player chooses the best action available to him given the (incomplete) information that he receives. A proposed solution of i-games in [Fer98] is to transform the original i-game into a c-game with imperfect information. Hence, Nash equilibrium is applied to the transformed game. Player's optimal bids are derived for the equilibrium condition.

Appendix F

Congestion Charges Calculation

Figure F.1 shows a 3-bus system. $G_{1,1}$, G_2, L_2 and $L_{2,3}$ entered as pool participants, and $G_{2,1}$ and $L_{2,3}$ enter as bilateral contract participants. Values of loads, generators, line flows, and LMPs are also shown Figure F.1.

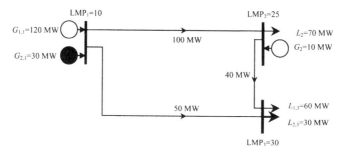

Figure F.1 Example of a 3-Bus System

F.1 CALCULATIONS OF CONGESTION CHARGES USING CONTRIBUTIONS OF GENERATORS

Figure F.2 shows the acyclic diagram of this example, and Figure F.3 gives detailed contribution of each generator to each line flow and each load.

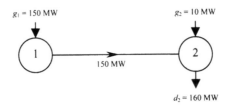

Figure F.2 Acyclic Diagram of Generator Contributions

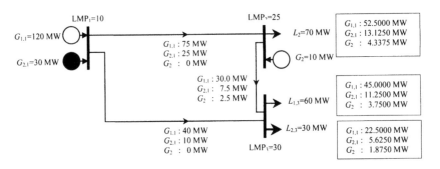

Figure F.3 Tracing Contributions of Generators

For this example, the absolute and relative contribution matrices are given by

$$A = \begin{bmatrix} 150.0 & 150.0 \\ 0.000 & 10.00 \end{bmatrix}, \quad R = \begin{bmatrix} 1.0000 & 0.9375 \\ 0.0000 & 0.0625 \end{bmatrix}$$

The charges for congestion are as follows:

- L_2: Pays   52.500 (25-10) = \$787.500   ($G_{1,1} \to L_2$)

  13.125 (25-10) = \$196.875   ($G_{2,1} \to L_2$)

  4.375 (25-25) = \$ 0.000   ($G_2 \to L_2$)

  Total = \$984.375

- $L_{1,3}$: Pays 45.000 (30-10) = \$900.000   ($G_{1,1} \to L_{1,3}$)

  11.250 (30-10) = \$225.000   ($G_{2,1} \to L_{1,3}$)

APPENDIX F

$$3.750\ (30\text{-}25) = \$\ 18.750\ (G_2 \rightarrow L_{1,3})$$
$$\text{Total} = \$1143.750$$

- $L_{2,3}$: Pays $22.500\ (30\text{-}10) = \$450.000\ (G_{1,1} \rightarrow L_{2,3})$
$$5.625\ (30\text{-}10) = \$112.500\ (G_{2,1} \rightarrow L_{2,3})$$
$$1.875\ (30\text{-}25) = \$\ \ 9.375\ (G_2 \rightarrow L_{2,3})$$
$$\text{Total} = \$571.875$$

Total congestion charges (TCC) = $2700

If we use Equation 10.28b for calculation of congestion charges:

$G_{1,1}$: $75\ (25\text{-}10) + 40\ (30\text{-}10) + 30\ (30\text{-}25) = 1125 + 500 + 150$
$$= \$2075$$

$G_{2,1}$: $25\ (25\text{-}10) + 10\ (30\text{-}10) + 7.5\ (30\text{-}25) = 375 + 200 + 37.5$
$$= \$612.5$$

$G_{1,1}$: $0\ (25\text{-}10) + 0\ (30\text{-}10) + 2.5\ (30\text{-}25) = \ \ 0 + \ \ 0 + 12.5$
$$= \$12.5$$
$$\text{Total} = \$2700$$

The congestion charges for individual lines are as follows:

Line 1-2: $75(25\text{-}10) + 25\ (25\text{-}10) + 0(25\text{-}10)$
$$= f_{1-2}\ (LMP_2 - LMP_1) = 100(25\text{-}10) = \$1500$$

Line 1-3: $40(30\text{-}10) + 10\ (30\text{-}10) + 0(30\text{-}10)$
$$= f_{1-3}\ (LMP_3 - LMP_1) = 50(30\text{-}10) = \$1000$$

Line 2-3: $30(30\text{-}25) + 7.5(30\text{-}25) + 2.5(30\text{-}25)$
$$= f_{2-3}\ (LMP_3 - LMP_2) = 40(30\text{-}25) = \$200$$
$$\text{Total} = \$2700$$

The payments for energy by pool's participants are as follows:

- L_2: Pays $\ \ 52.500\ (10) = \$525.000\ (G_{1,1} \rightarrow L_2)$

$$13.125\ (10) = \$131.250\ (G_{2,1} \rightarrow L_2)$$
$$4.375\ (25) = \$109.375\ (G_2 \rightarrow L_2)$$
$$\text{Total} = \$765.625$$

- $L_{1,3}$: Pays $45.000\ (10) = \$450.000\ (G_{1,1} \rightarrow L_{1,3})$
$$11.250\ (10) = \$112.500\ (G_{2,1} \rightarrow L_{1,3})$$
$$3.750\ (25) = \$\ 93.750\ (G_2 \rightarrow L_{1,3})$$
$$\text{Total} = \$656.250$$

Total payment for energy from pool's participants = $765.625 + $656.25 = $1421.875. Note that this total is less than what should be paid by the ISO to the pool's generators (120 × 10 + 10 × 25 = $1450) by $28.125. This $28.125 should come from $L_{2,3}$. The following shows how this is working:

Energy price at $G_{2,1}$ = 10 $/MWh

$G_{2,1}$ generates 30 MWh, for a total of $300

Now, if $L_{2,3}$ would pay for energy, its payment would be:

- $L_{2,3}$: Pays $22.500\ (10) = \$225.000\ (G_{1,1} \rightarrow L_{2,3})$
$$5.625\ (10) = \$\ 56.250\ (G_{2,1} \rightarrow L_{2,3})$$
$$1.875\ (25) = \$\ 46.875\ (G_2 \rightarrow L_{2,3})$$
$$\text{Total} = \$328.125$$

From this total, $300 would go to $G_{2,1}$ and the rest is $28.125. Note that $765.625 + $656.25 + $28.125 = $1450, which is the same amount the pool's generators are entitled for. The total energy payment (TEP) = $1450. Note that TEP should equal $\sum_i G_i\ \text{LMP}_i$ or (120 × 10 + 10 × 25 = 1450).

The total payments the ISO collects = TCC + TEP = $2700 + $1450 = $4150.

The ISO will pay G_1 and G_2 at LMPs where the generators exist, that is, the ISO will pay $1200 (or 120 ×10) to G_1 and $250 (or 10 ×25) to G_2. The total payment to generators is the same TEP collected by the ISO from the loads for energy. After paying TEP to generators, the ISO still has the TCC (or $2700). From this amount, the ISO will pay its award to FTR holders, and the extra money will be distributed to TOs based on certain criteria that take into consideration that some TOs have sold FTRs.

APPENDIX F

F.2 CALCULATIONS OF CONGESTION CHARGES USING CONTRIBUTIONS OF LOADS

Figure F.4 shows an example of the acyclic diagram and Figure F.5 gives in detail the contribution of each load to each line flow and each generator.

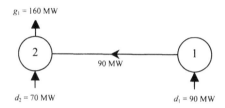

Figure F.4 Acyclic Diagram of the Load Contributions

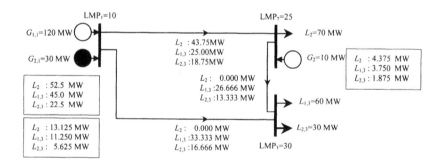

Figure F.5 Tracing Contributions of Loads

The absolute and relative contribution matrices are given by

$$A = \begin{bmatrix} 90.0 & 90.0 \\ 0.00 & 70.0 \end{bmatrix}, \quad R = \begin{bmatrix} 1.0000 & 0.5625 \\ 0.0000 & 0.4375 \end{bmatrix}$$

The charges for congestion are as follows:

- L_2: Pays 52.500 x (25-10) = $787.500 ($G_{1,1} \rightarrow L_2$)

$$13.125 \times (25-10) = \$196.875 \ (G_{2,1} \rightarrow L_2)$$
$$4.375 \times (25-25) = \$ \ \ 0.000 \ (G_2 \rightarrow L_2)$$
$$\text{Total} = \$984.375$$

- $L_{1,3}$: Pays $45.000 \times (30-10) = \$900.000 \ (G_{1,1} \rightarrow L_{1,3})$
$$11.250 \times (30-10) = \$225.000 \ (G_{2,1} \rightarrow L_{1,3})$$
$$3.750 \times (30-25) = \$ \ 18.750 \ (G_2 \rightarrow L_{1,3})$$
$$\text{Total} = \$1143.750$$

- $L_{2,3}$: Pays $22.500 \times (30-10) = \$450.000 \ (G_{1,1} \rightarrow L_{2,3})$
$$5.625 \times (30-10) = \$112.500 \ (G_{2,1} \rightarrow L_{2,3})$$
$$1.875 \times (30-25) = \$ \ \ 9.375 \ (G_2 \rightarrow L_{2,3})$$
$$\text{Total} = \$571.875$$

The total Congestion Charges (TCC) = $2700.00. Note that the TCC obtained using this alternative is the same as the TCC we obtained using the first alternative. If we use (10.28a) for the calculation of congestion charges:

L_2 : $43.75 \ (25-10) + 0 \ (30-10) + 0 \ (30-25) = 1125 + 500 + 150 = \$2075$

$L_{1,3}$: $25 \ (25-10) + 10 \ (30-10) + 7.5 \ (30-25) = 375 + 200 + 37.5 = \$612.5$

$L_{2,3}$: $0 \ (25-10) + 0 \ (30-10) + 2.5 \ (30-25) = 0 \ \ + \ \ 0 \ + \ 12.5 = \$12.5$

$$\text{Total} = \$2700$$

The congestion charges for individual lines are as follows:

Line 1-2: $43.75 \times (25-10) + 37.5 \times (25-10) + 18.75 \times (25-10)$
$= f_{1-2} \ (\text{LMP}_2 - \text{LMP}_1) = 100 \times (25-10) = \$1500$

Line 1-3: $0 \times (30-10) + 33.333 \times (30-10) + 16.666 \times (30-10)$
$= f_{1-3} \ (\text{LMP}_3 - \text{LMP}_1) = \ 50 \times (30-10) = \$1000$

Line 2-3: $0 \times (30-25) + 26.666 \times (30-25) + 13.333 \times (30-25)$
$= f_{2-3} \ (\text{LMP}_3 - \text{LMP}_2) = \ 40 \times (30-25) = \$200$

$$\text{Total} = \$2700$$

References

[Abd96] Abdul-Rahman, K., Shahidehpour, M., Agangic, M., and Mokhtari, S., "A Practical Resource Scheduling With OPF Constraints," IEEE Transaction on Power Systems, Vol. 11, No.1, pp.254-259, February 1996.

[Alo99a] Alomoush, M., and Shahidehpour, M., "Fixed Transmission Rights for Inter-Zonal and Intra-Zonal Congestion Management," IEE Proceedings - Generation Transmission and Distribution, Vol. 146, No. 5, pp. 465-477, September 1999.

[Alo99b] Alomoush, M., and Shahidehpour, M., "Impact of Wheeling Transactions on Zonal Congestion with FTR," Proceedings of the 1999 Large Engineering Systems Conference on Power Engineering, Nova Scotia, Canada, pp. 231-236, June 1999.

[Alo00a] Alomoush, M., and Shahidehpour, M., "Generalized Model for Fixed Transmission Rights Auction." Electric Power System Research, Vol. 54, No. 3, pp. 207-220, June 2000.

[Alo00b] Alomoush M., and Shahidehpour M., "Contingency Constrained Congestion Management with a Minimum Number of Adjustments in Preferred Schedules," International Journal of Electric Power and Energy Systems, Vol. 22, No. 4, pp. 277-290, May 2000.

[Ati97] Atiya, A., Talaat, N., and Shaheen, S., "An Efficient Stock Market Forecasting Model Using Neural Networks," Proceedings of International Conference on Neural Networks, Vol. 4, pp. 2112-2115, 1997.

[Aub82] Aubin, J., Mathematical Methods of Game and Economic Theory, North-Holland, Revised edition, 1982.

[Bab92] Baba, N., and Kozaki, M., "An Intelligent Forecasting System of Stock Price Using Neural Networks," Proceedings of International Joint Conference on Neural Networks (IJCNN'92), Vol. 1, pp. 371-377, 1992.

[Bac92] Bacha, H., and Meyer, W., "A Neural Network Architecture for Load Forecasting," Proceedings of International Joint Conference on Neural Networks (IJCNN'92), Vol. 2, pp. 442-447, 1992.

[Bak95] Bakirtzis, A., Petridis, V, Klartzis, S., Alexiadis, M., and Maissis, A., "A Neural Network Short Term Load Forecasting Model for the Greek Power System," presented at IEEE/PES SM-1995, Portland, OR, pp. 1-7, 95 SM 544-7 PWRS, 1995.

[Bar99] Baran, M., Banunarayanan, V., and Garren, K., "A Transaction Assessment Method for Allocation of Transmission Services," IEEE Transactions on Power Systems, Vol. 14, No. 3, pp. 920–928, August 1999.

[Bas99] Bastian, J., Zhu, J., Banunarayanan, V., and Mukerji, R., "Forecasting Energy Prices in a Competitive Market," IEEE Computer Application in Power, Vol. 12, No. 3, pp 40-45, July 1999.

[Bau93] Baumann, T., and Germond, A., "Application of the Kohonen Network to Short Term Load Forecasting," Proceedings of the Second International Forum on Application of Neural Networks to Power Systems (ANNPS'93), pp. 407-412, 1993.

[Bha00] Bhattacharya, K., "Strategic Bidding and Generation Scheduling in Electricity Spot-Markets", Proceeding of the International Conference on Electric Utility Deregulation and Restructuring and Power Technologies 2000, City University, London, 4-7 April 2000.

[Bia96] Bialek, J., "Tracing the Flow of Electricity," IEE Proceedings - Generation, Transmission and Distribution, Vol. 143, No. 4, pp. 313-320, July 1996.

[Bia97a] Bialek, J., "Elimination of Merchandise Surplus due to Spot Pricing of Electricity," IEE Proceedings - Generation, Transmission and Distribution, Vol. 144, No. 5, pp. 399-405, September 1997.

REFERENCES

[Bia97b] Bialek, J., "Topological Generation and Load Distribution Factors for Supplement Charge Allocation in Transmission Open Access," IEEE Transactions on Power Systems, Vol. 12, No. 3, pp. 1185-1193, August 1997.

[Bia98] Bialek, J., "Allocation of Transmission Supplementary Charge to Real and Reactive Loads," IEEE Transactions on Power Systems, Vol. 13, No. 3, pp. 749 -754, August 1998.

[Bjo00] Bjorgan, R., Song, H., Liu, C., and Dahlgren, R., "Pricing Flexible Electricity Contracts," IEEE Transaction on Power System, Vol. 15, No. 2, pp. 477-482, May 2000.

[Cai01] Website http://www.caiso.com: Sponsored by the California Independent System Operator, 2001.

[Cal00a] California Power Exchange, "California's New Electricity Market, Great Expectations: What Happens When New Markets Open?" http://www.calpx.com/news/publications/index.htm, 2000.

[Cal00b] Website http://www.calpx.com: Sponsored by the California Power Exchange, 2000.

[Cam01] Caminus Corporation, "WeatherDelta: A Weather Based Approach to Contract Pricing & Risk Management," http://www.caminus.com/news/pdf/WeatherDelta.pdf, 2001.

[Car82] Caramanis, M., Bohn, R., and Schweppe, F., "Optimal Spot Pricing: Practice and Theory," IEEE Transactions on Power Apparatus and Systems, Vol. PAS-101, No. 9, pp. 3234-3245, August 1982.

[Cas01] Castillo, E., Conejo, A., Pedregal, P., Garcia, R., and Alguacil, N., Building and Solving Mathematical Programming Models in Engineering and Science, Wiley inter-science, 2001.

[Cha93] Chaudary, S., Kalra, P., Srivastava, S., and Vinod, K., "Short Term Load Forecasting using Artificial Neural Networks," presented at ESAP'93, pp. 159-163, 1993.

[Cha95] Chattopadhyay, D., "An Energy Brokerage System with Emission Trading and Allocation of Cost Saving," IEEE Transactions on Power Systems, Vol. 10, No. 4, pp. 1939-1945, November 1992.

REFERENCES

[Che01] Chen, H., Canizares, C., and Singh, A., "ANN-Based Short-Term Load Forecasting in Electricity Markets", <u>IEEE/PES WM-2001</u>, Columbus, Ohio, January 2001.

[Coh84] Cohen, G., and Zhu, D., "Decomposition Coordination Methods in Large Scale Optimization Problems: The Non-Differentiable Case and The Use of Augmented Lagrangian," <u>Advances in Large Scale Systems, Theory and Application</u>, Vol. 1, pp. 203-266, JAI Press Inc., 1984

[Com98] Value at Risk for Energy, http://www.commodities-now.com/online/dec98/var.html.

[Deb01a] Deb, R., Albert, R., Hsue, L., and Brown, N., "How to Incorporate Volatility and Risk in Electricity Price Forecasting", http://www.energyonline.com/eolm/lcg_volatility.pdf, 2001.

[Deb01b] Deb, R., "Rethinking Asset Values in a Competitive Environment," http://www.energyonline.com/eolm/lcg_asset.pdf, 2001.

[Des88] Debs, A., <u>Modern Power Systems Control and Operation</u>, Kluwer Academic Publishers, 1988.

[Dju93] Djukanovic, M., Babic, B., Sobajic, D., and Pao, Y., "Unsupervised/supervised Learning Concept for 24-hours Load Forecasting," <u>IEE Proceedings-Generation, Transmission and Distribution</u>, Vol. 140, No. 4, pp. 311-318, 1993.

[Eht89] Ehtamo, H., Ruusunen, J., and Hamalainen, R., "A Hierarchical Approach to Bargaining in Power Pool Management," <u>IEEE Transactions on Automatic Control</u>, Vol. 34, No. 6, pp. 666-669, June 1989.

[Els93] El-Sharkawi, M.A., Marks II, R.J., Oh, S., and Brace, C.M., "Data Partitioning for Training a Layered Perceptron to Forecast Electric Load," <u>Proceedings of the Second International Forum on Application of Neural Networks to Power Systems (ANNPS'93)</u>, pp. 66-68, 1993.

[Ene98] "Why do Electricity Prices Spike," www.energyera.com/epasub/journal.htm, April 1998.

[Erk97] Erkman, I., and Ozdogan, A., "Short Term Load Forecasting using Genetically Optimized Neural Network Cascaded with a Modified Kohonen Clustering Process", <u>Proceedings of 12th</u>

REFERENCES

IEEE International Symposium on Intelligent Control, pp 107-112, 1997.

[Fan99a] Fang, R., and David, A., "Optimal Dispatch Under Transmission Contracts," IEEE Transactions on Power Systems, Vol. 14, No. 2, pp. 732-737, May 1999.

[Fan99b] Fang, R., and David, A., "Transmission Congestion Management in an Electricity Market," IEEE Transactions on Power Systems, Vol. 14, No. 3, pp. 877-883, August 1999.

[Fer96a] Ferrero, R., Rivera, J., Shahidehpour, M., and Ramesh, V., "Effect of deregulation on Hydrothermal Systems with Transmission Constraints," Electric Power Systems Research, Vol. 39, No. 3, pp. 191-197, September 1996.

[Fer98] Ferrero, R., Rivera, J., and Shahidehpour, M., "Application of Games with Incomplete Information for Pricing Electricity in Restructured Power Pools," IEEE Transactions on Power Systems, Vol. 13, No. 1, pp. 184-189, February 1998.

[Fte02] Website http://www.ftenergyusa.com/mwdaily/mdguide.asp

[Fud91] Fudenberg, D., and Tirole, J., Game Theory, Cambridge, Massachusetts: The MIT Press, 1991.

[Gro87] Gross, G., and Galiana, F., "Short-Term Load Forecasting," Proceedings of IEEE, Vol. 75, No. 12, December 1987.

[Gua97] Guan, X., Ni, E., Li, R., Luh, P., "An optimization-based algorithm for scheduling hydrothermal power systems with cascaded reservoirs and discrete hydro constraints," IEEE Transactions on Power Systems, Vol. 12, No. 4, pp. 1775–1780, Nov. 1997.

[Hao97] Hao, S., and Papalexopoulos, A., "Reactive Power Pricing and Management," IEEE Transactions on Power Systems, Vol. 12, No. 1, pp. 95-104, February 1997.

[Hap94] Happ, H., "Cost of Wheeling Methodologies," IEEE Transactions on Power Systems, Vol. 9, No. 1, pp. 147-156, February 1994.

[Hau92] Haurie, A., Loulou, R., and Savard, G., "A Two-player Game Model of Power Cogeneration in New England," IEEE Transactions On Automatic Control, Vol. 37, No. 9, pp. 1451-1456, September 1992.

[Hir01] Hirst, E., "Maximizing Generator Profits Across Energy and Ancillary-Services Markets," http://www.ehirst.com/PDF/asearningsmodel.pdf, 2001.

[Hob92] Hobbs, B., "Using Game Theory to Analyze Electric Transmission Pricing Policies in the United States," European Journal of Operational Research, Vol. 56, pp. 154-171, 1992.

[Hob00] Hobbs, B., Meier, P., Energy Decisions and the Environment: A Guide to the Use of Multicriteria Methods, Kluwer Publishers, July 2000.

[Hob01] Hobbs, B., Rothkopf, M., O'Neill, R., Chao, H., The Next Generation of Electric Power Unit Commitment Models, Kluwer Publishers, April 2001.

[Hul97] Hull, J., Options, Futures, and Other Derivatives, 3rd Edition, Prentice Hall, NJ, 1997.

[Ili97] Ilic, M., et al., "Toward Regional Transmission Provision and its Pricing in New England," Utility Policy, Vol. 6, No. 3, pp. 245-256, 1997.

[Jam69] Buchanan, J., Cost and Choice, Markham Publishing, pp. 42-43, 1969.

[Kam01] Kamat, R., and Oren, S., "Rational Buyer Meets Rational Seller: Reserve Market Equilibria under Alternative Auction Designs", February 2001.

[Kel01] Kelman, R., Barroso, L., and Pereira, M., "Market Power Assessment and Mitigation in Hydrothermal Systems," IEEE Transactions on Power Systems, Vol. 16, No. 3, pp. 354 –359, August 2001.

[Ker93] Kermanshahi, B., Poskar, C., Swift, G., Mclaren, P., Pedrycz W., Buhr, W., and Silk, A., "Artificial Neural Network for Forecasting Daily Loads of a Canadian Electric Utility," Proceedings of the Second International Forum on Application of Neural Networks to Power Systems (ANNPS'93), pp. 302-307, 1993.

[Kha92] Khadem, M., and Lago, A., "Short Term Load Forecasting Using Neural Networks," presented at Summer Workshop on Neural Network Computing for Electric Power Industry, Stanford, pp. 1-6, 1992.

REFERENCES

[Kha93] Khadem, M., "Application of Kohonen Neural Network Classifier to Short Term Load Forecasting," presented at IEEE/PES WM-1993, Columbus, Ohio, 1993.

[Kho95] Khotanzad, A., Hwang, R., Abaye, A., and Maratukulam, D., "An Adaptive Modular Artificial Neural Network Hourly Load Forecaster and Its Implementation at Electric Utilities," IEEE Transactions on Power System, Vol. 10, No. 3, pp. 1716-1722, August 1995.

[Kho97] Khotanzad, A., Rohani, R., Lu, T., Abaye, A., Davis, M., and Maratukulam, D. J., "ANNSTLF - A Neural-Network-Based Electric Load Forecasting System," IEEE Transactions on Neural Networks, Vol. 8, No. 4, pp. 835-846, July 1997.

[Kho98] Khotanzad, A., Rohani, R., and Maratukulam, D., "ANNSTLF - Artificial Neural Network Short-Term Load Forecaster - Generation Three," IEEE Transaction on Power System, Vol. 13, No. 4, pp. 1413-1422, November 1998.

[Kim95] Kim, K., Park, J., Hwang, K., and Kim, S., "Implementation of Hybrid Short-term Load Forecasting System Using Artificial Neural Networks and Fuzzy Expert Systems," IEEE Transaction on Power System, Vol. 10, No. 3, pp. 1535-1539, November 1995.

[Kir96] Kirby, B., and Hirst, E., "Unbundling Electricity: Ancillary Services," IEEE Power Engineering Review, Vol. 16, No. 6, pp. 5-6, June 1996.

[Kir97] Kirschen, D., Allan, R., and Strbac, G, "Contribution of Individual Generators to Loads and Flows," IEEE Transactions on Power Systems, Vol. 12, No. 1, pp. 52-60, February 1997.

[Kle97] Klein, J., "Interim Staff Report - Market Clearing Price Forecast, For the California Energy Market: Forecast Methodology and Analytical Issues," December 10, 1997. http://www.energy.ca.gov/electricity/97-12-10_MCP-FORECAST.pdf.

[Kor98] Koreneff, G., Seppala, A., Lehtonen, M., Kekkonen, V., Laitinen, E., Hakli, J., and Antila, E., "Electricity Spot Price Forecasting as a Part of Energy Management in Restructured Power Market," Proceedings of International Conference on Energy Management and Power Delivery (EMPD'98), Vol. 1, pp. 223-228, 1998.

[Kov94] Kovacs, R., and Leverett, A., "A Load Flow Based Method for Calculating Embedded, Incremental and Marginal Cost of Transmission Capacity," IEEE Transactions on Power Systems, Vol. 9, No. 1, pp. 272-278, February 1994.

[Lim96] Lima, J., "Allocation of Transmission Fixed Charges: an Overview," IEEE Transactions on Power Systems, Vol. 11, No. 3, pp. 1409-1418, August 1996.

[Liu00] Liu, Y., Alaywan, Z., Rothleder, M., Liu, S., and Assadian, M., "A Rational Buyer's Algorithm Used for Ancillary Service Procurement," IEEE Power Engineering Society Winter Meeting, Vol. 2, pp. 855-860, 2000.

[Lut57] Lute, D., and Raiffa, H., Games and Decisions - Introduction and Critical Survey, John Wiley & Sons, Inc. 1957.

[Ma98] Ma, H., and Shahidehpour, M., "Transmission Constrained Unit Commitment Based on Benders Decomposition," Electric Power and Energy Systems, Vol. 20, No. 4, pp. 287-294, April 1998.

[Mae92] Maeda, A., and Kaya, Y., "Game Theory Approach to Use of Non-Commercial Power Plants Under Time-of-Use Pricing," IEEE Transactions on Power Systems, Vol. 7, No. 3, pp. 1052-1059, August 1992.

[Mal94] Malliaris, M., "Modeling the Behavior of the S&P 500 Index: A neural Network Approach," Proceedings of the Tenth Conference on Artificial Intelligence for Applications, pp. 86-90, 1994.

[Mat99a] MathWorks Inc., Getting Started with MATLAB, 1999.

[Mat99b] MathWorks Inc., Neural Networks Toolbox User's Guide, 1999.

[Min00] Mina, J., and Xiao, J., "Return to Risk Metrics: The Evolution of a Standard," http://riskmetrics.com/research/techdoc/rmfinal.pdf, p.66.

[Moh95] Mohammed, O., et.al., "Practical Experience with an Adaptive Neural Network Networks Short Term Load Forecasting Systems," IEEE Transaction on Power System, Vol.10, No. 1, pp 254-265, 1995.

[Mor94] Morris, P., Introduction to Game Theory, New York: Spriger-Verlag, 1994.

REFERENCES

[Ner01] Website http://www.nerc.com: Sponsored by North American Electric Reliability Council, 2001.

[Neu47] von Neumann, J., and Morgenstern, O., The Theory of Games and Economic Behavior, 1947.

[Ng81] Ng, W., "Generalized Generation Distribution Factors for Power System Security Evaluations" IEEE Transactions on Power Apparatus and Systems, Vol. PAS-100, No. 3, pp. 1001-1005, March 1981.

[Ore01] Oren, S., "Design of Ancillary Service Markets", Proceedings of the 34th Hawaii International Conference on System Sciences, pp. 769-777, 2001.

[Orn01] Website http://www.ornl.gov: Sponsored by Oak Ridge National Laboratory, 2001.

[Pan00] Pan, J., Teklu, Y., Rahman, S., and Jun, K., "Review of Usage-Based Transmission Cost Allocation Methods under Open Access," IEEE Transactions on Power Systems, Vol. 15, No. 4, pp. 1218-1224, November 2000.

[Par98] Park, Y., Park, J., Lim, J., and Won, J., "An Analytical Approach for Transmission Costs Allocation in Transmission System," IEEE Transactions on Power Systems, Vol. 13, No. 4, pp. 1407-1412, November 1998.

[Pjm01] Website http://www.pjm.com: Sponsored by Pennsylvania-New Jersey-Maryland Interconnection, 2001.

[Qiu87] Qiu, J., and Shahidehpour, M., "A New Approach for Minimizing Power Losses and Improving Voltage Profile," IEEE Transactions on Power Systems, Vol. PWRS-2, No. 2, pp. 287-295, May 1987.

[Rah93] Rahman, S., Drezga I., and Rajagopalan, J., "Knowledge Enhanced Connectionist Models for Short Term Load Forecasting," Proceedings of the Second International Forum on Application of Neural Networks to Power Systems (ANNPS'93), 1993

[Rts96] Reliability Test System (RTS) Load Data, Version 1996, http://www.ee.washington.edu/research/pstca/rts/pg_tcarts.htm.

[Rud95] Rudnick, H., Palma, R., and Fernandez, J., "Marginal Pricing and Supplement Cost Allocation in Transmission Open Access,"

IEEE Transactions on Power Systems, Vol. 10, No. 2, pp. 1125-1132, May 1995.

[Rud97] Rudnick, H., Varela, R., and Hogan, W., "Evaluation of Alternatives for Power System Coordination and Pooling in a Competitive Environment," IEEE Transactions on Power Systems, Vol. 12, No. 2, pp. 605-613, May 1997.

[Ruu91] Ruusunen, J., Ehtamo, H., and Hamalainen, R., "Dynamic Cooperative Electricity Exchange in a Power Pool," IEEE Transactions on System, Man and Cybernetics, Vol. 21, No. 4, pp. 758-766, July/August 1991.

[Sch88] Schweppe, Bohn, F., R., Tabors, R. and Caramanis, M., Spot Pricing of Electricity, Kluwer Academic Publishers, 1988.

[Sha94] Shahidehpour, M., and Wang, C., Applications of Artificial Neural Networks in Multi-Area Generation Scheduling with Fuzzy Data, EPRI Publication TR-104219, July 1994.

[Sha98] Shahidehpour, M., and Ferrero, R., "Fuzzy Systems Approach to Short-term Power Purchases considering Uncertain Prices," chapter in Electric Power Applications of Fuzzy Systems, edited by M. El-Hawary, 1998.

[Sha99a] Shahidehpour, M., and Marwali, M., "Power Transmission Network," chapter in Wiley Encyclopedia of Electrical and Electronics Engineers, 1999.

[Sha99b] Shahidehpour, M., and Ferrero, M., "Electricity Supply Industry," chapter in Wiley Encyclopedia of Electrical and Electronics Engineers, 1999.

[Sha99c] Shahidehpour, M., and Yamin, H., "Risk Management using Game Theory in Transmission Constrained Unit Commitment within a Deregulated Power Market," chapter in IEEE/PES Tutorial on Applications of Gaming Methods to Power System Operation, 1999.

[Sha99d] Shahidehpour, M., and Alomoush, M., "Decision Making in a Deregulated Power Environment based on Fuzzy Sets," Chapter in Modern Optimization Techniques in Power Systems, edited by Y.H. Song, 1999.

[Sha00] Shahidehpour, M., and Marwali, M., Maintenance Scheduling in Restructured Power Systems, Kluwer Academic Publishers, May 2000.

REFERENCES

[Sha01] Shahidehpour, M., and Alomoush, M., <u>Restructured Electric Power Systems</u>, Marcel Dekker Publishers, May 2001.

[Shi89] Shirmohammadi D., et al., "Evaluation of Transmission Network Capacity Use for Wheeling Transactions," <u>IEEE Transactions on Power Systems</u>, Vol. 4, No. 4, pp. 1405-1413, October 1989.

[Shi91] Shirmohammadi, D. et al., "Cost of Transmission Transactions: An Introduction," <u>IEEE Transactions on Power Systems</u>, Vol. 6, No. 4, pp. 1546–1560, November 1991.

[Shi96] Shirmohammadi, D., Filho, X., Gorenstin, B., and Pereira, M., "Some Fundamental Technical Concepts about Cost Based Transmission Pricing," <u>IEEE Transactions on Power Systems</u>, Vol.11, No. 2, pp. 1002-1008, May 1996.

[Sil98] Silva, E., Mesa, S., and Morozowski, M. "Transmission Access Pricing to Wheeling Transactions: A Reliability Based Approach," <u>IEEE Transactions on Power Systems</u>, Vol. 13, No. 4, pp. 1451–1486, November 1998.

[Sin98] Singh, H., Hao, S., and Paralexopoulos, A., "Transmission Congestion Management in Competitive Electricity Markets," <u>IEEE Transactions on Power Systems</u>, Vol. 13, No. 2, pp. 672–680, May 1998.

[Sin99] Singh, H., <u>Game Theory Applications in Electric Power Market</u>, IEEE Tutorial # 99TP136-0, 1999.

[Ska01] Skantze, P., and Ilic, M., Valuation, <u>Hedging and Speculation in Competitive Electricity Markets: A Fundamental Approach</u>, Kluwer Publishers, October 2001.

[Sny92] Snyder, J., Sweat, J., Richardson, M., and Pattie, D., "Developing Neural Networks to Forecast Agriculture Commodity Prices," <u>Proceedings of the Twenty-Fifth Hawaii International Conference on System Sciences</u>, Vol. 4, pp. 516-522, 1992.

[Sri95] Srinivasan, D., Chang, C., and Liew A., "Demand Forecasting Using Fuzzy Neural Computation with Emphasis on Weekend and Public Holiday Forecasting," <u>IEEE Transaction on Power Systems</u>, Vol. 10, No.4, pp. 1897-1903, November 1995.

[Sto02] Stoft, S., <u>Power System Economics: Designing Markets for Electricity</u>, John Wiley & Sons, 2002.

[Str80] Stremel, J.P., Jenkins, R., Babb, R., and Bayless, W., "Production Costing Using the Cumulant Method of Representing the Equivalent Load Curve," IEEE Transactions on Power Apparatus and Systems, Vol. PAS-99, No. 5 pp. 1947-1956, September/October 1980.

[Str98] Strbac, G., Kirschen, D., and Ahmed, S., "Allocating Transmission System Usage on the Basis of Traceable Contributions of Generators and Loads to Flows," IEEE Transactions on Power Systems, Vol.13, No.2, pp. 527-534, May 1998.

[Szk99] Szkuta, B., Sanabria, L., and Dillon, T., "Electricity Price Short-Term Forecasting Using Artificial Neural Networks," IEEE Transactions on Power System, Vol. 14, No. 3, pp. 851-857, August 1999.

[Tor91] Torres, G., Traore, C., Mandolesi, F., and Mukhedkar, D., "Short Term Load Forecasting using a Fuzzy Engineering Tool," Proceedings of the First International Forum on Application of Neural Networks on Power Systems (ANNPS'91), pp. 36-40, 1991

[Uce01] Website http://www.ucei.berkeley.edu/ucei/datamine/datamine.htm, Sponsored by University of California Energy Institute (UCEI)

[Vaa01] Vaahedi, E., Chang, A., Mokhtari, S., Muller, N., and Irisarri, G., "A Future Application Environment for BC Hydro's EMS," IEEE Transactions on Power Systems, Vol. 16, No. 1, pp. 9–14, February 2001.

[Var97] Varaiya, P., and Wu, F., "MinISO: A Minimal Independent System Operator," Proceedings of the Thirtieth Hawaii International Conference on System Sciences, Vol. 5, pp 602-607, 1997.

[Voj96] Vojdani, A., Imparato, C., Saini, N., Wollenberg, B., Happ, H., "Transmission Access Issues," IEEE Transactions on Power Systems, Vol. 11, No. 1, pp. 41–51, February 1996.

[Wan95] Wang, S., Shahidehpour, M., Kirschen, D., Mokhtari, S. and Irisarri, G., "Short-Term Generation Scheduling With Transmission and Environmental Constraints Using an Augmented Lagrangian Relaxation," IEEE Transactions on Power Systems, Vol.10, No.3, pp. 1294-1301, August 1995.

REFERENCES

[Wan97] Wang, A., and Ramsay, B., "Prediction of System Marginal Price in the UK Power Pool Using Neural Networks," <u>Proceedings of International Conference on Neural Networks</u>, Vol. 4, pp. 2116-2120, 1997.

[Wil96] Willis, L., Finney, J. and Ramon, G., "Computing the Cost of Unbundled Services," <u>IEEE Computer Applications in Power</u>, Vol. 9, No. 4, pp. 16-21, October 1996.

[Woo96] Wood A., and Wollenberg B., <u>Power Generation, Operation and Control, Second Edition</u>, 1996, John Wiley & Sons, Inc.

[Yoo91] Yoon, Y., and Swales, G., "Predicting Stock Price Performance: A Neural Network Approach," <u>Proceedings of the Twenty-Fourth Annual Hawaii International Conference on System Sciences</u>, Vol. 4, pp. 156-162, 1991.

[Zob97] Zobian, A., and Ilic, M., "Unbundling of Transmission and Ancillary Services - Part I: Technical Issues and Part II: Cost-based Pricing Framework," <u>IEEE Transactions on Power Systems</u>, Vol. 12, No. 2, pp. 539–558, May 1997.

Index

A

A1, 355. *See* Control area, performance criterion, A1
A2, 355. *See* Control area, performance criterion, A2
Absolute percentage error, 42, 78. *See* APE
ac, 61, 66, 143, 166, 370, 376, 379, 380, 382, 383, 386-388, 400, 408, 412
Accuracy, 31, 47, 49, 55, 57, 63, 64, 70, 73, 75, 89, 111, 112, 113, 259
ACE, 355, 356. *See* Area control error
  measuring, 356
Acyclic
  diagram, 391, 489, 493
  state graph, 383
Adaptive, 15, 16, 30, 36, 58, 60, 71, 81, 89-91
  forecasting, 81, 89, 91
  learning, 71
  weight, 15, 36
Adequacy, 275, 357
Additional cost, 140, 180
Adjustment
  bid, 13
  factor, 259
A factor, 377, 386. *See* GSDF

Agent, 191
AGC, 12, 13, 19, 23, 312, 313, 316, 354-358, 366, 367. *See* Automatic generation control
  function, 354
  pricing, 358
  response, 356
  signal, 356
  system, 356
  unit revenue adequacy, 357
Aggregator, 5, 6, 8
Alberta, 63
Algorithm, 18, 26, 30, 31,195, 275, 296, 310, 348, 374, 379, 380, 381, 385, 390, 432, 433, 442
ALMP, 405, 422, 424, 428-431. *See* Average locational marginal price
Alternative, 370, 435, 494
  design, 312, 316-318, 320-322, 328, 330, 333, 334, 367
  definition, 78-80
  method, 77, 78, 81, 90, 91
  model, 15, 43-51
  technique, 16, 59
Analysis
  error, 16, 59, 60
  market power, 191
  microeconomic, 192
  performance, 16, 58, 59
  price difference, 106, 240

509

510 INDEX

price spike, 91
probability, 59
profit, 236
sensitivity, 15, 16, 53, 58, 169, 258, 259, 260
time series, 57
transaction, 19, 191, 195, 203
volatility, 16, 58, 59
Ancillary service, 2, 5-14, 18, 19, 116, 117, 150, 154, 164-171, 222, 223, 280, 311-324, 326, 331-342, 345, 346, 348-350, 353-355, 357, 360, 362, 364, 367, 372
   auction, 311, 312, 316, 320, 323, 326, 331, 360, 362, 364
   bid, 10, 165, 223, 315, 318, 323, 342, 345, 350
   definition, 312
   market, 10
   market design, 311, 320
      sequential approach, 10, 312, 334, 367
      simultaneous approach, 11, 312, 336, 354, 367
   non-spinning reserve, 10, 117, 141, 147-150, 154, 159, 160, 167, 169, 311, 313, 316, 335, 336, 338-342, 345-348, 351, 456, 459, 460
   regulation, 10, 13, 315, 316, 317, 335-337, 339-352, 355-357
   replacement reserve, 10, 316, 317, 335, 336, 339-341, 346, 348, 349, 352, 357
   requirement, 312
   scheduling, 318, 319
   spinning reserve, 10, 66, 115, 117-119, 141, 147-149, 154, 158, 159, 169, 173, 174, 223, 227-229, 231, 275, 277, 311, 313, 316, 335-352, 456, 457, 459, 460
   substitutability, 11, 335, 339
ANN, 15-17, 21, 25-27, 29-34, 36-38, 40, 42, 43, 45, 53-55, 57-60, 69-77, 81, 82, 84, 86, 88-91, 95, 105, 108-111, 259. *See* Artificial Neural Network
   application, 29
   architecture, 15, 26, 27, 29, 33
   feedback, 27
   feedforward, 27
      multi-layered, 30
      three-layer, 35
   input variable, 15, 32, 33
   recurrent, 30
   seasonal, 34
APE, 42, 45, 46, 47, 48, 49, 50, 51, 78, 79. *See* Absolute percentage error
Approximation, 195, 287, 377
   Gram-Charlier, 306
   linear, 143, 287, 296
   successive, 195
Arbitrage, 2, 7, 14, 18, 161-167, 169-171, 173, 174, 176, 177, 179-189, 259, 268
   application, 165
   bilateral contract, 166
   cross-commodity, 161, 163, 164, 166, 167
   emission allowance, 166, 182
   energy and ancillary service, 166
   example, 166
   gas, 166
   opportunity, 14, 18, 116, 161-166, 188, 189
   position, 163
   same-commodity, 161, 163
   spatial, 163
   steam, 166
   temporal, 163
   usefulness, 162
Arbitrageur, 162
Architecture, 5, 15, 26, 27, 33, 35, 43, 44, 50-53
Area control error, 355. *See* ACE
ARMA, 24, 25. *See* Autoregressive moving average
Artificial Neural Network, 15. *See* ANN
Asset

INDEX

market, 162
price, 162
valuation, 15, 19, 112, 117, 233, 235, 236
ATC, 421. *See* Available transfer capability
Auction, 6, 11, 12, 14, 19, 246, 312, 315-318, 320, 322, 323, 331-339, 343-348, 350, 351, 353, 354, 358, 360, 362, 367, 412-417, 420, 454
  10-minute, 358
  ancillary service, 311, 312, 316, 320, 323, 326, 331, 360, 362, 364
  centralized, 11, 412
  example
    10-bid, 326, 331
    4-bid, 323
  FTR, 412
  marginal pricing, 347, 348, 350, 351, 353, 354, 367
  primary, 11
  rational buyer, 335, 336, 338, 340, 346-354, 367
  secondary, 12
  sequential, 312, 317, 320, 323, 334, 335, 354
Auctioneer, 11, 412, 413, 415, 416
Augmentation, 18, 141, 276, 288, 289, 462, 463
Automatic generation control, 12, 23, 312. *See* AGC
Autoregressive moving average, 24. *See* ARMA
Auxiliary principle, 143
Availability, 8, 13, 18, 38, 75, 215, 216, 233, 234, 267, 268, 271, 304, 315, 412, 454
Available transfer capability, 421. *See* ATC
Average locational marginal price, 405. *See* ALMP

B

Back propagation, 31, 32, 36
Backup supply, 314

Balancing, 12, 13, 22, 312, 317, 318
  market, 12
Bargaining, 192, 195, 202-204
  game, 203
Base Load Forecast, 44. *See* BLF
Benders
  cut, 18, 276, 287, 296, 298, 300-302, 304
    formulation, 296
  decomposition, 18, 275, 276, 285-287, 303, 310
    application, 287
  formulation
    standard form, 286
Bialek, 380, 381, 383, 386, 389, 390, 392, 393
Bid
  adjustment, 13
  ancillary service, 10, 165, 223, 315, 318, 323, 342, 345, 350
  capacity, 10, 316-318, 320, 321, 326
    reservation, 316, 317, 328, 334, 357
  curve
    slope, 201
  decremental, 13, 422
  energy, 10, 12, 146, 147, 236, 316-318, 320, 322, 324, 326, 327, 334, 357, 359, 367
  incremental, 13
  purchase, 247
  sale, 247
  format
    two-part, 357
Bidding
  curve, 242
  information, 215
  pattern, 16, 58, 59, 67, 70
  price, 146, 147, 233, 236, 238, 242, 248, 258-267, 274
    coefficient, 258, 260
  quantity, 145, 146, 247, 248, 258
  slope, 146
  strategy, 14, 17, 22, 63-66, 70, 116, 145-147, 231

512 INDEX

  based on PBUC, 145
  price, 258, 339, 346, 349, 351
  quantity, 258
  simulation, 258
Bilateral
  agreement, 197
  contract, 4, 7, 18, 63, 64, 117,
    145, 146, 165, 171-174, 203,
    227, 236, 246-249, 258, 259,
    262, 263, 267, 268, 312, 354,
    357, 366, 367, 371, 413, 416,
    432, 434-439, 443-447, 450,
    454-456, 477, 489
Black start capability, 314
Blackout, 314
BLF, 44. *See* Base Load Forecast
Block diagram, 33
Broker, 5, 8, 453
Brokerage, 195
Bulk, 7, 9, 311, 314
Bus
  load, 25, 70, 74, 75
  price, 70-73, 139
  price profile, 72
By-product, 454

C

CAISO, 11, 316, 334-336, 338. *See*
  California ISO
California ISO, 6, 316, 334
Capacity
  payment, 13, 318, 320, 321, 322,
    326, 328-331, 333, 334, 357,
    367
  reservation, 319
Case study, 81, 150, 154, 261, 296,
  306, 322
Centralized, 4, 9, 11, 203, 213, 412
CDF, 244, 257. *See* Cumulative
  distribution function
C factor, 378, 379, 386. *See* GLDF
CFD, 2. *See* Contracts for
  Differences
c-game, 217, 219, 486, 487. *See*
  Complete game
Change Load Forecast, 44. *See* CLF

Characteristics function, 214, 484,
  485
Charge
  allocation, 20, 371
  loss, 434, 441, 451, 452
  wheeling, 66
Chronological simulation, 65
CLF, 44. *See* Change Load Forecast
Coalition, 191, 192, 197, 198, 212,
  213, 485, 486
  counter-, 212
  grand, 191, 196, 198, 212-214,
    231, 486
COB, 63
Cogeneration, 186, 187, 195
Collective rationality, 485
Collusion, 192, 194, 197, 213
Comment, 50, 88, 386
Commodity, 11, 18, 57, 161, 162,
  163, 164, 166
Common, 383-385, 391, 433, 439,
  440, 447-451, 486
  root, 383, 384
Competition, 1, 2, 4, 5, 6, 8-11, 13,
  19, 116, 162, 191, 194, 200, 202,
  213, 215, 230, 231, 234, 311, 315,
  334, 335, 369, 371, 434, 435, 453
  imperfect, 213, 259
  perfect, 202, 213, 259, 261
Complete game, 217, 225, 232, 486.
  See c-game
Compliance rating, 356
Composite reliability evaluation,
  303
Conclusion, 160, 188, 230, 273, 310,
  367, 453
Conditional probability, 112, 113,
  217, 219, 226, 228
  distribution, 112
Confidence level, 234, 237, 245,
  246, 272
Congestion, 2, 7, 9, 11, 12, 60, 61,
  63, 70-72, 86, 88, 116, 139, 150,
  154, 231, 319, 323, 326, 327,
  331-334, 359-362, 364, 366, 367,
  369-371, 393, 394, 399, 400, 408-

INDEX

411, 413, 421, 422, 424, 425, 428-432, 434, 435, 439, 441, 445, 447, 451-454, 489-491, 493, 494
  charge, 11, 61, 370, 371, 394, 408-410, 422, 425, 429-431, 434, 441, 452, 453, 454, 491, 494
    calculation, 489
  cost, 413, 432
  credit, 11, 409, 411, 422, 431, 454
  index, 71
  management, 19, 369. *See also* Transmission, congestion, management
    inter-zonal, 406, 422, 425, 427, 431
    intra-zonal, 422, 431
    zonal, 421
  minimization, 285
Constraint
  alleviation, 5
  contingency, 290, 292
  coupling, 119
  energy and reserve, 124, 141
  EUE, 287
  relaxed, 121
  steady state, 290, 292
  transmission flow, 287
  voltage, 287
Contingency, 22, 70, 211, 275, 290, 292, 294, 295, 298-302, 310, 311, 315, 376, 414, 456, 458
  constraint, 290, 292
  n-1, 276
  subproblem, 295
Contract path, 373
  method, 370, 373
Contract protection mode, 434
Contracts for Differences, 2. *See* CFD
Contribution, 371, 372, 374, 376, 378-384, 387, 388, 392, 401, 431-434, 439-441, 447, 449-454, 457, 489, 490, 493
  matrix

absolute, 385, 449
relative, 385, 449
Control area, 19, 57, 311-314, 354-357, 366, 367, 421, 430
  performance criterion
    A1, 355
    A2, 355
Control Performance Standard 1, 355. *See* CPS1
Control Performance Standard 2, 355. *See* CPS2
Convergence, 32, 128, 141, 143, 288, 289
  criterion, 128
  LR, 288
Convex, 143, 289
Convexity, 288, 462
Cooperation, 194, 197, 198, 208, 212, 213
Cooperative game, 195, 196, 212, 213, 231
Coordinated multilateral trade model, 6
Coordination, 5, 196, 197, 203, 231, 313, 453, 485
Cost
  allocation
    method, 386, 392
    usage-based, 372
  curve, 152, 179, 199, 200, 207, 242
  increment, 198, 199
  marginal, 4, 61, 64, 145, 146, 193-195, 198, 199, 201, 202, 205-207, 210, 212, 214, 217, 220-222, 227, 243, 248, 258, 262, 264, 267, 271, 305, 337, 338, 348, 352, 393, 394
  minimization, 115
  procurement, 11, 336-338, 344-348, 350, 353, 354, 367
  production, 117, 195, 204, 207, 210, 215, 237, 275, 277, 288, 298, 300-303
  running, 372
  social, 11, 336, 337, 347-353

Cost-based unit commitment, 115, 116
Counter-coalition, 196, 212-214, 484, 485
Counter-flow, 376, 452
  method, 376
Counterparty risk, 63
Coupling constraint, 119
CPS1, 355. *See* Control Performance Standard 1
CPS2, 355. *See* Control Performance Standard 2
CPU time, 296, 302
Credit, 376, 393, 409, 411, 412, 421, 428, 429, 431, 442, 453
Cross-commodity arbitrage, *see* Arbitrage, cross-commodity
Cumulant method, 305
Cumulative distribution function, 244. *See* CDF
Curtailment, 20, 84, 371, 431-439, 444, 447, 454
  weight, 445
Curve fitting, 90, 91
Customer, 9

D

Data pre-processing, 16, 58, 60, 81-83, 90, 91. *See also* Pre-processing
Day-ahead, 9, 12, 17, 115, 275, 312
  forward market, 12, 312
  schedule, 12, 17, 115, 275, 312
dc, 290, 291, 303, 373, 374, 377, 379, 380, 382, 383, 386-388, 418, 425
Decentralized, 203, 207
Decision
  making, 189, 202, 367
  variable, 143, 195, 289
Decomposability, 289
Decomposition, 1, 18, 119, 127, 143, 275, 276, 285, 286, 288, 289, 294, 376, 379
  Benders, 18, 275, 276, 285-287, 303, 310

Decremental bid, 422
Defuzzified, 31
Delivery factor, 397. *See* DF
Depreciation, 189
Deregulation, 22, 230
Derivative, 121, 131
Deviation
  instructed, 358
  uninstructed, 358
  frequency, 312, 354, 356, 366, 367
DF, 397, 399. *See* Delivery factor
D factor, 377, 386. *See* GGDF
Differentiable, 143, 289
Direct access, 9
Directed graph, 383
DISCO, 5-8, 246, 315, 329, 333, 334, 357, 367, 432, 434
Discretization, 177, 180, 342
Dispatch
  economic, *see* Economic, dispatch
  transmission constrained, 65
Disposition, 93, 99, 103, 106, 108
Distributed, 1, 409, 451, 452, 485, 492
Distribution factor, 66, 371, 376, 379, 381
  method, 376, 388
  topological, 380, 381
Domain, 383, 384, 433
Dominate, 483
Downstream-looking algorithm, 380
Droop characteristics, 356
  governor, 356
Dual variable, 6, 10, 305, 394
Duality gap, 128, 288
  relative, 128
Dynamic
  programming, 119, 288, 289, 340, 341
    single-unit, 120
  scheduling, 314
  braking resistor, 314

INDEX

E

Economic
  dispatch, 4, 125-128, 133, 165, 173, 174, 212, 213, 290, 397
    security-constrained, 299
  operation, 2, 22
  signal, 5, 453
Efficiency, 7, 20, 164, 169, 233, 235, 335, 369
  market, 200, 215, 230, 231
Efficient allocation, 485
8-bus system, 422
ELDC, 305. *See* Equivalent load duration curve
Electric Reliability Council of Texas, 164. *See* ERCOT
Electricity
  market, 1, 2, 4-7, 10, 12, 19, 23, 112, 164, 369. *See* Market
    hierarchy, 1
    model
      bilateral, 4
      hybrid, 4
      PoolCo, 4
    overview, 371, 453
    operation, 1
    structure, 1
  price, 15-17, 23, 57-61, 63-65, 69-80, 91, 93, 105, 108, 111, 112, 164, 165
    forecasting, 57, 61. *See* Price forecasting
    probability distribution, 59
    volatility, 61
      reason, 62
  pricing, 60
Embedded cost, 372-375, 386
Emission allowance, 18, 182-186
Energy
  balancing, 317
  bid, 10, 12, 146, 147, 236, 316-318, 320, 322, 324, 326, 327, 334, 357, 359, 367
  imbalance, 311, 313
  interchange, 430
  market, 9, 64, 150, 249, 270
  forward, 6, 12, 315, 318, 323, 367
  real-time, 315, 317
  payment, 317, 318, 320, 324, 326-328, 330, 331, 333, 334, 357, 492
  trading, 9, 12, 312
Equality constraint, 402, 415, 435, 436, 462
Equilibrium, 192, 193
  Nash, 193, 194, 203, 217, 221-223, 228, 231, 232, 487
  Stackelberg, 195
Equivalent load duration curve, 305. *See* ELDC
ERCOT, 164. *See* Electric Reliability Council of Texas
Error analysis, 16, 59, 60
EUE, 18, 276, 279, 280, 287, 303-306. *See* Expected unserved energy
  definition, 303
  minimization, 276, 287
Example, 67-68, 80, 92, 96, 98, 109, 111, 113, 118, 164-165, 166-188, 192-194, 207, 212-214, 218-222, 225-230, 240-258, 268-270, 280-283, 306-310, 342-347, 349-354, 387-412, 417-420, 424-431, 443-453, 489-494. *See also* Numerical example. *See also* Case study
Exempt Wholesale Generator, 7. *See* EWG
Exhaustive search, 340
Expected unserved energy, 18, 303, 460. *See* EUE
Expected value, 59, 98, 99, 103, 106, 108, 109, 111-113, 163, 225, 234, 236, 238, 304, 461
Ex-post
  price, 359, 360, 362, 364, 366
  pricing, 317
External constraint, 213
Extraction, 11, 370, 406, 408, 409, 412-414, 421, 425

F

FACTS, 314
Feasibility, 125, 286, 307, 309, 421, 425
Feasible, 11, 212, 310, 341-343, 345, 346, 414, 415, 421, 425, 454
Federal Energy Regulatory Commission, 2. *See* FERC
FERC, 2, 8, 311, 313, 314, 315, 318, 355
   Order 888, 2, 355
Financial
   instrument, 2
   transmission right, 408. *See* FTR
Firm
   point-to-point service, 413, 421
   transmission
      reservation, 421, 428
      right, 2, 371, 408. *See* FTR
      service, 372, 374, 413, 421
5-unit System, 150, 469
Fixed transmission right, 408, 421. *See* FTR
Flowchart, 129
FOR, 33, 222, 236, 271, 272, 303, 304, 306, 307, 313, 387. *See* Force outage rate
Force outage rate, 65, 271, 276, 303, 315. *See* FOR
Forecast
   single-point, 108
   point, 108, 111
Forecasting
   accuracy, 16, 21, 54, 55, 58, 60, 63, 86
   adaptive, 81, 89, 91
   duration, 23
   electricity price, 57
   information, 15
   load, 14-16, 22-24, 31-33, 36, 52, 53, 57, 63, 64, 79, 116, 358
   method
      artificial intelligence-based, 21
      evolutionary programming, 21, 69
      expert system, 21, 30, 69
      fuzzy system, 21, 31, 69
      regression, 21, 69
      state-space, 21, 24, 25, 69
      statistical, 17, 21, 59, 105, 110, 111, 236, 274
   model, 21, 259
   performance, 16, 30, 55, 59, 60, 84, 85, 88, 89
   price, 2, 14, 16, 57-59, 63-65, 69-71, 73, 76, 79, 81, 88, 91, 95, 108, 116, 268
   strategy, 91
ForePrice, 16, 17, 58, 59
Formulation, 17, 117, 122, 126-128, 132, 139-141, 150, 154, 160, 186-188, 275, 276, 286, 287, 293, 294, 303, 304, 307, 310, 320, 321, 340, 341, 348, 414, 435, 436
Forward
   market
      day-ahead, 12, 312
      hour-ahead, 12, 312
   price, 113
Framework, 16, 58, 59, 236, 238, 240, 246, 267, 268, 274, 336, 443
Free mode, 434
Frequency deviation, 312, 354, 356, 366, 367
FTR, 2, 20, 371, 393, 408-421, 424, 425, 428-434, 439, 442, 451-454, 492
   auction, 412, 414, 419, 454
   award, 414, 420
   credit, 409, 412, 432, 442, 453
   feasibility, 425
   feasible, 415
   holder, 20, 371, 409, 413, 416, 430-432, 442, 451, 452, 454, 492
   limit, 415
   marginal dispatched, 415
   market price, 415
Fuel
   consumption, 17, 152, 174, 177, 186, 187, 233, 455, 456
   cost, 189, 277, 290

INDEX

price, 151, 152
  variation, 150, 151
Fulfillment, 10
Function
  cost, 455
  emission, 455
  fuel consumption, 455
Futures contract, 163
Fuzzified, 31
Fuzzy, 31
  expert system, 30, 31
  logic, 30
  system, 21, 31, 69

G

Game
  basic probability distribution, 225
  complete, *see* Complete game
  cooperative, 195, 196, 212, 213, 231, 485
  core of, 485
  hyper-, 195
  incomplete, *see* Incomplete game
  Nash, 195
  non-cooperative, 196, 212, 213, 231, 484, 485
    Stackelberg, 195
  N-player, 483
    cooperative, 485
  parlor, 192
  power transaction, 195
  probability distribution, 216
  theory, 19, 191-193, 195, 196, 230, 483
  two-participant, 194
  zero-sum, 192
Gaming, 7, 15, 116, 226-229, 259, 367
  methodology, 218, 225
Gas, 18, 63, 65, 162, 164, 166, 174, 176-178, 180, 181
GENCO, 2, 5-10, 14, 15, 17, 21, 63, 64, 67, 115, 116, 118, 140, 160, 165, 166, 171, 174, 177, 180, 233, 236, 237, 240, 242, 246, 247, 248, 258, 262, 315, 316, 318, 322-334, 357, 360-367, 432, 434
Generalized generation distribution factor, 377. *See* GGDF
Generalized load distribution factor, 378. *See* GLDF
Generating company, 2, 8, 14. *See* GENCO
Generation
  asset valuation, 17, 59, 112, 116, 233-237, 268, 273, 274
  capacity valuation, 233, 234, 267-274
  scheduling, 15, 21, 23, 151, 222, 231, 233
Generation shift distribution factor, 377. *See* GSDF
Generator
  availability, 66
  bidding strategy, 66
  outage, 16, 58, 59, 66-68, 70, 71, 76
  site, 66
Generator shift factor, 397. *See* GSF
Genetic algorithm, 30, 31
GGDF, 377-379, 388, 389, 392, 393. *See* Generalized generation distribution factor
GLDF, 378, 379. *See* Generalized load distribution factor
Global optimum, 288
Gram-Charlier approximation, 306
Grand coalition, 191, 196, 198, 212-214, 231, 486
Greedy algorithm, 348
Gross
  demand, 381
  power, 381, 382, 390
GSDF, 377, 388, 389. *See* Generation shift distribution factor
GSF, 399. *See* Generator shift factor, 399

H

Hard constraint, 115, 116

Heat rate, 18, 65, 154, 164, 165, 277
  incremental, 151
Hedging, 2, 334, 367, 432
  tools, 2
Henry Hub, 63
Heuristics, 27, 32
Hidden layer, 26, 27, 30, 32, 34, 35, 43, 70
Hierarchical, 335
Historical price forecast, 240, 254
Homogeneous, 61, 68, 70, 72
Hopfield network, 27
Hour-ahead
  forward market, 12, 312
Hydro, 21, 66
Hydroelectricity, 62
Hydrothermal, 22
Hyper-game, 195

I

IEEE, 223, 224
Identical unit, 150
i-game, 215, 217, 219, 486, 487. *See* Incomplete game
Imbalance market, 366, 367
Impact, 16, 22, 32, 53-56, 58, 64-66, 70, 71, 81, 82, 84, 86, 88, 89, 94, 108, 116, 150, 151, 153, 169, 173, 233, 236, 259-261, 264, 266, 267, 270-274, 299, 376, 377, 379, 387, 403
Impacting factor, 81, 89, 111
Imperfect competition, 213, 259
Imputation, 485
Incentive compatibility, 335
Incomplete game, 215, 217, 219, 486, 487. *See* i-game
Increment, 198, 199, 264, 294, 342, 352, 393, 436, 458
Incremental
  bid, 422
  cost, 116, 145, 146, 198-201, 206, 212, 276
  heat rate, 151
  profit, 169, 171
Independent, 1, 5

power plant, 7. *See* IPP
system operator, 1. *See* ISO
Individual rationality, 485
Industrial organization theory, 192
Inequality constraint, 401, 402, 415, 462
  double-sided, 465
Infeasibility, 293, 304, 305, 309
  cut, 304, 305, 309. *See also* Benders, cut, infeasibility
Inflow, 380, 382-384, 448, 449
Injection, 11, 291, 370, 406, 408, 409, 412-415, 421, 424, 425, 436
Input
  factor, 53, 108
  layer, 26, 27, 32, 33, 35, 43, 45, 70
in-service, 276
Instructed deviation, 358
Interest, 189, 306
Interconnected Operations Services Working Group, 314
Inter-zonal, 67, 71, 154, 406, 407, 421, 422, 425-431, 454
  congestion management, 406, 422, 425-427, 431
  line, 67, 406, 407, 422, 425, 427-430
Intra-zonal, 421, 422, 425, 431, 454
  congestion management, 422, 431
  line, 425
IPP, 7. *See* Independent power plant
ISO, 1, 4-7, 9-12, 14, 15, 17, 20, 60, 61, 64, 115, 116, 195, 196, 222, 223, 229, 232, 246, 275, 311, 312, 315, 317, 318, 320-331, 333-335, 337, 338, 357-364, 366, 367, 371, 412, 413, 415, 416, 424, 432-436, 441, 442, 445, 447, 451, 452, 454, 492. *See* Independent system operator
  California, 6, 316, 334
  MaxISO, 6, 10
  MinISO, 5, 9
  PJM, 6

INDEX

J

Jacobian matrix, 292, 458

K

Kirschen, 383, 386, 391-393, 431, 433, 439, 442
Kohonen
  map, 30
  network, 27, 28, 30, 31

L

Lagrangian
  augmentation, 288, 462
    application, 141
    term, 141, 464, 465
  function, 120, 121, 123, 128, 129, 131, 141, 288, 462, 463
  multiplier, 119, 120, 122, 137, 285, 288, 289, 349, 459
  relaxation, 18, 119, 276, 288, 289. *See* LR
  method, 18
  term, 120, 129
Large-scale system, 70, 125
Learning
  adaptive, 71
  strategy, 29
  supervised, 29, 30
  unsupervised, 29, 30
Leasing, 18, 188, 189
LFC, 355. *See* Load frequency control
Liability, 409, 428
Line
  limit, 16, 58, 59, 66, 67, 70-73, 396-398, 401-404, 406, 415, 416, 419, 420, 431
  outage, 16, 58, 59, 66, 67, 70, 72, 300, 456
Linear, 15, 16, 25, 30, 44, 59, 77, 136, 137, 141, 143, 174, 195, 201, 205, 287, 290, 294, 296, 306, 310, 321, 340, 359, 376, 436, 438, 444, 484
  approximation, 143, 287, 296

programming, 195, 287, 340
sensitivity factor, 290, 310. *See* LSF
Linearization, 143, 276, 289, 290, 292, 383
Link, 383-385, 391, 433, 448, 449
Liquidity, 162
List of symbols, 117, 277, 455
LMP, 10, 20, 60, 61, 64, 66, 154, 371, 393, 394, 396, 397, 399, 403-412, 422, 424, 425, 427, 430-434, 441, 442, 445-447, 452-454, 489, 492. *See* Locational marginal price
  application, 405
  calculation, 61
Load
  balance, 10, 204, 275, 290, 304, 459
  distribution, 65, 66, 70, 297
  flow, 66, 376, 387. *See also* Power flow
    dc, 290, 291, 377, 383, 387, 425
  fluctuation, 72, 356
  following, 312, 314
  forecasting, 14-16, 22-24, 31-33, 36, 52, 53, 57, 63, 64, 79, 116, 358
    application, 21
    category, 23
    long-term, 22
    model
      ARMA, 24
      dynamic, 24
      load shape, 24
      peak load, 24
      state-space, 24, 25
    performance, 45, 51
    short-term, 21, 23
      ANN, 25
    very short-term, 23
  level, 17, 58, 59, 91-93, 95, 105, 111, 113, 304
    high, 93, 94, 99
    low, 93

medium, 93, 95
pattern, 16, 22, 32, 58, 59, 66, 67, 70, 74
  affecting factor, 22
shape, 22-24, 29, 195
variation, 303, 310, 413
Loading level, 280, 282, 283, 284
Load frequency control, 355. *See* LFC
Load serving entity, 11, 409. *See* LSE
Local generation, 165, 171, 173, 258, 262, 263
Locational marginal price, 393. *See* LMP
Logarithm, 83, 100
Long-term
  planning, 5
  price forecasting, 63
Loop flow, 370
Loss, 10, 20, 61, 180, 193, 204, 206-211, 216, 234, 235, 246, 298, 301, 311, 314, 371, 379-382, 387, 389, 393, 394, 397, 399, 401, 402, 404, 431, 433-436, 441, 447, 451, 452, 454
  charge, 434, 441, 451, 452
  marginal, 61, 394, 431, 447
Lossless, 205, 207, 209, 389, 391, 405, 424
Loss of Load Probability, 10. *See* LOLP
LP, 290. *See* Linear programming
LR, 288. *See* Lagrangian relaxation
LSF, 290, 291. *See* Linear sensitivity factor

M

Machine learning, 30, 31
MAIN, 63
Maintenance, 5, 7, 8, 65, 315, 372, 459
  schedule, 5, 65, 315
MAPE, 42, 43, 45-52, 54, 55, 58, 74, 75, 78-80, 83-87, 89, 90. *See* Mean absolute percentage error

definition, 80
new, 80, 81, 90
traditional, 78-80
Mapping, 54, 85, 91, 108, 109, 111, 414
MAPS, 65
Marginal
  cost, 4, 61, 64, 145, 146, 193-195, 198, 199, 201, 202, 205-207, 210, 212, 214, 217, 220-222, 227, 243, 248, 258, 262, 264, 267, 271, 305, 337, 338, 348, 352, 393, 394
  loss, 61, 394, 431, 447
  pricing, 336, 338, 347, 348, 349-353, 354, 367, 393
  unit, 394, 396
Marginal pricing auction
  formulation, 348
  solution, 348
Market
  ancillary service, 2, 7, 10, 13, 166, 312, 342, 357
  balancing, 12
  coordinator, 19, 191, 196, 197, 201, 231
  decision, 15, 21
  design, 15, 316, 317, 334, 354, 367
  dynamics, 4
  energy, 9, 64, 150, 249, 270
  entity, 5, 8
  forward, 12
  liquidity, 162
  manipulation, 63
  model, 246, 247
  monitoring, 2
  operation, 2, 22
    component, 14
  operator, 1, 5, 22
  participant, 1, 4, 5, 7, 9, 10, 11, 14, 22, 60, 63-65, 116, 194, 197, 234, 259, 336, 338, 347, 357, 393, 409, 412, 413, 421, 422, 431, 432, 434

price, 13, 17, 19, 71, 105, 112, 116, 125, 139, 146, 148, 150, 151, 154-159, 166, 167, 169, 171, 173, 182, 186, 194, 201-203, 211, 213, 215, 216, 233-236, 238, 241-244, 247, 248, 254, 256, 257-264, 266-274, 312, 415
  forecasted, 111, 233, 238, 254, 256, 258, 259, 262-264, 266, 459, 460
  simulation, 236, 241, 254, 270, 271
real-time, 9, 12, 13, 312, 358, 367
risk, 2
rule, 1, 5, 246
settlement, 9, 234, 236, 237, 242, 243, 246-248, 268, 274, 336, 337, 358, 360-364, 366, 411, 432
  rule, 246
  simulation, 242
structure, 4-6, 17, 160, 246, 274, 310, 312, 355, 369, 372
transmission, 9, 11
volatility, 2, 61, 264, 266, 267, 271
Market-clearing price, 6, 9, 60, 246, 318, 320, 342. *See* MCP
Market power, 2, 5, 7, 13, 14, 19, 63, 64, 191, 195, 211, 335, 354
  analysis, 191
Marketer, 5, 8, 17, 59, 64, 67
Matching process, 318, 321, 359
Master problem, 18, 275, 276, 286, 287, 295, 298, 301, 303-306, 309
  formulation, 287
Mathematical
  model, 15, 25, 65, 412, 454
  derivation, 461
MATLAB, 31, 32, 69
  toolbox, 31, 32
  training function, 31
Maximization, 120, 192, 202, 334
Maximum sustained ramp rate, 119, 456. *See* MSR

Max-min
  criterion, 214
  strategy, 214
  value, 484, 485. *See also*
    Characteristics function
MaxISO, 6, 10
MCP, 6, 9, 60, 61, 64-66, 86-90, 246, 247, 318-336, 338-340, 342, 346, 347, 349, 350, 352, 353, 357, 359-367, 413-415, 420. *See*
Market-clearing price
  calculation, 60
  unconstrained, 81
Mean absolute percentage error, 38, 42, 58. *See* MAPE
Mean reversion, 61
Mechanism, 9, 20, 336, 338, 357, 369, 371, 372, 376, 409, 412
Membership degree, 30
Merit order, 317, 348
Metering, 13, 314
Methodology, 18, 117, 119, 140, 141, 160, 192, 216, 218, 231, 382
Microeconomic analysis, 192
Minimization, 11, 120, 293, 294, 305, 334, 338, 341, 348, 462
  problem
    equality constrained, 462
    inequality constrained, 462
Minimum down/up time, 65. *See*
  Minimum ON/OFF time
Minimum ON/OFF time, 10, 155, 173, 273, 457, 469
MinISO, 5, 9
Mismatch, 355
Module, 16, 31, 43, 59, 72, 77, 236-238, 268
  performance analysis, 59
  price forecasting, 58
  price simulation, 58
  volatility analysis, 59
Modified
  AC-OPF, 422
  IEEE 30-bus system, 223, 477
Monopoly, 21

Monte Carlo simulation, 109, 257, 267
MSR, 118, 119, 277, 456. *See* Maximum sustained ramp rate
Multilateral, 371, 432, 434-439, 443, 444, 447, 454
Multiplier updating, 124, 133, 134
MVA-mile method, 376
MW-mile method, 370, 373, 374, 376, 392, 439, 440

N

n-1 contingency, 276
Nash
  bargaining
    method, 202, 204
    model, 203
    problem, 202
    scheme, 195
    solution, 204
  equilibrium, 192, 194, 203, 217, 221-223, 228, 231, 232, 487
    solution, 203
  game, 195
  John, 192
National Climatic Data Center, 40. *See* NCDC
NCDC, 40. *See* National Climatic Data Center
NERC, 314, 315, 355, 356. *See* North American Electric Reliability Council
Network
  configuration, 66, 373
  constraint, 286, 298
    linearization, 290
  integration service, 413
  stability service, 314
  structure, 87
  violation
    minimization, 276
Neuron, 25-28, 30, 32-35, 43-45, 70, 87
  axon, 25, 26
  dendrite, 25
  soma, 25, 26
  synapse, 25, 26
No-arbitrage pricing, 162
  theory, 162, 163
Non-competitive, 19, 191, 231
Non-convexity, 288
Non-cooperative
  game, 195, 196, 212, 213, 231, 484, 485
  payoff region, 483. See also Payoff, region
Non-discriminatory, 2, 5, 7, 20, 371, 412, 434, 435, 453, 454
Non-homogeneous, 61, 68, 70, 72, 151
Non-linear, 15, 24, 25, 59, 205
Non-separable, 143, 289
Non-spinning reserve, see Ancillary service, non-spinning reserve
Normal
  distribution, 109, 238, 256, 259, 264, 270, 274
  form, 483
North American Electric Reliability Council, 314. *See* NERC
Numerical example, 167, 240, 342, 371, 432, 443
NYMEX, 63

O

O&M, 65. *See* Operation and maintenance
  fixed cost, 65
  variable cost, 65
Oak Ridge National Laboratory, 314. *See* ORNL
OASIS, 18, 315. *See* Open Access Same Time Information System
Objective function, 119, 120, 202, 203, 205, 277, 285, 287, 288, 290, 293, 294, 338, 340, 415, 434, 436
Obligation, 19, 115, 163, 171, 311
118-bus system, 67, 72, 73, 479
Open access, 5, 7, 313, 314, 453
Open Access Same Time Information System, 315. *See* OASIS

INDEX

Operating cost, 7, 275, 276, 356
Operation and maintence, 65
OPF, 61, 67, 70, 108, 436. *See* Optimal power flow
Opportunity cost, 318, 321, 322, 326, 334, 367, 415
Optimal
   power flow, 6, 61, 231, 379. *See* OPF
   solution, 143, 169, 178, 180, 183, 204, 287, 289, 403
   strategy, 180, 181, 183, 221
Optimality, 121-124, 132, 143, 289
   condition, 121, 123, 131, 137
Optimization, 11, 31, 115, 122, 151, 165, 180, 203, 205-208, 222, 276, 288-290, 300, 302, 321, 348, 359, 394, 427, 445, 447
Optimum
   global, 288
Option valuation, 112
ORNL, 314. *See* Oak Ridge National Laboratory
Outage, 7, 8, 63, 66-68, 299, 300, 302, 303, 356, 358
   generator, 16, 58, 59, 66-68, 70, 71, 76
   equipment, 303, 310
   line, 16, 58, 59, 67, 70, 72, 295, 298, 300, 302, 456, 458
Outflow, 380, 382-384, 449
Output layer, 26, 27, 30, 32, 34, 35, 43, 70
Over-generation, 355
Over-priced, 18, 161, 162

P

P&L distribution, 234. *See* Profit & Loss distribution
Pairwise-contract, 432
Palo Verde, 63
Pancaking, 370
Paradigm, 115, 116, 162, 336
Parallel flow, 370
Pareto optimal, 483, 485
Parlor game, 192

Partitioning, 29, 30, 43, 44
Pattern
   bidding, 16, 58, 59, 67, 70
   load, 16, 22, 58, 59, 67, 74
   flow, 203
Payment
   capacity, 13, 318, 320-322, 326, 328-331, 333, 334, 357, 367
   combined, 320, 321
   energy, 317, 318, 320, 324, 326-328, 330, 331, 333, 334, 357, 492
Payoff, 191, 192, 194, 196-198, 200-208, 210, 212-215, 217, 220-222, 227-231, 483-486
   conditional, 217, 220, 222, 231
   expected, 217
   function, 202, 204, 215, 217, 483, 486
   low-risk, 162
   matrix, 213
      expected, 220
   net, 210
   pair, 483
   region, 483
PBUC, 14, 17, 18, 111, 115-117, 119, 130, 139, 145, 150, 156, 160, 165-167, 169, 174, 178, 180, 182, 185-189, 222, 223, 236, 237, 242, 248, 250, 257, 258, 262, 263, 268, 318, 455. *See* Price-based unit commitment
   constraint
      system
         emission, 118
         energy and reserve, 118
         fuel, 118
      unit
         fuel, 119
         generation, 118
         minimum ON/OFF time, 119
         ramping, 119
   emission-constrained, 182
   flowchart, 129
   formulation, 117

PDF, 244. *See* Probability density functio
Penalty
  factor, 247, 248, 462
  function, 462
  term, 288, 289, 304
  variable, 287, 293, 294, 304
Percentage Error, 78
Perceptron, 27, 29, 30
  multi-layer, 27, 31
  single-layer, 27
Perfect competition, 202, 213, 259, 261
Performance
  analysis, 16, 58, 59
  evaluation, 81, 89
  load forecasting, 45, 51
Pessimism–optimism criterion, 214
PG&E, 65
Phase shifter, 18, 276, 279, 280, 285, 287, 291
  adjustment, 276
Physical constraint, 271
Piecewise linear, 165, 207, 290
PJM, 6, 10, 11, 17, 61, 62
  ISO, 6
Planning
  long-term, 5
Point forecast, 108, 111. *See also* Single-point forecast
Point-to-point, 11, 411, 421
PoolCo, 4, 315
Pool protection mode, 434
Postage-stamp method, 370, 372
Post-processing, 84, 100
Post-processor, 30
Power
  exchange, 6, 195, 259, 313
  flow
    ac, 376, 379, 380
    dc, 373, 379, 380, 418
    full ac, 379
  market, 2, 4, 5, 7, 1-19, 57, 59-61, 63, 81, 92, 96, 100, 161-163, 166, 171, 188, 192, 194, 195, 197, 198, 211-213, 222, 230, 231, 233-235, 268, 275, 280, 369
  type, 9
  pool, 17, 63, 64
  system, 1, 2, 10, 12, 14, 15, 21, 22, 23, 56, 65-67, 88, 111, 195, 200, 203, 211, 230, 311, 389, 392, 453
    centralized, 203
    decentralized, 203
    monopolistic, 21
    restructured, 11, 13, 15, 19, 21, 369, 377, 434, 453, 486
    security, 15, 22
    transaction, 195
Preferred schedule, 20, 371, 406, 409, 422, 424, 454
Pre-processing, 30, 82, 83, 95, 100
Price
  coefficient, 122, 212
  curve, 16, 58, 61, 65, 96, 152, 177-180, 185, 199
  difference, 105-108, 239, 240, 241, 255
    analysis, 106, 240
  discrepancy, 116, 161, 162
  driver, 16, 58, 59, 67, 111
  electricity, 1, 16, 57, 58, 60, 61, 63, 164
  ex-post, 359, 360, 362, 364, 366
  forecast level, 17, 59, 95, 97, 98-100, 102, 105, 111
    extra-high, 102, 103
    high, 98, 99, 102, 103
    low, 98, 102
    medium, 98, 102
  forecasting, 2, 14, 16, 57-59, 63-65, 69-73, 76, 79, 81, 88, 91, 95, 108, 109, 112, 116, 268
    affecting factor, 64
    ANN factor, 70
    application, 111
    category, 63
    long-term, 63
    short-term, 16, 57, 63, 64, 112, 268

INDEX

indicator, 9, 371
market, *see* Market price
market clearing, 6, 9, 60, 246, 318, 320, 342, 394, 413
pair, 98
probability distribution, 17, 58, 59, 91, 93-95, 98-100, 103, 105, 106, 108, 109, 111, 113, 236
procurement, 337, 338, 348
reversal, 335, 336, 347, 349, 350, 351, 353, 354, 367
signal, 121
simulation, 16, 58, 67, 254, 267
spike, 16, 17, 58, 59, 82-84, 91-93, 95, 97, 98, 100, 102, 103, 105
  analysis, 91
  at different load levels, 91
  at different price forecast levels, 95, 100
spot, 4, 13, 163, 164, 193, 194, 195, 197, 198, 200, 202, 206, 207, 213, 216, 222, 228, 229, 231, 236, 238, 240
variation, 150
Price-based unit commitment, 17, 115. *See* PBUC
Price-taking, 335
Pricing
  congestion-based, 370
  demand substitution, 337
  electricity, 18, 60, 163
  ex-post, 317
  flow-based, 370
  marginal, 336, 338, 347-354, 367, 393
  mechanism, 162
  methodology, 370
  no-arbitrage, 162
  pay-as-bid, 336, 337
  rule
    multi-part, 375
    MW-mile, 375
    usage-based, 376
  scheme, 202, 370
  substitution, 336
  supply substitution, 337, 338
  transmission, 12, 14, 20, 369-372, 379, 393, 431, 432, 453
  uniform, 336
Primal, 307, 309
Prioritization, 431, 432, 434, 435, 445
Priority, 20, 371, 434
  procedure, 434
    contract protection mode, 434
    free mode, 434
    pool protection mode, 434
pro forma tariff, 1, 311
Probability, 16, 17, 58, 59, 91-93, 95, 97-100, 102, 103, 105, 106, 108, 109, 111-113, 215-219, 225, 226, 228, 231, 234-236, 238, 241, 243-245, 254, 256, 257, 263, 264, 267, 274, 461
  analysis, 59
  density function, 244. *See* PDF
  distribution, 17, 58, 59, 91, 93, 95, 98-100, 103, 105, 106, 108, 109, 111-113, 215-219, 225, 231, 236, 238, 243, 254, 256, 257, 274, 461
  conditional, 112
  exponential, 238
  lognormal, 238
  normal, 109, 238, 256, 259, 264, 270, 274
  of price, 91, 94, 98, 100, 103
  triangular, 238
  theory, 192
Processor, 31
Procurement
  cost, 11, 336-338, 344-348, 350, 353, 354, 367
  price, 337, 338, 348
Production cost, 117, 195, 204, 207, 210, 215, 237, 275, 277, 288, 298, 300-303
Profit & Loss distribution, 234. *See* P&L distribution
Profit analysis, 236

Profitability, 19, 167, 169, 235
Profit-based unit commitment, 115
Proportional sharing principle, 380, 382
Pure strategy, 483
PX, 6, 9, 11, 195. *See* Power exchange

Q

Quadratic programming, 125, 127
Quick start capability, 46, 65, 457

R

Ramp rate, 10, 151, 167, 169, 233, 272, 275, 280, 283, 290, 321, 357, 359, 456
  down, 469
  up, 457, 469
Ramping, 12, 65, 119, 126, 173, 276, 278, 280, 282-285, 290, 360, 362, 364
  constraint, 119, 173, 276, 278, 280, 285
  delay, 280, 282, 283
  down, 282-285
  model, 280
  up, 280, 283-285
Ranking process, 317, 320, 322, 323, 325, 334
Rational buyer auction, 335, 336, 338, 340, 346-354, 367
  formulation, 339
  solution, 340
Reactive supply and voltage control, 313
Real power
  interruption, 457, 458
  loss replacement, 314
Real-time
  market, 9, 12, 13, 312, 358, 367
  utilization factor, 320
Re-dispatch, 312, 406, 409, 422, 425, 430

Reference bus, 377-379, 386, 387, 397-400, 402, 403, 413, 415, 416, 419, 420, 444, 447
Regulated, 2, 7, 8, 10, 115
Regulation, *see* Ancillary service, regulation
Regulation and frequency response, 313, 355
Reinforcement, 369, 372
Relative duality gap, 128
Relaxation, 119, 120, 288
Reliability, 1, 7-9, 12-14, 18, 31, 118, 275, 280, 303-305, 310, 311, 314, 315, 319, 354, 375, 397
  constraint, 18, 275, 280
  index, 303
  maximization, 285
  requirement, 118, 305, 310, 319
Reliability Test System, 38. *See* RTS
Replacement reserve, *see* Ancillary service, replacement reserve
Reserve
  non-spinning, *see* Ancillary service, non-spinning reserve
  replacement, *see* Ancillary service, replacement reserve
  spinning, *see* Ancillary service, spinning reserve
Response capability, 233, 272
Restructure, 1, 2, 5, 7, 9, 10, 11, 13-19, 21, 57, 59, 61, 115, 116, 160, 161, 195, 230, 231, 235, 280, 310, 311, 326, 336, 354, 355, 366, 367, 369, 370, 371, 376, 434, 453, 454, 483, 486
Retail access, 1
RETAILCO, 5, 7, 8
Revenue, 20, 116, 117, 162, 163, 189, 192, 233, 237, 243, 244, 262, 330, 357, 369, 420, 424, 430, 431, 440
  expectation, 20, 369
Risk
  analysis, 112, 233
  counterparty, 63

INDEX

management, 2, 15, 17, 59, 63, 232
market, 2
minimization, 19, 191
Riskless, 162, 163
Rolled-in, 372
Root
 common, 383, 384
 node, 384
RTS, 38, 467. *See* Reliability Test System

S

Same-commodity arbitrage, *see* Arbitrage, same-commodity
Scarcity, 335
Schedule
 commitment, 111, 298, 305, 307
 day-ahead, 12, 17, 115, 275, 312
 unit commitment, 18, 275, 276, 287, 298, 299
Scheduling
 ancillary service, 318, 319
 generation, 15, 21, 23, 151, 222, 231, 233
 process, 290
Scheduling and dispatch, 312
Scheme, 15-17, 20, 58, 61, 83, 84, 109, 195, 317, 369, 370, 371, 393, 421, 422, 428, 431, 432, 454
Screening, 290
SCUC, 14, 18, 115, 116, 275, 276, 285-287, 290, 293, 296, 298-303, 455. *See* Security-constrained unit commitment
 constraint
  emission, 275
   regional, 279
   system, 278
  fuel, 278
  load balance, 10, 275, 290, 304
  minimum ON/OFF time, 275
  network flow, 275
  ramp rate, 275
  reactive power
   generation, 279, 280

operating reserve, 279
real power generation, 280
system
 operating reserve, 277
 spinning reserve, 277
 voltage, 279, 280
thermal unit minimum starting up/down time, 278
transformer tap, 279
transmission flow, 276, 279, 280, 290, 303
unit generation, 118, 278
Seasonal network, 15, 34, 35
Security, 2, 5, 6, 7, 10, 14, 17, 21, 22, 65, 115, 116, 207, 210, 231, 275, 290, 299, 303, 310, 376, 415
Security-constrained unit commitment, 10, 115, 275. *See* SCUC
Self-provision, 314
Sensitivity, 15, 16, 53, 54, 58, 67, 71-73, 156-159, 169, 238, 258-260, 270, 288, 291, 372, 376, 379
 analysis, 15, 16, 53, 54, 58, 67, 71, 156-159, 169, 258-260, 270
 index
  ac flow, 379
 study, 58, 67
Separable, 120, 143, 289
Sequential, 10, 19, 66, 312, 317, 318, 320, 323, 334, 335, 339, 354, 367
 approach, 10, 312, 334, 367
 auction, 312, 317, 320, 323, 334, 335, 354
Settlement, 9, 242, 246-248, 268, 336, 337, 358, 360-364, 366, 411, 432
 10-minute, 361
 rule, 246, 247
SFT, 421, 424, 425, 426. *See* Simultaneous feasibility test
Shapley value criterion, 195
Short-term

load forecasting, 15, 21, 23, 29, 30, 33, 38
price forecasting, 16, 57, 63, 64, 112, 268
Similarity index, 78
Simplex multiplier, 287, 296
Simulation
  bidding strategy, 258
  chronological, 65
  example, 67
  market price, 236, 241, 254, 270, 271
  market settlement, 242
  Monte Carlo, 109, 257, 267
  statistical model, 236, 274
  strategy, 66
Simultaneous, 11, 18, 19, 161, 312, 335, 336, 339, 354, 367, 421
  approach, 11, 312, 336, 354, 367
  feasibility test, 421. *See* SFT
Single-point forecast, 108
Single-unit dynamic programming, 120. *See* Dynamic programming, single-unit
6-unit system, 166, 167, 187, 476
Slope
  bid curve, 201
Social
  cost, 11, 336, 337, 347-353
  welfare, 6, 10
Solution methodology, 17, 117, 119, 134, 139, 150, 154, 160, 186, 187, 188
Spark spread, 164, 165
Spatial arbitrage, *see* Arbitrage, spatial
Spike, 16, 17, 58, 59, 82-84, 91-93, 95, 97, 98, 100, 102, 103, 105
Spinning reserve, *see* Ancillary service, spinning reserve
Spot
  market, 4, 6, 63, 145, 163, 166, 167, 169, 173, 194, 201-203, 215, 216, 233, 236-238, 241, 242, 254, 259, 260, 262, 263, 409, 422, 424, 429-431

price, 4, 167, 169, 173, 194, 201-203, 215, 216, 233, 236-238, 241, 242, 254, 259, 260, 262, 263
  actual, 236, 238, 240
  forecasted, 236, 238
  simulation, 238
price, 4, 13, 163, 164, 193, 194, 195, 197, 198, 200, 202, 206, 207, 213, 216, 222, 228, 229, 231, 236, 238, 240
pricing, 206
Stabilizer, 314
Stackelberg
  equilibrium, 195
  non-cooperative game, 195
Standard deviation, 61, 109, 238, 256, 259
Starting point, 146, 147, 312
Start-up
  cost, 10, 155, 273, 277, 459
  emission, 457
  fuel, 457
Statistical, 17, 21, 59, 105, 110, 111, 236, 238, 254, 256, 274
Steady state, 276, 290, 292, 294, 295, 298, 300-302, 356, 458
  constraint, 290, 292
  subproblem, 294
Steam, 18, 164, 166, 186-188
Step calculation, 124, 133-135
STLF, 15, 21, 22, 29-31, 33. *See* Short-term load forecasting
Stopping criterion, 42
STPF, 16, 57. *See* Short-term price forecasting
Strategy
  bidding, 14, 64, 70, 145-147, 233, 242, 248, 258-261, 264, 266
  forecasting, 91
  training, 91
  pure, 483
Subgradient method, 124, 133
Subproblem, 18, 119, 125, 275, 276, 286-290, 293-296, 298, 300-304, 307, 309, 310

INDEX

contingency, 295
formulation, 293, 303
reactive, 276
steady state, 294
transmission, 294, 295
voltage, 295
Substitutability, 11, 335, 339
Substitution, 335-337, 339, 341, 342, 346, 347, 351, 352, 354, 367
Successive approximation, 195
Supplemental service, 12, 311, 317, 357
Synchronized, 119, 147, 316

T

Tap
　changer, 276
　　adjustment, 276
　transformer, 18, 287
Tariff, 5, 313-315, 370
　pro forma, 1, 311
TC, 313, 374. *See* Transmission customer
TCC, 491, 492, 494. *See* Transmission Congestion Contract
Telemetering, 314
Temporal arbitrage, *see* Arbitrage, temporal
10-minute
　auction, 358
　requirement, 315
　settlement, 361
Testing
　process, 45
　set, 40-42, 44, 45, 49
Thermal, 21, 210, 397, 401, 402, 410, 421, 425
3-bus system, 296, 310, 417, 489
30-bus system, 223, 326, 477
36-unit system, 150, 154, 472
Threshold, 26, 128
Tie-line, 312, 354, 356, 357, 366, 367
Time series analysis, 57

TO, 290, 303, 415, 431, 452, 462. *See* Transmission owner
TP, 313. *See* Transmission provider
Tracing, 20, 342, 346, 371, 372, 376, 380, 381, 383, 391, 392, 431, 433, 439, 442, 447, 454
　contribution, 20, 371
　flow-based, 20, 371, 454
　method, 376, 379, 392
　　Kirschen's, 383, 386, 391-393, 431, 433, 439, 442
　　Bialek's, 380, 381, 383, 386, 389, 390, 392, 393
Training, 16, 26, 30-36, 38, 40, 42-46, 53, 55, 61, 69, 70, 72-74, 81-91, 111
　process, 30, 32, 33, 35, 36, 38, 42, 45, 84
　set, 40, 41, 74
　strategy, 91
　vector, 42, 43, 81, 84-86, 89
Transaction
　analysis, 203
　price, 198, 199, 204-206, 207-210
TRANSCO, 5-7, 18, 231, 315
Transfer capability, 13, 66
Transformer tap, 458, 459
Transmission
　capacity, 4, 6, 10, 154, 195, 373-375
　　unscheduled, 374
　　unused, 374
　charge, 20, 202, 369, 370, 371-373, 385, 392, 431, 454
　　calculation, 439
　　service, 431
　　usage, 431
　congestion, 2, 6, 10, 20, 57, 60, 61, 116, 139, 154, 322, 323, 370, 371, 393, 409, 428, 429, 431, 453, 454
　　charge, 393, 409, 429, 431
　　calculation, 441
　　contract, 2
　　cost, 61, 431
　　credit, 428

impact, 71, 361
management, 19, 369. *See* Congestion management,
price, 6, 10
simulation, 139
constraint, 13
control, 1
cost, 311, 372, 374, 379, 386, 431
   allocation, 372, 387
   embedded, 372
   fixed, 372, 373, 375, 377, 380, 453
   service, 431
credit, 432
customer, 313, 371, 413. *See* TC
facility, 7, 373, 375, 376
flow constraint, 276, 290, 303
grid, 1, 8, 370
line, 20, 216, 223, 276, 297, 303, 304, 315, 326, 369, 371, 392, 399, 408
loss, 193, 204, 209, 371, 432, 433, 436, 441, 447
market, 9, 11
network, 6, 11, 19, 116, 259, 303, 304, 369, 372, 373, 377-379, 414, 431, 432, 453
network model
   dc, 303
   transportation, 303
owner, 1, 5, 14, 22, 415, 431, 451, 452
owner, 1, 5, 14, 22, 415, 431, 451, 452. *See* TO
ownership, 1
path, 61, 370, 371, 373, 408
pricing, 12, 14, 20, 369, 370-372, 379, 393, 431, 432, 453
provider, 313, 434. *See* TP
right, 11, 393, 408, 412
security, 4, 5, 17, 18, 286, 300, 302
service, 8, 11, 314, 355, 373, 376, 386, 413, 421, 431, 453
   firm, 372, 374, 413, 421
subproblem, 294, 295

system, 2, 5, 6, 7, 9, 12, 19, 66, 204, 311, 314, 371, 373-375, 383, 386, 393, 394, 432, 435
tariff, 5, 313, 314, 370
usage, 11, 372, 414
   charge
      calculation, 439, 451
violation, 5, 276, 290, 298, 299, 432
Transportation model, 303
Tuple, 483, 485
Two-participant
   bargaining problem, 204
   game, 194

U

UC, 17, 115. *See* Unit commitment
Unavailability, 314
Unbundled service, 2
Unbundling, 2, 10, 116, 311, 313, 453
Uncertainty, 218, 225, 234, 303, 310, 355
Unconstrained MCP, 81
Under-priced, 18, 161, 162
Uniform pricing, 336
Uninstructed deviation, 358
Unit characteristics, 248, 469, 472, 476
Unit commitment, 2, 6, 10, 14, 17, 18, 22, 59, 115, 116, 125-128, 160, 226, 229, 231, 275, 276, 279, 285-289, 292-294, 296, 298, 299, 301, 303, 310. *See* UC
   cost-based, 115, 116
   price-based, *see* Price-based unit commitment
   profit-based, 115
   schedule, 18, 275, 276, 287, 298, 299
   security-constrained, *see* Security-constrained unit commitment
   status, 125, 279
   traditional, 14, 275, 287
   transmission-constrained, 285

INDEX

Unloaded, 119, 147, 155
Unused transmission capacity method, 374
Unsynchronized, 119
Updating multiplier, 124. See Multiplier updating
UPLAN-E, 65
Upstream-looking algorithm, 380
Usage charge, 7, 20, 371, 382, 432, 434, 451
Utility function, 192, 193
Utilization factor
   real-time, 320

V

Value at Risk, 19, 234. See VaR
Value of Lost Load, 10. See VOLL
VaR, 19, 234-238, 240, 245, 246, 257, 261-274. See Value at Risk
   application, 235
   calculation, 234, 236, 237, 246, 257, 268
   definition, 235
Variable fuel price, 140
Vertically integrated
   structure, 7, 9, 354
   utility, 5, 355
   monopoly, 453
Violation
   coupling constraint, 119
   line flow, 406, 425
   minimization, 275, 290, 303
   transmission, 5, 290, 298, 299, 432
   transmission flow, 276, 298, 299
Volatility, 2, 16, 57-59, 61, 63, 91, 108, 162, 257, 259, 264, 266, 271
   analysis, 16, 58, 59
   annualized, 63
   coefficient, 259
   market, 61, 264, 266, 267
VOLL, 10. See Value of Lost Load
Voltage
   constraint, 276, 290, 292, 293, 298, 301-303, 458
   subproblem, 295

W

Weather, 15, 22, 24, 30, 31, 33, 36, 38-40, 47, 50, 53-55
   cloud cover, 33
   humidity, 15, 16, 23, 25, 32, 33, 53-56
      relative, 56
   light intensity, 23
   precipitation, 23
   temperature, 15, 16, 22, 24, 25, 31-34, 37-40, 43-45, 47, 50, 51, 53, 55, 56
      average, 31
      effective, 16, 55, 56
   thunderstorm, 23, 33
   wind, 15, 16, 23, 32, 53-56
Weight, 29, 31, 36, 84, 86, 89, 382, 432, 444, 447
   adaptive, 36
Weighting factor, 294, 312, 317, 318, 324, 328, 334, 367
Wheeling, 66, 373, 374, 379
Willingness, 10, 20, 371, 432, 434, 438, 445, 454
Withdrawal, 370, 414, 415
Working group
   Interconnected Operations Services, 314
WTI, 63

Z

Zero-sum game, 192
ZMCP, 60, 61, 64, 66, 86, 87, 88, 89. See Zonal market clearing price
   calculation, 61
Zonal
   boundary
      defining, 422
      determining, 405
   congestion management, 421
   market clearing price, 60. See ZMCP
Zone definition, 405

Printed in the USA/Agawam, MA
March 15, 2021

771596.008